Immunology

Immunology

Authors

Arbind Acharya & Sanjay Kumar

Co-*authors*

Rishi Kant Singh, Sandeep Kumar
Munendra Singh Tomar, Naveen Kumar,
Anand Kumar Patel, Lokesh Yadav, Kaushalendra,
Alok Shukla

2024

Daya Publishing House®

A Division of

Astral International Pvt. Ltd.

New Delhi – 110 002

ISBN: 978-93-5461-950-2 (HB)

Publisher's Note:

Every possible effort has been made to ensure that the information contained in this book is accurate at the time of going to press, and the publisher and author cannot accept responsibility for any errors or omissions, however caused. No responsibility for loss or damage occasioned to any person acting, or refraining from action, as a result of the material in this publication can be accepted by the editor, the publisher or the author. The Publisher is not associated with any product or vendor mentioned in the book. The contents of this work are intended to further general scientific research, understanding and discussion only. Readers should consult with a specialist where appropriate.

Every effort has been made to trace the owners of copyright material used in this book, if any. The author and the publisher will be grateful for any omission brought to their notice for acknowledgement in the future editions of the book.

Published by : **Daya Publishing House®**
A Division of
Astral International Pvt. Ltd.
– ISO 9001:2015 Certified Company –
4736/23, Ansari Road, Darya Ganj
New Delhi-110 002
Ph. 011-43549197, 23278134
E-mail: info@astralint.com
Website: www.astralint.com

Digitally Printed at : **Replika Press Pvt. Ltd.**

Preface

The study of Immunology has assumed tremendous importance with the progress in different disciplines of Biosciences. Varied aspects of Immunology have now been incorporated in the university curriculum of a number of subjects of Biosciences and Biomedical Sciences. Therefore main aim to write down this book is to present knowledge in lucid and easy manner. This book is dedicated to provide the present knowledge of basic mechanism of Immunology especially to Undergraduate and Postgraduate students. An attempt to arrange diverse areas of Immunology study has been made in a coherent manner. Therefore, this book provides a clear summary of the principle and emerging scope of Immunology in a concise and easily manageable format. However, much advancement in the field of molecular biology has been laid down and this advancement has undoubtedly thrown a light to decipher many of unrebuild scientific truth and proper understanding of the immune system in the vertebrates and still it is continually added. It is remarkable to us that new paradigm continues to be established in the field of Immunology and a unifying principle emerges from analysis of several complex molecules and their role in immune recognition and effectively combats to the disease. The lure to study Immunology clearly comes from its direct relation to human health and diseases.

It is not before the work of Louis Pasteur on rabies vaccine in 1885; Immunology becomes the research area of interest. After that, continuous endeavor of the immunologists gave present dimension to understand the various diseases. It has always attracted the people that how our body discriminates the self from non-self through the immune system and starts a battle against non-self or foreign. It is due to this, somebody called our body as a battle field, where regular war between foreign body and our immune system takes place. It is self and non-self discrimination of our immune system that has led down a thwart on tissue grafting, and thus origin of Transplantation Immunology. Perfectly, immune system has developed a patrolling mechanism that keeps eyes to see or vigil whether any intruding activity is done or not. If yes, a second mechanism triggers that effectively eliminate it out. It is very interesting that some people are prone to infection and malignancies, whereas some are not. Now-a-days, due to least side effects, medicine has relied on Immunology to search out cure for many of the diseases, and therefore, much impetus has been started to give on immunotherapeutic strategies to various diseases such as allergy, infection, auto-immunity and malignancies, and many of them are adopted. Vaccination before and after birth are well and enough to understand, how Immunology has become important for health and diseases. No wonder, one day, vaccination strategies would be developed against most dreaded diseases such as cancer and AIDS.

We have tried best to make easier this book. Therefore, rather than breaking a general subject into too many separate chapters, its various aspects are dealt with together so that their interrelationship can be appreciated. However, this is an initiation and we feel a lot of possibility for its improvement. Therefore, we solicit suggestions from readers, students and teachers so that the book may further improve in future editions.

We would like to put a special word of appreciation and thanks to Astral International Pvt Ltd, New Delhi for his interest to publish our book and giving us so much time.

Arbind Acharya

Sanjay Kumar

Contents

THE AUTHORS

Prof. Arbind Acharya, the principal author of the book, is currently serving as a Professor and Head of the Department of Zoology. With over three decades of dedicated research and teaching experience in the field of tumor immunology and cancer biology, Prof. Acharya is a prominent figure in his area of expertise. Throughout his career, he has effectively supervised numerous students at both Ph.D. and P.G. levels, guiding them to the successful completion of their theses. Prof. Acharya's contributions extend beyond academia as he has led and successfully concluded several major research projects, receiving recognition and funding from various esteemed agencies. His work has resulted in the publication of more than 70 research papers in internationally renowned journals, solidifying his position as a respected authority in the field. During this tenure, Prof. Acharya has undertaken research visits to numerous countries through faculty exchange programs, fostering global partnerships and enriching the academic landscape. Furthermore, Prof. Acharya's commitment to education is evident in his efforts to introduce immunology into the syllabi of both PG and UG levels within the Department of Zoology, ensuring that future generations of students benefit from a comprehensive and up-to-date curriculum.

Dr. Sanjay Kumar, who obtained his Ph.D. from the esteemed Department of Zoology at Banaras Hindu University under the guidance of Prof. Arbind Acharya, has been a significant contributor to the book. Following the completion of his Ph.D., Dr. Kumar further honed his expertise by engaging in Post Doctoral research at both Banaras Hindu University and Jawaharlal Nehru University. His research acumen and dedication of more than 10 years are evident through the publication of multiple research papers in highly reputable international journals during his doctoral and post-doctoral studies. In the collaborative effort of preparing the book, Dr. Kumar's contributions are of paramount importance, and it is second only to the principal author, Prof. Arbind Acharya. His expertise and insights have undoubtedly enriched the content, making it a comprehensive and authoritative resource in the field of tumor immunology and cancer biology.

CO-AUTHORS

Dr. Rishi Kant Singh earned his Ph.D. in Cancer Biology from the Department of Zoology, Banaras Hindu University, Varanasi, India. During his Ph.D. he has published several papers in highly reputed international journals. Currently, he is working as a Post-Doctoral Fellow to understand the molecular mechanism of B cell differentiation and B cell acute lymphoblastic leukemia at the Indian Institute of Technology, New Delhi, India. He has contributed significantly to book preparation and provided valuable suggestions to enhance the overall quality of the book.

Dr. Sandeep Kumar did his Ph.D. in Zoology from the Department of Zoology, Institute of Science, Banaras Hindu University under the supervision of Prof. Arbind Acharya. As an avid immunology enthusiast, he is an aspiring author determined to contribute to the field of immunology through my writing. His passion for this subject drives him to explore the intricate mechanisms of the immune system, its role in maintaining health, and its significance in combating diseases. With a background in biology, his contribution and valuable suggestion have proven invaluable in enhancing the overall quality of the book.

Dr. Munendra Singh Tomar did his Ph.D. in Zoology from the Department of Zoology, Institute of Science, Banaras Hindu University under the supervision of Prof. Arbind Acharya. Currently, Dr. Tomar is working as a Research Associate at the University of Colorado, USA. During his Ph.D. he has worked in tumor immunology targeting NK cell activity during non-Hodgkin's lymphoma progression and publishing several research papers in international journals of high repute. His knowledge of tumor immunology helps to make the book more comprehensive and impactful.

Dr. Naveen Kumar is a highly accomplished researcher who holds a Ph.D. in Zoology from the prestigious Banaras Hindu University in Varanasi, Uttar Pradesh, India. He completed his Master of Science in Zoology at Kurukshetra University, Haryana, India, in 2017, and qualified for UGC NET-JRF in the same year. During his Ph.D. he published his research work in reputed international journals. Currently, Dr. Kumar is working as a scientist at Taurmed Technologies, BioNEST, in the central discovery center of Banaras Hindu University. He has contributed significantly to book preparation and helped to make the book more impactful.

Mr. Anand Kumar Patel is currently undertaking his Ph.D. in Zoology at the prestigious Department of Zoology, Institute of Science, Banaras Hindu University, under the guidance of Prof. Arbind Acharya. His journey in academia began with an M.Sc. in Zoology from the same department. After qualifying for CSIR NET-JRF, he joined the department in 2020 with a determined focus on pursuing his doctoral research. Mr. Patel has made substantial contributions to the preparation of the book. His dedication, expertise, and insights have proven invaluable, enhancing the overall quality and depth of the content.

Mr. Lokesh Yadav is currently pursuing his Ph.D. in Zoology from the Department of Zoology, Institute of Science, Banaras Hindu University, under the supervision of Prof. Arbind Acharya. He did his M.Sc. from the Department of Molecular and Human Genetics, Banaras Hindu University. Upon qualifying for the CSIR JRF-NET, he joined the department in 2020 to pursue his doctoral research. Significantly, Mr. Yadav has made substantial contributions to the preparation of the book. His substantial contribution and valuable suggestion have proven invaluable in enhancing the overall quality of the book.

Dr. Kaushalendra did his Ph.D. from the Department of Zoology, Banaras Hindu University. Currently, he is working as an Assistant Professor at the Department of Zoology, Pachhunga University, Aizawl, Mizoram. During his Ph.D. he has published several papers in highly reputed international journals and has a very good understanding of his area of expertise. During book preparation, his substantial contribution makes it more comprehensive and impactful.

Dr. Alok Shukla received his Ph.D. in Zoology from the Department of Zoology, Institute of Science, Banaras Hindu University, under the supervision of Prof. Arbind Acharya. During his Ph.D., he published several research articles on the molecular pharmacology of anticancer medications against non-Hodgkin's lymphoma. Dr. Shukla's substantial contribution to the book provided unique insights and knowledge into its content, making it more comprehensive and impactful.

CHAPTER 1

OVERVIEW OF DEFENCE MECHANISM IN VERTEBRATES

The system through which an enormous variety of cells and molecules capable of recognizing and eliminating invading pathogenic microorganisms, developed, has been known as the immune system. The immune system is an exclusively adaptive and dynamic defense mechanism of vertebrates to eliminate unlimited types of foreign pathogenic bodies, invading the organism.

The immune system function in two ways, first it recognizes one foreign pathogen from another, and at the same time discriminates between foreign molecules and the bodie's own cells and proteins, secondly it conceives an effector response pattern consisting of cells and molecules to participate in the process of elimination or neutralization of the foreign invaders .In this way the immune system works on the principle of recognition and response, generated in the body of the host organism, suited to a particular pathogen.

The recognition pattern induced by the development of defensive cells and molecules, through the immune system, is further established as a memory response. The memory response is virtually an amplified immune reaction caused by later exposure of the organism to the same or identical foreign pathogen. Obviously, since the immune system has already been induced in the body of an organism, suited to a particular foreign pathogen, later exposure to the same foreign pathogen trigger the effector response, resulting in the elimination or neutralization of the foreign pathogen. And thus the disease likely to be caused by invasion of a particular pathogen is averted.

The chapter deals with the components –cells, and molecules that make the immune system and the mechanisms to protect the host body against foreign pathogens.

Historical Background

The English word "immunity" has been derived from the Latin word "immunis" which means exemption of the host body from contracting diseases. The factor of immunity was actually observed in 430 B.C. as reported by Thucydides, the great historian of the Peloponnesian War. It was recorded that person who has suffered from the plague could nurse the sick without contracting the disease second time and this observation leads to the notion of immunity. Eventually, attempts came into being of inducing immunity.

In the 15th century Chinese and Turks practiced to induce immunity against smallpox either by deep inhaling through their nostrils, the dried crusts of smallpox pustules or inserting it into small cuts on the skin. The technique of inserting the crust through the cuts on the skin was however reformed by the English Physicist Edward Jenner as inoculation of the cowpox pustules. Experimentally, it was found that a boy of eight years old, who was inoculated with the fluid from a smallpox pustule, did not develop smallpox.

Jenner's technique of inoculation soon spread throughout Europe and the most important and crucial phase evolved in the technique of vaccination (from the Latin "vacca" meaning cow). Louis Pasteur succeeded in growing the bacterium; he later injected the cultured bacterium of Cholera in to his chicken and became seek but soon recovered. Then he again injected the same chickens with the attenuated strain of bacterium and he wonderrd to found that the chicken survived and as completely protected from the disease. He called this attenuated strain as vaccine. Pasteur experiment with anthrax bacillus (Bacillus anthracis) on a group of sheep that did not die, where as the unvaccinated animals died. This marked the beginning of the science of immunology in 1885. Later Pasteur administered the first vaccine to a boy against rabies and the boy survived.

However, the mechanism of immunity could not be detailed by Pasteur, when the vaccination he found, had attained a status. The mechanism of immunity was reported by Von Behring, the nobel laureate in medicine in 1901. Behring and Kitasto, followed by other researchers demonstrated that an active component of the immune system could neutralize toxin, precipitate toxins and agglutinate bacteria. Elvin Kabat, demonstrated that active molecules in the immunoglobulin fraction (serum) called antibodies mediate immunity. The antibodies flood the body fluids (then called humors). And thus the concept of humoral immunity came to be recognized.

Elie Metchnikoff by his experiments concludes that certain blood cells which he termed phagocytes were more active in immunized animals than in non-immunized animals, causes immunity. This observation led to the concept of cell mediated (consisting of blood monocytes and neutrophils). Later in 1950s the lymphocytes were identified as the cell responsible for both cellular and humoral immunity. Further experiment proved the roles of both the cells and humors in immunity.

Early Theories of Immunity

The earlier concern of the immunologists was the specificity of the antibody molecules that accounted for its binding – the antigen (the general term for a substance that binds with a specific antibody). Experiment, showed that antibodies have a capacity for an almost unlimited range of reactivity. Ehrlich in 1900 explaining the action of serum antibodies proposed that cells in the blood contain a wide variety of side chain receptors that could react with infectious agents and inactivate them. That interaction between an infectious agent and a cell bound side chain receptor would release the side chain in a corresponding specificity. That is to say that an antigen would select the appropriate side chain receptor to render immunity. This theory is known as the selective theory.

Prior to information about DNA and RNA, another theory was developed by which antigen played a central role in determining the specificity of the antibody molecule. This theory postulates that a template around which antibody would fold, and take instructions from its configuration. By 1960 this postulate was disapproved, and the selective theory was redefined in the form of the clonal selection theory. According to this theory an individual lymphocyte expresses membrane receptor that are specific for a particular antigen. An antigen binds to a specific receptor and as a result, a cell is activated, causing it to proliferate into a clone of cells that have the same immunological specificity as the parent cell. This theory has held its postulation in the realm of modern immunology.

Component of Immune System

There are two types of mechanisms to combat pathogens – innate immunity (by birth) and adaptive (induced) immunity. The former operates through phagocytic cells, such as macrophages, and the latter through lymphocytes and the antibodies and the other molecules they produce. Just after the host's exposure to a pathogen, the first line of defense is brought into action by the innate immune system, clearing the pathogens within a few days. In case the invading pathogens elude the innate response system the adaptive immune system is triggered with a high degree of specificities as well as the remarkable property of memory. Both the immune response functions as a highly interactive and cooperative system, producing total protection from infectious disease. The innate response system to invading pathogens is called non-specific and the adaptive response system is called specific (operating with a high degree of specificity to the invading microorganisms).

Table 1-1: Nobel Prizes for Immunologic Research

Year	Recipient	Country	Research
1901	Emil von Behring	Germany	Serum antitoxin
1905	Robert Koch	Germany	Cellular immunity to tuberculosis
1908	Elie Metchnikoff Paul Ehrlich	Russia Germany	Role of phagocytosis (Metchnikoff) and antitoxin (Ehrlich) in immunity
1913	Charles Richet	France	Anaphylaxis
1919	Jules Bordet	Belgium	Complement-mediated bacteriolysis
1930	Karl Landsteiner	United States	Discovery of human blood group
1951	Max Theiler	South Africa	Development of yellow fever vaccine
1957	Daniel Bovet	Switzerland	Antihistamines
1960	F. M. Burnet Peter Medawar	Australia Great Britain	Discovery of acquired immunological tolerance
1972	Rodney R. Porter Gerald M. Edelman	Great Britain United States	Chemical structure of antibody
1977	Rosalyn R. Yalow	United States	Development of radioimmunoassay
1980	Baruj Benacerraf George Snell Jean Dausset	United states United States France	Major Histocompatibility Complex
1984	Cesar Milstein Georges F. Köhler Niels K. Jerne	Great Britain Germany Denmark	Monoclonal antibody
1987	Susumu Tonegawa	Japan	Gene rearrangement in antibody Production
1991	E. Donnall Thomas Joseph Murray	United States United States	Transplantation immunity
1996	Peter C. Doherty Rolf M. Zinkernagel	Australia Switzerland	The specificity of the cell-mediated immune response

Non-specific or Innate Immunity

The innate immunity operates through barriers- anatomic, physiologic, phagocytic, and inflammatory.

Anatomic Barriers

The skin is the most effective barrier to the penetration of microorganisms. The outer layer of the skin (epidermis) consists of dead cell and is filled with a waterproofing protein called keratin. The layer below the epidermis is called the dermis. It is composed of connective tissue, blood vessels, sebaceous glands, and sweat glands. The sebaceous gland is associated with the hair follicles and produces an oily secretion called sebum. The sebum consists of lactic acid, fatty acids which maintain the pH the of skin between 3 and 5. This low pH level inhibits the penetration or entry of microorganisms, provided the skin has no cut, or breaks to serve as a route of infection. However, the intact skin is penetrated by biting insects (e.g., mosquitoes, mites, ticks, fleas, and sand flies etc.) if these harbor pathogenic organisms when they intromit into the body as they feed.

Table 1-2: Means of Non-specific host defense in vertebrates.

Type	*Mechanism*
Anatomic barriers	
Skin	Mechanical barrier retards the entry of microbes. Ancidic environment (pH 3-5) retards the growth of microbes.
Mucous membrane	Normal flora competes with microbes for attachment sites and nutrient. Mucous entrap foreign micro-organisms. Cilia propel microorganisms out.
Physiological barriers	
Temperature	Normal body temperature inhibits the growth of some pathogens. Fever response inhibits the growth of some pathogens.
Low pH	Acidity of stomach contents kills most ingested pathogens.
Chemical mediators	Lysozyme cleaves bacterial cell walls. Interferon induces an ntiviral state in uninfected cells. Complement lyses microorganisms or facilitates phagocytosis.
Phagocytic/endocytic barriers	Various cells internalize (endocytosis) and break down foreign macromolecules. Specialized cells (blood monocytes, neutrophil, tissue macrophages) internalize (phagocytosis), kill, and digest whole microorganisms
Inflammatory barriers	Tissue damage and infection induce leakage of vascular fluid, containing serum proteins with antibacterial activity, and influx of phagocytic cells into the affected area.

Another line of innate barrier to infection is composed of saliva, tears, and mucous secretion, by which the potential invaders are washed away or entrapped in the antibacterial and antiviral substances contained in mucous (viscous fluid secreted by epithelial cells of mucous membranes). The conjunctivae, alimentary, respiratory, and urinogenital tracts are lined by mucous membranes. However, some organisms e.g., influenza virus and the organisms that causes gonorrhoea have a surface molecule by which they bind to the mucous membranes and escape the innate immune system.

Physiological Barriers

The physiological barriers are composed of body temperature and soluble factors. High body temperature inhibits the growth of bacteria. Low pH of the stomach causes the decaying of bacteria. Lysozymes (a hydrolytic enzyme found in mucous secretion and in tears) interferon (a group of proteins produced by virus infected cells) and complements (a group of serum proteins that circulate in the blood in an inactive state) are the soluble factors serving as physiologic barriers to the growth of microorganisms. Lysozymes cleave the peptidoglycan layers of the bacterial cell wall. Interferon binds to the nearby sells inducing a generalized antiviral states. Complements are converted into active state, through a variety of specific and non-specific immunologic mechanisms, damaging the membranes of pathogenic organisms.

Phagocytic Barriers

The antibacterial mechanism generated by the phagocytic system is known as Phagocytosis, a cell's plasma membrane expands around the particular material to form large vesicles called phagosomes (Fig. 1-1). Phagocytosis is one type of endocytosis. Phagocytosis is conducted by cell types, capable of other forms of endocytosis consisting of blood monocytes, neutrophils, and tissue macrophages. The other forms of endocytosis are receptor-mediated endocytosis wherein extracellular molecules are internalized after binding by specific cellular receptors and 'pinocytosis' whereby cells take up fluid from the surrounding medium and its molecules.

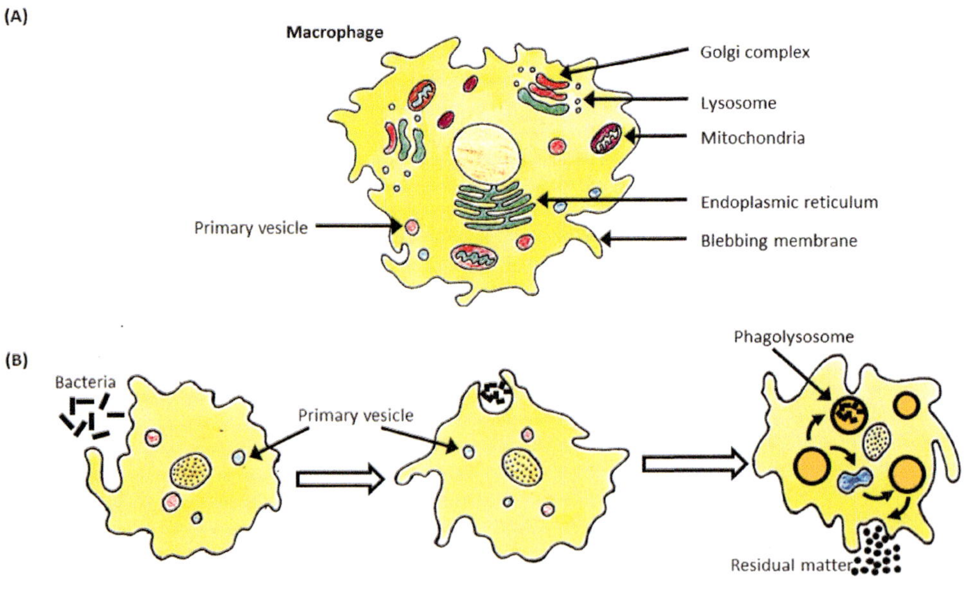

F-1-1. Diagrammatic structure of macrophage showing ruffling membrane. b. Mechanism of phagocytosis. I, showing attachment of foreign particles by pseudopodia of macrophage; II, engulfment of foreign particles and formation of phagosome; III, fusion of phagosome with primary lysosome; IV, digestion of ingested foreign particles; and V, exocytosis of undigested materials.

Inflammatory Barriers

A tissue may be damaged either by a wound or by an invading pathogen. This damaged tissue induces a complex sequence of events. These events are known as inflammatory response of the host body. The inflammation is characterized by any of these symptoms - redness (*rubor*), swelling (*tumor*), heat (*calor*), pain (*dolor*) and loss of function (*functio laesa*) - as reported in the first century and second century AD.

The following are the major events seen under inflammatotory response of the body; triggered by bacterial infection (fig. 1-2).

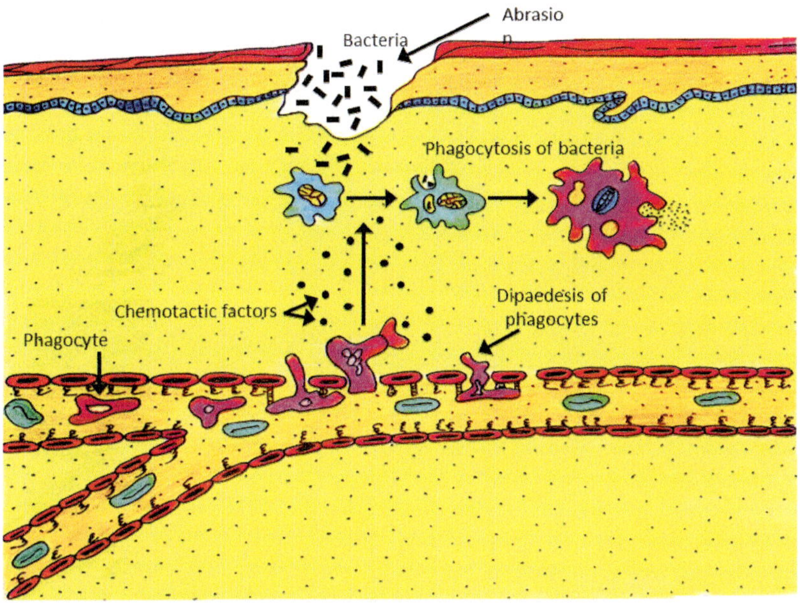

F-1-2. Generalized diagram of inflammatory response. Stepwise events take place during the bacterial infection at the site of tissue abrasion. This includes the release of various vasoactive and chemotactive factors at the site of tissue abrasion, which facilitates the increase of blood flow to the abrasion site and enhances the phagocytic action.

- Blood vessels dilate and nearby capillaries engorged with blood, resulting in redness (erythema) of tissue and rise in the temperature of tissues.
- Tissue swelling (edema) occurs as a result of accumulation of exudates. The accumulation is caused by the increased capillary permeability letting an influx of fluid and cells into the tissues.
- The increased capillary permeability also facilitates the flowing in of phagocytes into the t issues. The influx of phagocytes into the tissues develop into a complex series of events – adherence of the cells to the endothelial wall of the blood vessels, emigration into the tissue, and finally migration to the

site of inflammation (inflammatory response). This inflammatory response is called chemotaxis. Pus formation at the site of inflammation is caused by accumulation of dead cells, digested material, and fluid.

The inflammatory response involves a complex action of various chemical ingredients. The liver in response to tissue damage produces an acute phase protein called C-reactive protein (C- polysaccharide cell wall components found in the damaged tissue, resulting in the increased clearance of the pathogen through complement mediated lysis or increased Phagocytosis.

Histamine produced by a variety of cells through vasodilatation and increased capillary permeability, is also a major chemical participating in inflammatory response. Kinins, present in the blood plasma in an inactive form are activated by tissue injury and function as histamines. This very chemical causes pain in the pathogen affected area or site stimulating the host to protect the injured area.

Fibrin, the main component of blood clotting, enters into the inflammatory response system as insoluble strands, to prevent the spread of infection. As a result of the above chemical mediation the tissue repair process is increased and the regeneration of new tissues set in.

Specific or Adaptive Immunity

The reaction of the adaptive immune system is specific to the antigen challenging the host body. The mode of action of the adaptive immune system is to increase the state of immune reactivity by recognition of molecules and developing an immunological memory. As a result of this immunological memory, the immune system gets adopted to the response to an antigen, when encountered for a second time. However, the interaction between the innate and adaptive immune system continue to eliminate the invading pathogens.

Cells of Immune System

The Immunological Memory developed in the Adaptive Immune System is a complex function of Plasma Cells and Antigen Presenting Cells.

Plasma Cells

Lymphocytes are one of the white blood cells, found in the bone marrow. Lymphocytes are of two types B-lymphocytes and T-lymphocytes. B-lymphocytes leave the bone marrow and live in lymphoid organs as unique antigen binding receptor on its membrane (fig. 1-3. a) called antibodies. Antibodies are glycoprotein. Any antibody molecule consist of two identical heavy and light polypeptide chains each of which is held together in a pair by a disulfide bond forming a Y-shaped cleft within which antigen binds. B-lymphocytes or B-cells when encountering the antigen that matches its membrane bound antibody causes the cell to divide into memory B-cells and effector B-cell called plasma cells. The plasma cells are capable of secreting 2000 molecules of the antibody per second and constitute the humoral branch of immune system specialized in the recognition of antigens.

T-lymphocytes after leaving the bone marrow migrate into the thymus and mature therein, forming on its membrane a unique antigen-binding molecule called T cell receptor. This T- cell receptor can recognize an antigen that is bound to a cell-membrane protein called major histocompatibility complex (MHC) molecules.

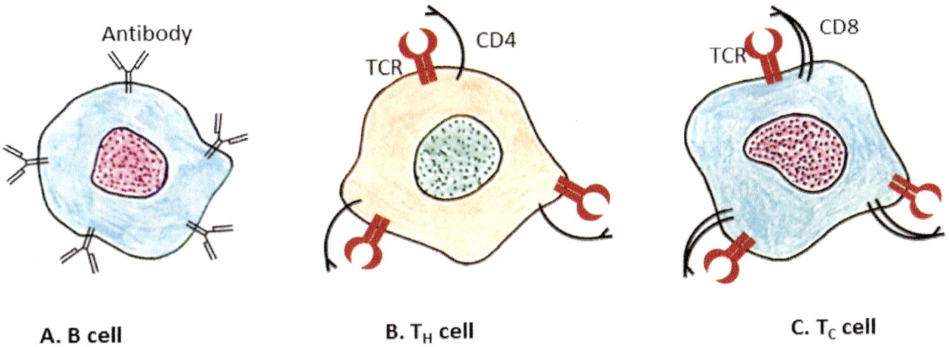

F-1-3. Expression of the membrane-bound molecule on the lymphocytes. A. B cells express and secretes antibodies, B. T helper (T_H)- cells express CD4 molecules for recognition of membrane bound MHC-molecule, and C. T-cytotoxic (T_c)-cells express CD8 molecules for recognition of membrane-bound MHC-molecules.

The MHC molecules are of two types in a chain pattern like B-cells, class I MHC molecules are found in all nucleated cells consisting β_2 micro globulin. Class II MHC molecule is antigen, presenting cells consisting of α and β glycoprotein chain. On encounter with antigen the T cell proliferates and differentiates into memory T cells and effector T cells. These cells may be identified by the presence of CD4 and CD8 membrane glycoprotein (fig 1-3. b, c). T cells have two subpopulations called T-helper (T_H) and T-cytotoxic (Tc). A T_H cell is interacting with an antigen turns in to an effector cell and secretes cytokines. The secreted cytokines activate B cells, T cells, macrophages, and various other cells to work in the immune response. A Tc cells that recognizes an antigen – MHC class I molecule complex proliferation and differentiates into an effector cell called a cytotoxic T-Lymphocyte (CTL). The CTL is very potent in killing any foreign antigen.

Antigen Presenting Cells

T- Lymphocytes, by their subpopulation known as T_H (T-helper) recognized only antigen that is displayed together with class II MHC molecules on the surface of antigen presenting cells (APCs). APCs are specialized cells consisting of macrophages, B lymphocytes, and dendritic cells. The antigen-presenting cells have two distinguishing properties; expression of class II MHC molecule on their membranes, and delivery of a co stimulatory signal for activating T_H cells. The T_H cells recognize and interact with MHC class II – antigen complex molecule on the membrane of the antigen presenting cells (Fig. 1-4).

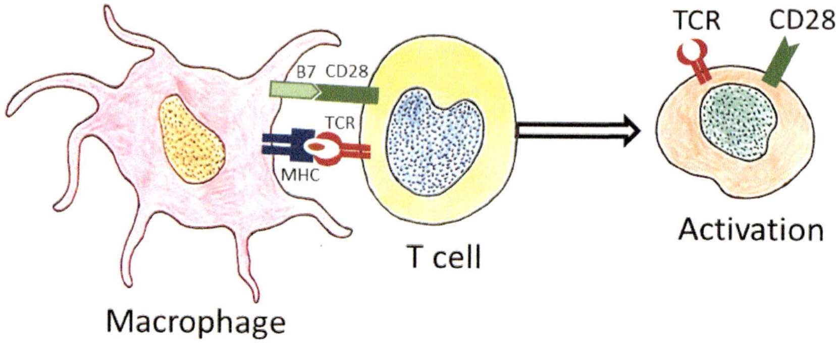

F-1-4. Diagrammatic representation of macrophage-T cell interaction. Interaction activates the T cells to release effector mediators.

Interaction Between Non-specific and Specific Immunity

In the process of elimination of the pathogen, the innate (non-specific) and adaptive (specific) system works in cooperation with each other. An encounter with microbes generates macrophages to give danger or warning signals. These signals are recognized by antigen specific T cells, to bring into action the adaptive immune system. Macrophages on encounter with microbes secret immunoregulatory hormones called cytokines to combat the particular intracellular pathogens. The innate immune system acts in cooperation with the adaptive immune system. Cytokines produce antibodies, which bind to the microbes and kill them. This interaction poses an awful resistance to bacterial infection.

Humoral and Cell Mediated Immune Response

The immune response elicited by the administration of serum antibody to a non-immunized organism is known as humoral immunity and it can be transferred by the administration of serum antibodies, which is collected from an immunized organism. The response elicited by the direct involvement of important immune cells is called cell-mediated immunity. It can be only transferred by the administration of T cells from an immune individual to a non-immune individual. Humoral immunity works in the interaction of B cells with antigen and their subsequent proliferation and differentiation into antibody-secreting plasma cells (fig 1-5). Antibody affects the humoral response by binding to antigen and neutralizing it for its elimination. An antigen coated witha an antibody can be eliminated by formation of clusters that are more readily ingested by phagocytic cells, preventing the antigen from binding to host cells.

Whereas cell mediated immune response relies on the T cells including T_H and T_C-cells both acting as effector cell population to antigenic challenge. The T_H cells containing cytokines, as secretes, activate various phagocytic cells, and facilitate the killing of microorganisms. This process of elimination of antigen is known as a cell mediated immune response. By this process various virus-infected cells and tumor cells are cleared of microorganisms. T_C-cells after activation kills the microorganisms or antigen bearing cells directly.

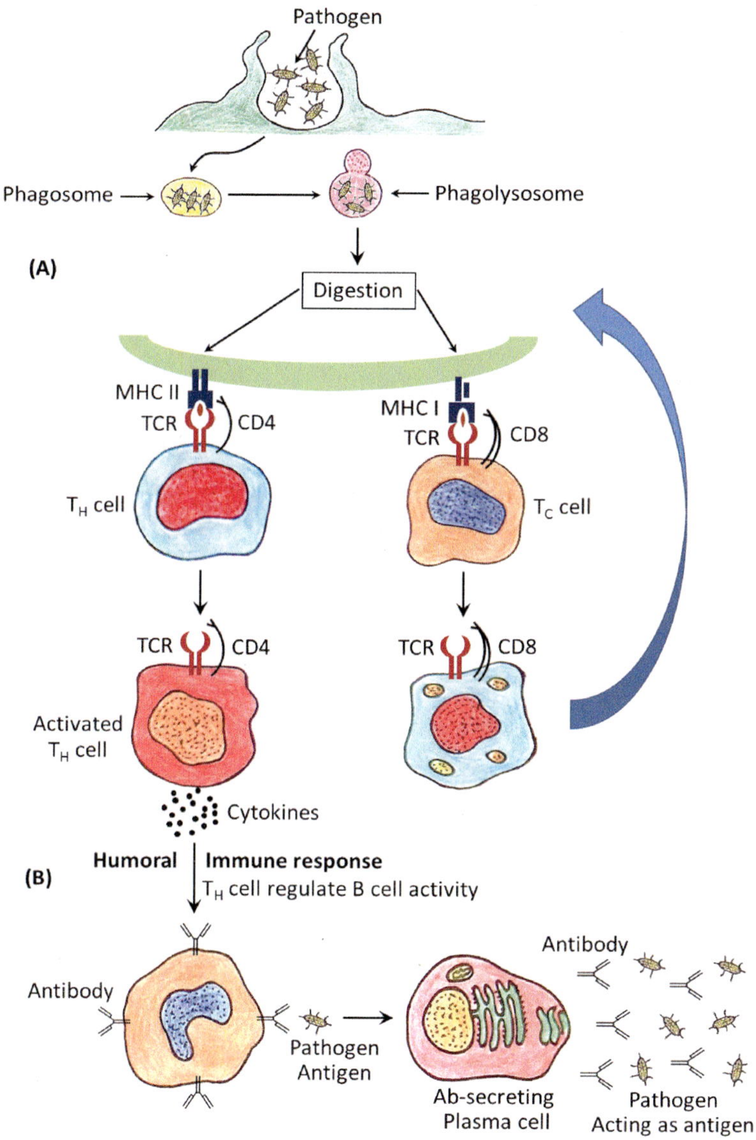

F-1-5. Generalized diagrammatic representation of cell-mediated and humoral immune response. In cell-mediated immune response (a), expression of endogenous and exogenous antigens by MHC I and MHC II molecules recognized by T_C-cells and T_H-cells respectively, and In humoral immune response (b), B cells interact with antigen by membrane-bound antibody and differentiate into antibody secreting plasma cells.

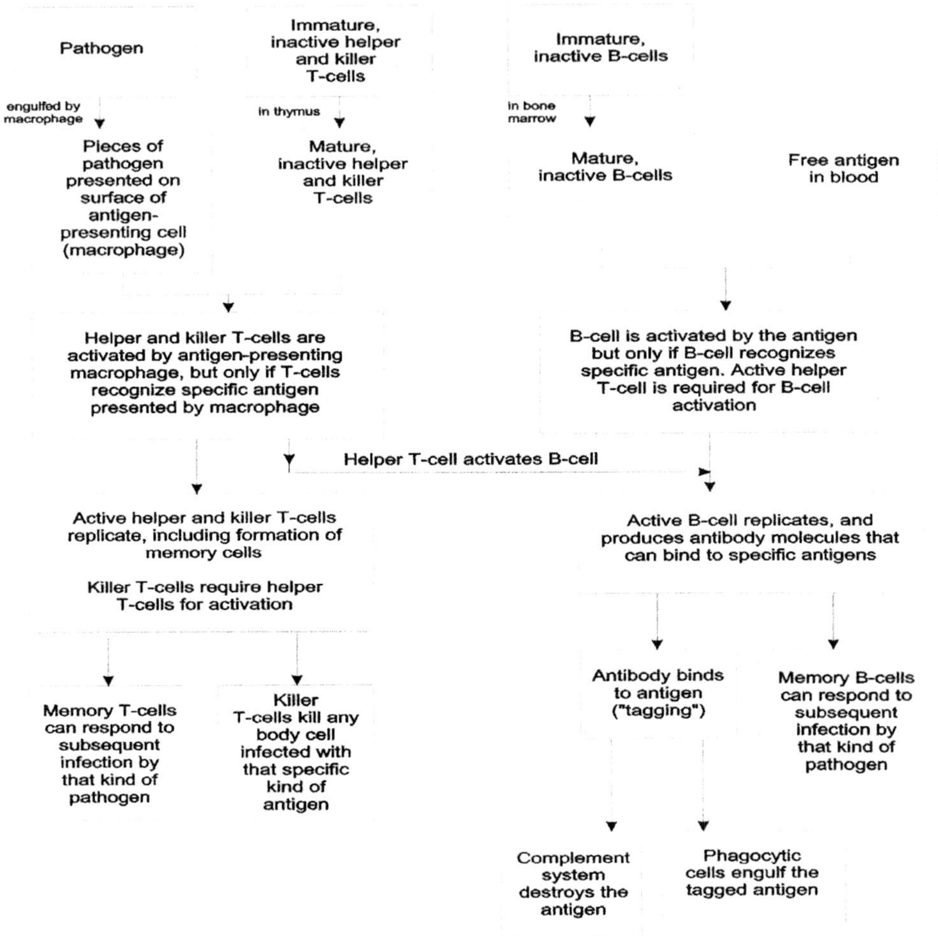

Fig. 1. 6. Overview of Immune system. Both humoral and cell-mediated immune responses are interlinked, and influenced by each other.

Antigen Recognition by Immune System

Different and distinct cell membrane molecules account for the recognition of antigens in the immune system. B and T lymphocytes recognize discrete sites on the antigen called antigenic determinants or epitopes. Epitopes are the immunologically active regions on a complex antigen. B cells (a component of the humoral immune response) recognize a wide variety of epitopes displayed on the surface of bacteria or viral particles, as well as on soluble proteins, glycoprotein polysaccharides, or Lipopolysaccharides that have been released from invading pathogens. T cells (a component of the cell-mediated immune response) recognize protein epitopes on self-cells and cancerous cells. The epitope is displayed with class I MHC and class II MHC molecules.

Development of Lymphocytes Specificity and Diversity

Protein antigens differ from each other by their amino acid content. Gene segments encoding the antibody molecule generate the specificity of B-cells by their random rearrangements. As a result of this process, each mature B cell has a single functional gene encoding the antibody heavy chain and a single functional gene encoding antibody light chain. Thus the cell synthesizes and displays antibodies with one specificity on its membrane. All 10^5 antibody molecules on a given B-lymphocyte have identical specificity giving each B-lymphocyte and the clone of daughter cells so created a distinct specificity for a single epitope on an antigen. This is why the mature B-lymphocyte is said to be antigenically committed. The random gene rearrangement during B-cell maturation in the bone marrow generates a great number of different antigen specificities. As a result B-cells multiply, each expressing a unique antibody, estimated to exhibit collectively more than 10^8 different antigen specificities. The self–components of any B cells are estimated by a selection process in the bone marrow, reducing the mature B cells proliferation. As a result, self active antibodies (autoantibodies) are not produced.

Each T lymphocyte cell expresses about 10^5 receptors on a cell and its clonal progeny have identical specificity for antigens. Through a selection process in the thymus, the enormous potential diversity is later decreased and only T cells with antigen-recognizing receptors are allowed to mature. A detailed study on this aspect of lymphocyte specificity and diversity will be carried on in chapters 9 and 10.

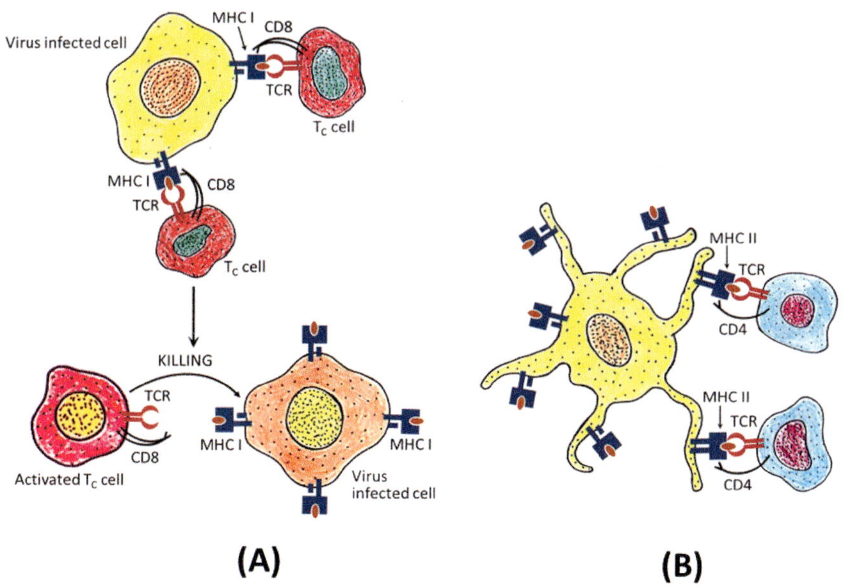

F-1-7. Endogenous and exogenous antigens presented by MHC class I and MHC class II are recognized by cytotoxic T cells expressing CD8 molecules, and T-helper cells expressing CD4 molecules respectively.

Role of Membrane Bound Glycoprotein Major Histocompatibility Complex (MHC)

There are two classes of membrane bound glycoprotein encoded by the major histocopatibility genetic complex. The T_H cells generally recognize antigen combined with MHC class II molecules and T_E cells generally recognizes MHC class I molecules (fig 1-7). The MHC have multiple loci, and encode the above two classes of MHC molecules. There are three genetic loci in human in each class –in class 1; A, B, C and in class II; DR, DP, and DQ. There are two genetic loci in mice in each class, in class I, K and D and in class II, IA and IE. Both class I and class II MHC gene are highly polymorphic; that is within a species each gene exhibits many different forms called alleles is inherited from each parent for each locus. The ability to present an antigen to T lymphocytes influenced by the particular set of alleles inherited by an individual. Different alleles display wide variation in their amino acid sequences depending on the distal region (farthest from the membrane). These distal region form a cleft within which the antigen peptide (MHC molecule can bind to a spectrum of antigenic peptides derived from the intracellular degradation of antigen molecules) sits and is presented to T lymphocytes.

Antigen Processing and Presentation

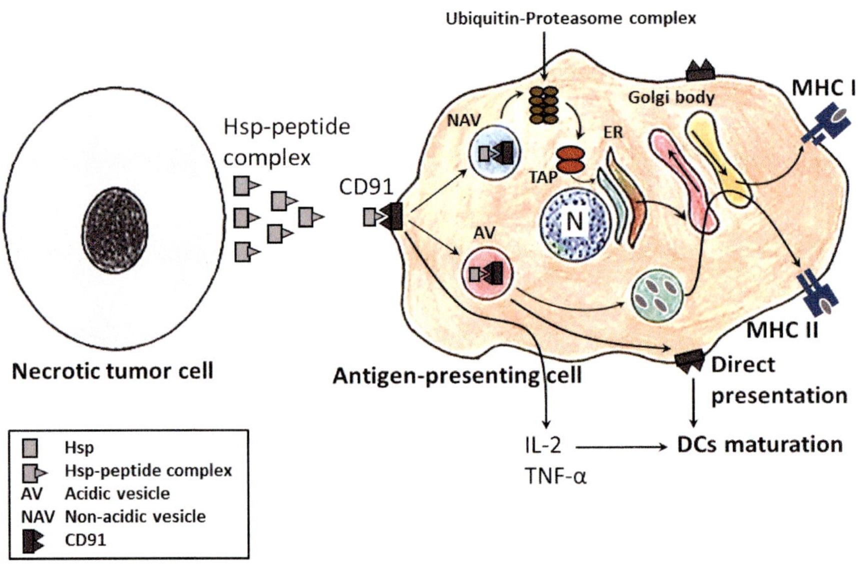

Fig. 1-8. Diagrammatic representation of antigen processing and presentation. Note the involvement of ubiquitin-proteasome complex for the processing of antigen to present on cell surface through MHC I whereas the antigen presentation through MHC II especially involves the formation of acidic vesicle. Heat shock proteins (HSP), a molecular chaperone, play a crucial role in the antigen processing and thus antigen presentation. Adapted from P. Deepak, S. Kumar, and A. Acharya. 2006. European Journal of Inflammation 4(1): 1-10.

A foreign protein antigen is converted into small antigenic peptides to form complex with class I or class II MHC molecules. The degradation or conversion of proteins is called the processing and presentation of antigens. The Processing and presentation of a particular antigen is determined by the route it takes to enter a host cell.

Class II MHC molecules are expressed within the endocytic processing pathway and those peptides produced by degradation of antigen in this pathway bind to the cleft within the class II MHC molecules and exposed to the cell surface. T cells recognize antigen combined with class II MHC molecules. Fig. 1-8 depicts the whole mechanism involved in antigen processing and presentation to the cell surface. This is why the T cells displaying CD4 are said to be class II MHC restricted cells and function as T_H cells.

Thus, the exogenous antigen is processed and presented to the T cells. Unlike the exogenous antigen is produced within the host cell itself. The endogenous antigen is converted into peptide fragments, bind to class I MHC molecules and exported to the cell membrane. T cells recognize antigen combined with class I MHC molecules. This is why T cells displaying CD8 are said to be class I MHC molecule restricted cells and with their specific receptors eliminate the antigen (MHC class I complex).

Clonal Selection of Lymphocytes

The random gene rearrangement determines the specificity of each T and B lymphocytes long before its encounters with antigen. The process of rearrangement produces antigen reactive clones if T- and B-lymphocytes. An antigen binds to a specific T- or B-cells, and stimulates it to divide repeatedly into a clone of cells with identical antigen specificity as the original cell. This process of binding of an antigen is called clonal selection (fig. 1-9). Immunologic recognition memory is complex reaction patterns of clonal selection. During clonal selection, the number of lymphocytes specific for an antigen is greatly amplified. The memory cells have a long life span than the naïve lymphocytes from which they arise. An antigen when comes into contact with a naïve immunocompetent lymphocyte primary response is produces. Later contact of the host with an antigen produces a more rapid and intensive response called secondary response.

Within 5-7 days naïve antigen T_H cells activate B cells. Subsequent proliferation and differentiation of the activated B cells takes place in about 14 days and plasma cells produced in this process, begin to die. The primary and secondary responses, as above, occur in the humoral branch of the immune system. In the cell-mediated branch of immune system, the recognition of an antigen MHC complex by a specific mature T lymphocyte induces clonal proliferation into various T cells with effector functions (T_H cells and T_C) and into memory T cells. Under cell-mediated response it has been observed that skin grafts are rejected more rapidly and vigorously, on a second time, than at first grafts. This faster graft rejection is the result of expanded memory cells.

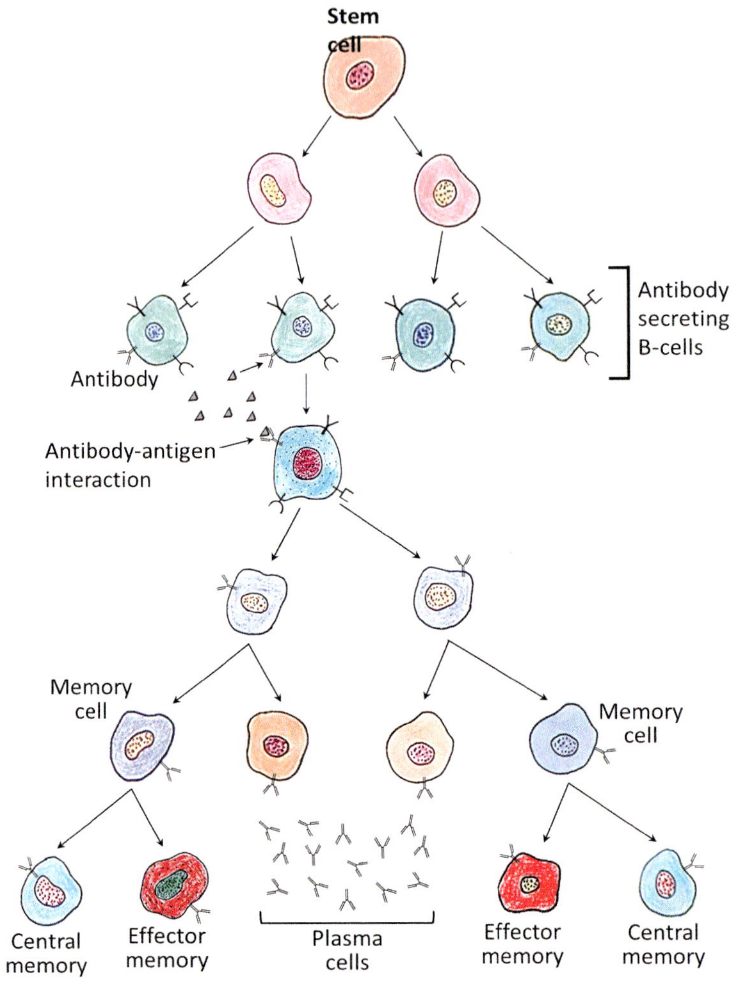

F-1-9. Diagrammatic representation of Ehrlich's side chain theory of clonal selection, in which a B cell expresses antibodies not-specific to particular antigen, binds with antigen to further clonal expansion of B cells specific for that particular antigen.

A specific immunologic response is generated as an interaction among several different types of cells, consisting of T_H cells (affecting humoral immunity and cell-mediated immunity).

Generation of the Humoral Immunity and Cell Mediated Responses

When mature B-cells encounter an antigen, its activation, proliferation and differentiation are triggered. The various stages of B-cells division and differentiation are stimulated by cytokines secreted by T_H cells. The activated B-cells undergo a series of cell division over about five days, differentiation into a population of both antibody-secreting plasma cells and memory cells. Cytokines secreted by T_H cells activate various T effector cells to cause cell-mediated responses. A detailed study of this immune system will be carried in a later chapter.

Immunity in Non-Vertebrate and Multicellular Organisms

Seemingly, immunologic responses present an evolutionary spectacle in organisms of many non- vertebrate phyla and multicellular organisms. However, the signal transduction pathways used to evoke immunity are strikingly similar, since some of the key element of this signaling pathway in vertebrates, insect and even plant has been found to contain evolutionary conserved molecule. In vertebrates, antibodies and lymphocytes play an important role under immune system, whereas in insect and plants, peptides exert a strong antibacterial or antifungal activity.

Consequences of Inadequate Immune Response

Inadequate immunity may cause several diseases viz., immunodeficiency, allergy, asthma, graft rejection, autoimmune diseases, Antigen like plant pollen, food or animal danders have been reported to cause allergy and asthma. Charles Richet (1902) immunized dogs against the toxins of the aquatic organism physalia, who when exposed to sub lethal doge of the toxins reacted to Extinct or protect themselves from the toxic effects called phylaxis and the unimmunised dogs fail to react defensively, a condition known as anaphylaxis. For his discovery of anaphylactic response Richet was awarded Noble Prize in 1913.

Allergic reactions in human are, however, not fatal. In the anaphylactic response or allergy one immunoglobulin called IgE is bound to its specific antigen (allergen) release substances and so when an allergic individual is exposed to an allergen (allergic antigen) symptoms like sneezing,wheezing difficulty in breathing (asthma) dermatitis , hives (skin eruption) and in more sensitive case inflammation of airways may develop significantly.

A fatal reaction, known as graft-versus-host disease is caused by the transplantation of cells or an organ from another individual. In it, the immune system of the host reacts to the transplanted cell or an organ treating as non-self. However, the disease can be potentially treated by drugs. The administration of drugs to control graft–versus-host reaction may result in suppression of all immune function in the host, so that the host is no longer protected by its immune system and becomes susceptible to infectious diseases. Thus suppression of immunity has remain a challenge to the immunologists even now.

The immune system malfunction, wherein the host loses its sense of self and non-self easing immune attack upon the host is another scenario of inadequate immunity. The condition is know as autoimmunity. This causes debilitating of immune reactions result in multiple serosis, rheumatoid arthritis etc. Genetic abnormality is caused by defects in innate or specific immunity or due to damage induced by chemical, physical or biological agents to the innate or specific immune system. This condition is known as immune deficiency. The diseases caused by immune deficiency, depend on the number of affected components. AIDS (Acquired immune deficiency syndrome) is a complex of such an immune deficiency disease in which retrovirus human immunodeficiency virus, or HIV cause a collapse of immune system, for which no effective drugs is known so far.

CHAPTER 2

CELLS AND ORGANS OF IMMUNE SYSTEM

The human body consists of many different organs and tissue, which play an important role in the immune system. There are two main types of organs, the primary lymphoid organs and the secondary lymphoid organs. The primary lymphoid organs provide the appropriate micro environment for the development and maturation of lymphocytes. The secondary lymphoid cells capture antigen from tissue or vascular spaces and thus are sites where lymphocyte interact with them. The blood vessels and lymphatic system connect this organ to develop into a functional whole. In the lymphoid organs, there are various white blood cells or leukocytes, which participate in the immune system, by clearing antigen. The vital lymphocytes especially T-lymphocytes secrete various protein molecules called cytokine. These molecules play an important role in the immunoregulatory response in the immune system. A detailed study of the formation of blood cells and functions of lymphoid organs follows in this chapter.

Hematopoiesis

The generation of stem cells starts right from the embryonic stage of the human body. This process is known as hematopoiesis, through which red and white blood cells begin to form in the embryonic yolk sac during the first week of development. In the third month of gestation, stem cells, in this case, hematopoietic stem cells (HSC) migrate from the yolk sac to the fetal liver and from there to the spleen, in a span of third to seven month of gestation. Stem cells can differentiate in-to other cell types by cell division.

Every functionally specialized mature blood cell is divided from the same type of stem cell. A hematopoietic stem cell is pluripotent to differentiate in various ways and thereby generate erythrocytes, granulocytes, monocytes, mast cells lymphocytes and megakaryocytes. These stem cells are few, normally less than one HSC per 5 x 10^4 in the bone marrow. The study of HSC is however different because of their scarcity and even hard to grow *in vitro*. This is why, little is known about how they proliferate, differentiate and regulate. Hematopoietic stem cells are at stable level throughout the adult life because of their capacity for self-renewal. HSC show enormous proliferation capacity at the level of increased demand for hematopoiesis. The enormous capacity of stem cells to proliferate and differentiate has been seen in mice whose hematopoietic system are fully destroyed by a lethal dose of X-rays (950rads; one rad represents the absorption by an irradiated target of an amount of radiation corresponding to 100ergs/gm of target). Unless infused with normal bone- marrow cells from a genetically identical mouse, such irradiated mice die within 10 days of exposure to rad. The infusion of bone marrow cells (0.01%-0.1%) completely restores the hematopoietic system.

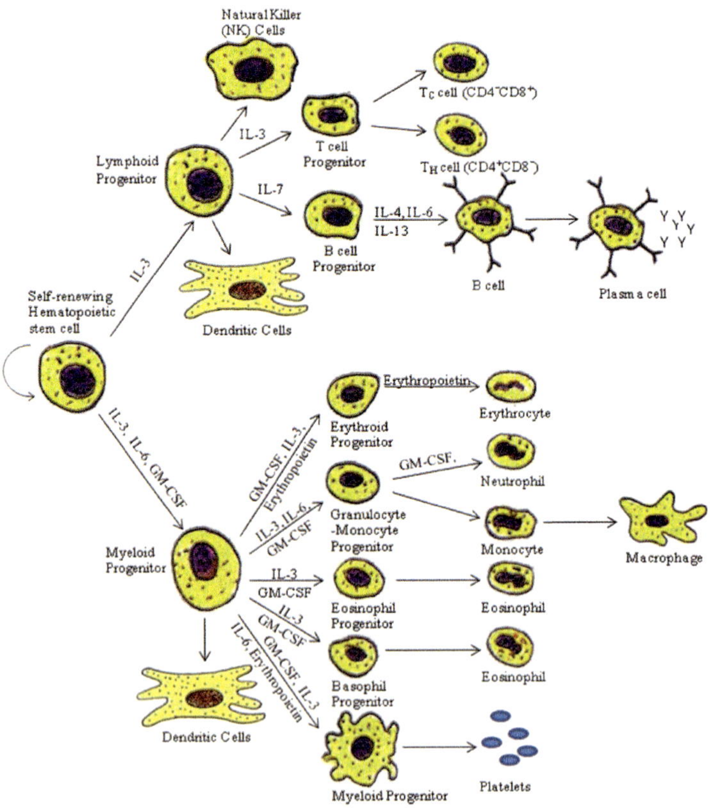

Fig-2-1. Proliferation and differentiation of various hematopoietic cells by different cytokines. Bone-marrow stromal cells are the major source of cytokines, which activate macrophage and T helper cells and induce hematopoietic activity resulting in the rapid expansion of the population of different blood cell types.

In the early stage of hematopoiesis, pluripotent stem cell differentiates either as a common lymphoid progenitor or myeloid stem cells. A progenitor cell or a particular stem cell is controlled by the microenvironment prevailing in it, in the process of differentiation. In the presence of cytokines progenitor cells proliferate and differentiate into the corresponding cell type – erythrocyte, leukocyte or platelet-generating cells. Fat cells, endothelial cells, fibroblast and macrophages are called stromal cells, which affects the differentiation of hematopoietic stem cells (fig. 2-1).

Factor Stimulating hematopoietic System

The cell culture system has various growth factors responsible for the survival, proliferation, differentiation and maturation of hematopoietic cells. Cytokines stimulate the formation of hematopoietic cell colonies in bone marrow cultures. A family of acidic glycoprotein which induces the formation of distinct hematopoietic cell lines has also been found among the cytokines. The glycoproteins are among the colony stimulating factors (CSFs), besides multilineage, CSF also known as IL-3 (interleukin-3), macrophage CSF (M-CSF), granulocyte CSF (G-CSF) and granulocyte–macrophage CSF (GM-CSF). EPO (erythropoietin) produced by the kidney is another glycoprotein (hematopoietic cytokine), which induces the terminal development of erythrocytes and regulates the production of red blood cells. Cytokines play an important role in the differentiation cells and account for hematopoiesis (fig.2-1).

Regulatory Mechanism of Hematopoiesis

Differentiation of stem cells and their descendents is regulated by different set of lineage-determining and lineage-specific genes at appropriate times and in the correct order. Thus hematopoiesis is dependent on genes. Studies of mice suggest that if mice fail to produce red cells or particular white blood cells, the protein specified gene is always non-existent. This phenomenon is known as targeted disruption, which blocks the production of the protein that it encodes. Table 2-1 shows transcription factors, which account for hematopoietic lineages. The absence of this factor or under functioning leads to the death of the infected organism, at an early age.

Table 2-1: Transcription Factor Essential for Hematopoiesis

Factors	*Cell types produced*
BM11	Cells of Myeloid and Lymphoid origin
GATA-1	Erythrocytes
GATA-2	Erythrocytes and cells of Myeloid and Lymphoid origin
Ikaros	Cells of Lymphoid origin
PU.1	Erythrocytes and cells of Myeloid and Lymphoid origin
Oct-2	B cells and differentiation of B cells into plasma cells

The steady state level in human beings depends on the production of mature blood cells equal to their loss (this accounts for especially in the process of aging). Hematopoiesis, is a continuous process that generally maintain a steady state level of blood cells in the organism. Hematopoiesis is regulated by complex mechanism to ensure steady state level of the various blood cells by increasing tenfold to twenty fold in defense against hemorrhage or infection. This regulation of hematopoiesis is accomplished through following process or factors;

- Production of cytokines with hematopoietic activity by activated T cells and macrophages.
- Removal of some cells by the controlled induction of cell death or apoptosis.
- Regulation of expression of receptors for hematopoietically active cytokines in stem cell and progenitor cells.
- Control of the level and types of cytokines.

Any one or a combination of the above regulatory factors can affect rate of cell production and differentiation.

Programmed Cell Death or Apoptosis

Apoptosis is a process through which a cell actively participates in bringing about its own death. The phenomena of apoptosis are known as programmed cell death, as distinctive morphological changes take place in the death of a mature cell (fig.2-2). Another process through which a cell dies is known as "necrosis". In necrosis cell dies by injuries.

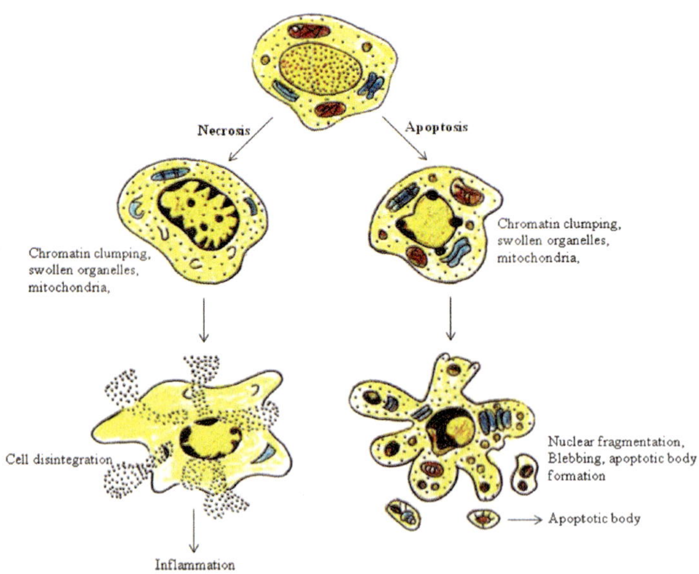

Fig.-2-2. Generalized diagrammatic representation of necrosis and apoptosis that lead to the death of injured cells, results in the release of cell contents, which induces the inflammatory response.

Apoptosis is markedly different from necrosis. In apoptosis the intracellular contents including proteolytic and other lytic enzymes, cationic proteins and oxidizing molecules are not released into surrounding tissue. In necrosis, the injured cell swells and bursts, and the intracellular contents are released to the surrounding tissue causing inflammation - a damaging response or danger signal.

Hematopoiesis produces leukocytes that after living a characteristic life span die by programmed cell death. If programmed cell death fails to occur, a leukemic state may develop. Proper numbers of hematopoietic progenitor cells are also maintained by programmed cell death.

Table 2-2 shows genes that regulate apoptosis. A normal adult has about 5L of blood with 2000 lymphocytes/mm^3, for a total of about 10^{10} lymphocytes. During acute infection, the total lymphocyte count rises to 40-150 x 10^9. At this stage, as the antigen levels subside, apoptosis signals and triggered causing the death of lymphocytes (fig. 2-5).

Table 2-2: Genes involved in the regulation of Apoptosis

Genes	*Apoptosis*
bcl-2	-
bcl-XL	-
bcl-XS	+
bax	+

Cells of Immune System

The immune system is composed of lymphocytes, which constitutes 20-40 percent of the body's white blood cells and 99% of the cells in the lymph (Table 2-3). Lymphocytes by continually migrating into the tissue and lymphoid organs integrate the immune system. The lymphocytes are of three main characteristics- B cells, T cells and natural killer cells represented by their action and cell membrane components. Fig. 2-3 shows the sizes of various lymphoid cells and their function.

B cells mature in the bone marrow are different from other lymphocytes by their synthesis and display of membrane bound antibodies which serve as receptors for antigens. Interaction between the antigen and the membrane-bound antibodies on a mature naive B cell, as well as interaction with T cells and macrophages selectively induces the activation and differentiation of B cell clones, of corresponding specificity. The B cells in this process generate plasma cells and memory cells, of varied life span against antigens.

T cells matured in the thymus like B cells, have membrane receptors for antigens. T cell receptor recognizes only antigen that is bound to particular classes of self-molecules. T cells recognize an antigen only when it is bound to a self-molecule encoded by genes within the Major Histocompatibility complex (MHC). Thus the cell is restricted to binding antigen displayed on self cells together with MHC molecules, presenting on the viruses infected cells, cancer cells, and grafts.

The decaying effect of these self-molecules altered as infected cells, cancer cells are countered by T cell, and the normal functioning of the body continue unaffected. Membrane molecules of T cells recognize antigen-bound cells by their functional agent CD4 and CD8. CD4 membrane molecules recognizes antigen bound to class II MHC and CD8 membrane molecule and CD8 membrane molecules recognizes antigen bound to class I MHC molecules.

Table 2-3: Normal adult blood-cell counts

Cell type	*Cells/mm³*	*%*
Red blood cells	5.0×10^6	
Platelets	2.5×10^5	
Leukocytes	7.3×10^3	
Neutrophil		50-70
Lymphocyte		20-40
Monocyte		1-6
Eosinophil		1-3
Basophil		<1

$CD4^+$ T cells generally function as T helper (T_H) cells and are restricted to class II MHC molecules. $CD8^+$ T cells are generally function as T cytotoxic (T_C) cells and are restricted to class I MHC molecules. The ratio of $CD4^+$ and $CD8^+$ T cells in the normal human body is approximately 2:1 in the peripheral blood, which may be changed by immunodeficiency diseases, autoimmune diseases and other disorders. T_H cells on recognizing an antigen class II MHC molecules begin to secrete various cytokines which activate B cells, T_E cells and other cells to help immune response. Tc cells interact with an antigen class I MHC molecules on the surface of an altered self-cells (virus infected cell or tumor cell) and proliferate to develop into an effector cell called a cytotoxic T lymphocyte (CTL). CTLs in contrast to T_H cells secrete few cytokines but eliminate altered self-cells to a great extent. Thus T_H and T_C cells help the immune system of the body.

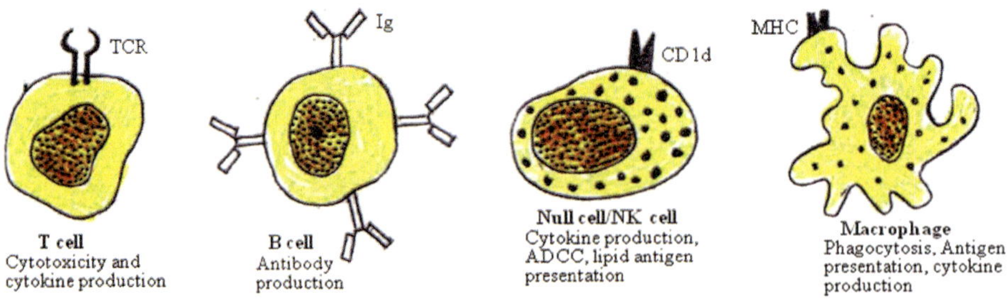

Fig. -2-3. Different types of lymphoid cells and their primary functions

Null cells are a small group of lymphocytes in the peripheral blood. These cells are large granular lymphocytes consisting 5-10% of the lymphocytes in human peripheral blood. Null cells do not produce antigen-binding receptors, so they lack precise immunologic specificity and memory. In 1976, it was reported that certain null cells display cytotoxic activity against a wide range of tumor cells. Because of their cytotoxic activity null cells are also called as natural killer (NK) cells. Null cells display an important role in host defense to tumor cells and cells infected with some, though not all, viruses. NK cell can recognize potential target cells by distinguishing a notable reduction in the displayed class I MHC molecules and an unusual profile of surface antigens displayed by tumor cells or cells infected by some viruses. By attaching to the antibodies made by the immune system, the NK cells destroy the targeted cells. This activity of NK cell is known as antibody-dependent cell-mediated cytotoxicity (ADCC). This activity has been discussed in the following chapter, in terms of available details. NK cells are a supplement to the production of cytokines required to support antibody production by cells and the expansion of cytotoxic T cells.

Macrophages and Phagocytosis

Monocytes circulating in the blood stream enlarge and migrate to the tissues forming specific tissue macrophages. Macrophages are dispersed throughout the body in a particular tissue becoming fixed macrophages and those remaining motile are called free or wandering macrophages, serving different functions in different tissues. There are a variety of potential activators of macrophages in the course of an immune response, viz., phagocytosis of particular antigens, cytokines secreted by activated T_H cells (IFN-γ) or components of bacterial cell walls. Interferons gamma (IFN-γ) is one of the most potent activators of macrophages. Because of greater phagocytic activity, activated macrophages are a potential eliminator of pathogens, including virus-infected cells, tumor cells and intracellular bacteria. Fig 2-4 shows the morphology of a monocytes and a macrophage.

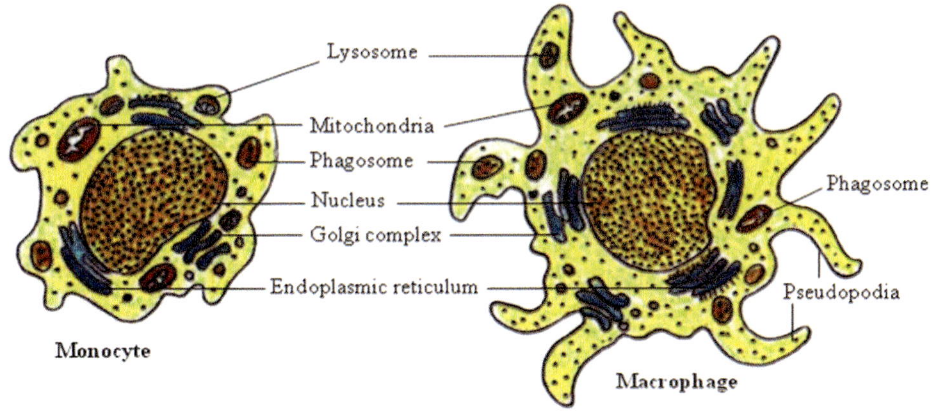

Fig.-2-4. Differentiation of macrophages from monocytes with a distinct morphology.

Macrophages in their first step of phagocytosis are attracted to a variety of substances generated in an immune response and then the antigens adhere to the macrophage cell membrane. This adherence induces membrane protrusions (called pseudopodia) and encloses the material within a membrane-bounded structure called a phagosome. A phagosome enters the endocytic processing pathway and fuses with a lysosome. This process is called phagolysosome. Lysosomes contain enzymes, which digest the ingested material, and get eliminated by a process called exocytosis. Fig 2-5 shows the Phagocytosis of antigens.

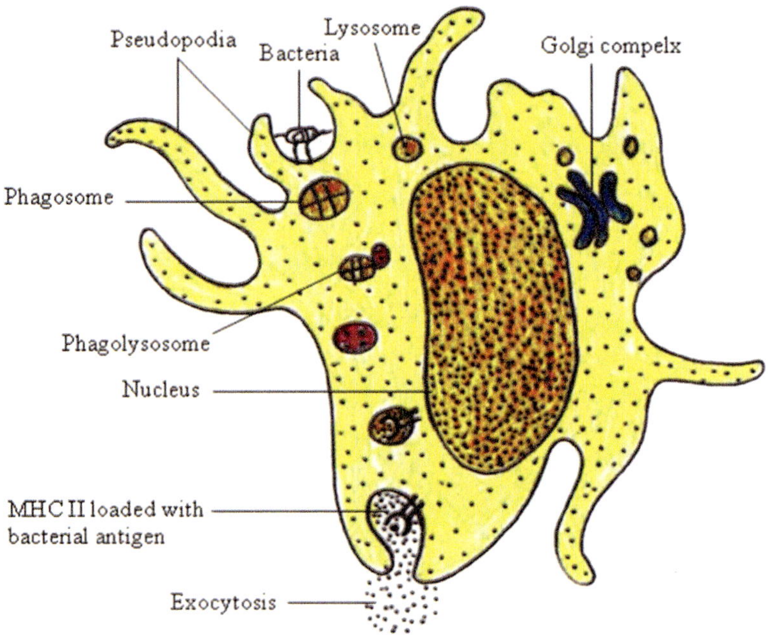

Fig.-2-7. Process of phagocytosis and presentation of antigen through MHC class II molecules by macrophages (APCs).

Destruction of Phagocytosed Microorganisms

Activated macrophages cause destruction of phagocytosed microorganisms, through some oxygen-dependent and oxygen-independent elimination mechanism. During Phagocytosis, a metabolic process known as the respiratory burst occurs in activated macrophages. This activates a membrane-bound oxidase that catalyses the reduction of oxygen to super oxide anion – a reactive oxygen intermediate. It is highly toxic to ingested microorganisms. Besides, super-oxide anion other powerful oxidizing agents including hydroxyl radicals and hydrogen peroxide are also generated by the respiratory burst phenomena. Nitric oxide gas is also a potent antimicrobicides. Nitric oxide and substances derived from it are potent antimicrobicides.

L-arginine $+O^2$ + NADPH$\rightarrow$NO + L –citrulline + NADP

Activated macrophages produce a group of defensins consisting of antimicrobial and cytotoxic peptides. The molecules of defensins are cysteine-rich cationic peptide containing 29-35 amino acid residues. A peptide has six invariant cystein, and for a circular defensin peptide to kill a variety of bacteria including *Staphylococcus aureus, Streptococcus pneumoniae, Haemophilus influenzae, Escherichia coli,* Pseudomonas aeruginosa. Besides, activated macrophages secrete a cytokine known as tumor necrosis factor α (TNF-α) a cytotoxic agent for some tumor cells. Due to lysosomal contents, phagocytosed microorganisms are killed in the phagosome itself. However, some microorganisms develop a resistant variety e.g., *Salmonella typhymurium, Mycobacterium avium, Mycobacterium tuberculosis, Mycobacterium leprae, Candida albicans* etc. These all are intracellular pathogens and are resistant to the contents of lysosomes. Some pathogens survive by escaping from phagosome and proliferate within the cytoplasm of infected macrophages.

Secretion of Cytotoxic Factors

Essential to the development of immune response are secretion of activated macrophages. By inflammatory reaction the immune system eliminates foreign pathogens. For the development of inflammatory response, the activated macrophages secrete cytokines such as interleukin-1 (IL-1) and interleukin-6 (IL-6), which activate lymphocytes and TNR-α to provide fever by affecting the thermoregulatory center in the hypothalamus. The inflammatory response is further continued by the production of complement protein in the liver. The hydrolytic enzymes also contained within the lysosome of macrophages are also secreted when the cells are activated, which in some cases contributes to tissue damage, like TNF-α which can kill a variety of cells.

Granulocytes

On the basis of cellular morphology and cytoplasmic staining characteristics granulocytes classified as Neutrophils, Eosinophils and Basophils are phagocytic cells, like macrophages. They are produced in the bone marrow by hematopoiesis and respond to many types of infections through circulation in the peripheral blood to affected tissues.

Neutrophils employ both oxygen dependent and oxygen independent pathways to generate antimicrobial substances. Fig. 2-6 shows the typical morphology of granulocytes, content of neutrophiles rich in reactive oxygen intermediate and reactive nitrogen intermediates.

Eosinophils are also phagocytic cells like neutrophils and can migrate to tissues through peripheral blood, and serve as defense against parasitic organisms. Basophils are non-phagocytic granulocytes that function by releasing pharmacologically active substances from their cytoplasmic granules and play a major role in certain allergic responses.

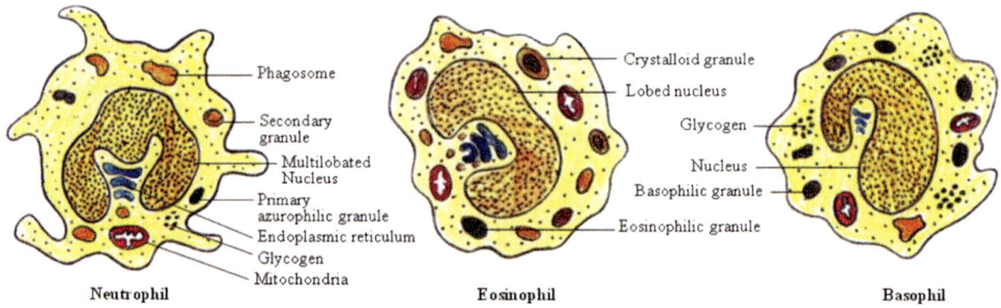

Fig.-2-6. Typical nuclear morphology of neutrophils (polymorphonuclear cells), eosinophils and basophils, and their cytoplasmic granules.

Mast Cells

Mast cells are found in wide variety of tissues, including the skin, connective tissues of various organs and mucosal epithelial tissue of the respiratory, gastro-urinary and digestive tracts, have large number of cytoplasmic granules that contain histamine and other pharmacologically active substances. They play an important role in the development of allergies.

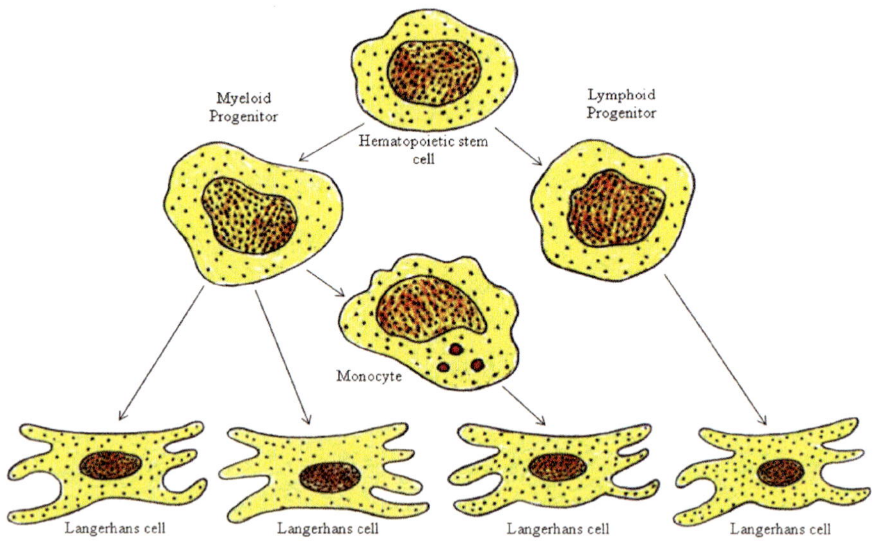

Fig.-2-7. Possible pathways of dendritic cell development. Langerhans cells, interstitial cells, and myeloid dendritic cells originated from a lymphoid precursor of hematopoietic stem cells, whereas lymphoid dendritic cells originated from a lymphoid precursor of hematopoietic stem cells.

Dendritic cell resembling the dendrites of nerve cells, have been isolated for in vitro study by employing the techniques to isolate enzyme and gentler dispersion. These cells have been classified by their location as follows;

- Circulating – consisting 0.1% of the blood leukocytes.
- Interstitial – existing in organs (heart, lungs, liver, kidney and gastrointestinal tract)
- Langerhans – found in skin and mucous membrane
- Interdigitating – populating T cells area of secondary lymphoid tissue and the thymic medulla.

Since these cells constitutively express high level of both class II MHC molecule members of the co-stimulatory B7 family are more potent antigen–processing cells than macrophages and B cell. The path taken by these cells in their development are still under investigation. However, two possible developmental pathways are shown in Fg. 2-7. Another type of dendritic cell, the follicular dendritic cells do not express class II MHC molecules and therefore do not function as antigen-presenting cells for T_H cell activation. The follicular dendritic cells have been reported to contribute to the development of memory B cells.

Organs of the Immune System

The organs and tissues of the body play an important role in the immune system, by their diverse functions. By their functions they can be distinguished as primary and secondary lymphoid organs. The lymphocytes circulate through these organs and in this process they trap the antigen as also provide sites for maturation of lymphocytes. These structurally and functionally diverse lymphoid organs and tissues are interconnected with the blood vessels so provide immunity to the entire organism. The thymus and bone marrow are the primary (or central) lymphoid organs where maturation of lymphocytes takes place. The mucosal-associated lymphoid tissues (MALT) and gut associated lymphoid tissues (GALT) – lymph nodes, spleen etc. are various secondary (or peripheral) lymphoid organs. Lymphocytes are capable of mounting an immune response only after they are matured in a central lymphoid organ. T cells mature in the thymus and B cells mature in the bone marrow. Fig.2-8 shows the human and murine lymphoid system.

Central lymphoid organs, as already stated are the thymus and bone marrow. Thymus- is a flat, bilobed organ situated above the heart. Each lobe is surrounded by a capsule and is divided into lobules. Each lobule consists of an outer compartment or cortex. The cortex is densely packed with immature T cells called thymocytes. The inner compartment or medulla is sparsely populated with thymocytes.

The process through which the immature T cells mature in the thymus is not authoritatively understood. However, it is held that after progenitor T cells formed in hematopoiesis enter the thymus and a rapid proliferation of thymocytes sets in and mature thymocytes migrate from cortex to medulla. The maturation thus continues and finally, the matured T cells leave the thymus through post-capillary venules. It has also been shown in some studies that a small subpopulation of cortical thymocytes leaves the thymus without entering the medulla.

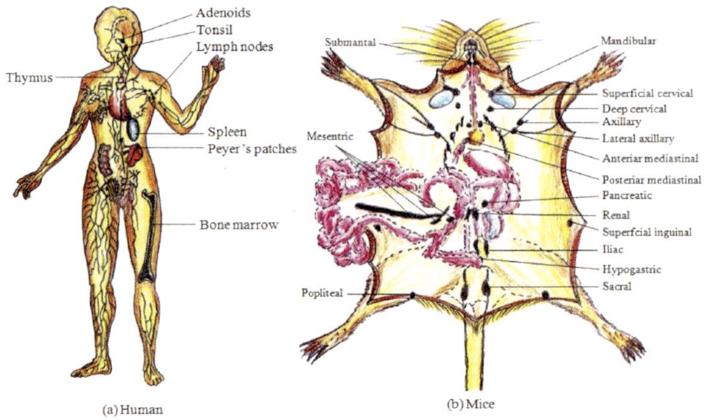

Fig.-2-8. Generalized diagrammatic representation of various organs of humans (a), and of mice (b).

The growth and maturation of thymocytes take place in a three-dimensional stromal cell network composed of epithelial cells, interdigitating dendritic cells, and macrophages. Fig. 2-9 shows the cells involved in the maturation of thymocytes. Any developing thymocyte is charged with a two-step selection process. If it is unable to recognize self-MHC molecule, is subjected to programmed cell death, and that which recognizes self-MHC molecule plus foreign antigen, is allowed to mature. The programmed cell death is initiated by an external agent or inability of a thymocyte to recognize self-MHC molecules as also an antigen, is subjected to further studies. However, it has been known for some time that cortical thymocytes, especially in rodents, begin to die by induction of endogenous glucocorticoids (or corticosteroids). In a study with human and mice, it has been found that cell-mediated immunity (as provided by matured thymocytes) is decreased by experimental removal of the thymus as also with aging. At the time of birth and in infants, the average weight of thymus is 70g but atrophy decreases its weight in elderly humans, making them more susceptible to infectious diseases, when the weight is only 3g. The Experiment has shown that mice who received a 33 month old thymus showed a significantly lower improvement in immune function than a mice who received the newborn thymus.

The prominent lymphoid organ – bone marrow, is the site for the maturation of B cells. Unlike birds, and rabbits that have a gut associated lymphoid tissue as the primary site of B cell maturation, there are regions of bone marrow in mammals, cattle and spleen for maturation of B cells. In the process of maturation, stromal cells within the bone marrow interact directly with the B cells and secrete various cytokines that are required for the maturation process. Like thymocytes, B cells also undergo a selection process, whereby they are eliminated with self-reactive antibody receptors. A detailed discussion of the contribution of bone marrow to the immune system has been made in a further chapter.

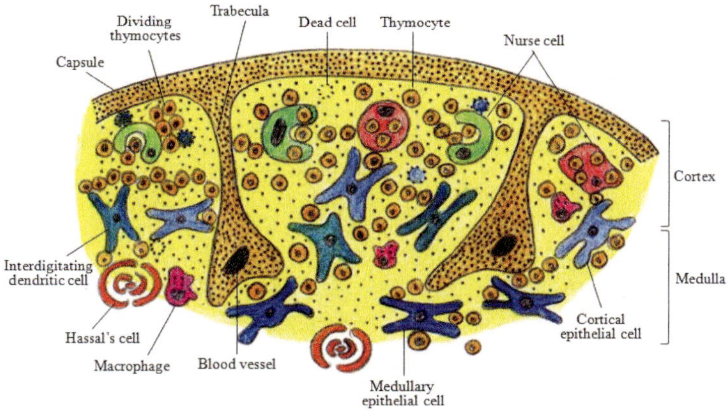

F-2-9. Diagrammatic cross-sectional view of thymus showing medulla and cortex. the presence of "Hassal's corpuscles" is the characteristic feature of the thymus, along with the macrophages, thymocytes and, some dead cells.

The fluid component portion of the blood is called plasma. Blood circulating under pressure seeps through the thin wall of the capillaries into the surrounding tissues. A major quantity of plasma returns to the blood through capillary membranes. This quantity of blood is known as interstitial fluid. This interstitial fluid does not completely exhaust on the circulation and the remaining interstitial fluid, called lymph, flow from the spaces in connective tissues into a network of larger collecting vessels called lymphatic vessels.

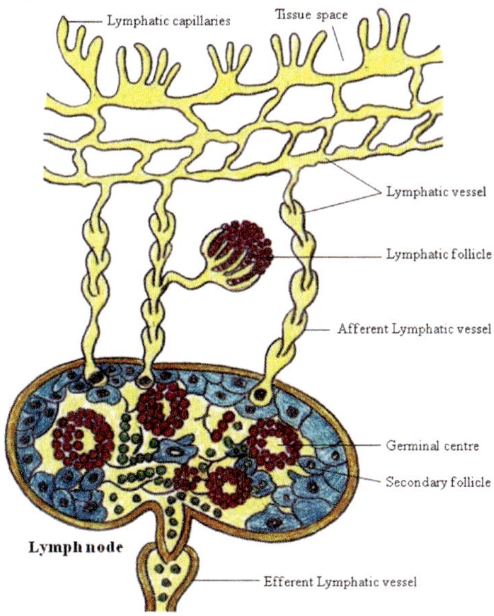

Fig.-2-10. Connection of lymphatic vessels or capillaries with intestinal tissues of lymph nodes.

The largest lymphatic vessels called the thoracic duct empties the blood into the left subclavian vein near the heart. This entire mechanism is called lymphatic system. This system ensures that the blood flows in only one direction. When a foreign antigen enters the tissues it is picked up by the lymphatic system (draining all the tissue of the body) and carried to various organized lymphoid tissues such as lymph nodes. The lymph nodes trap the foreign antigen and interact by activation. Fig. 2-10 is illustrative of the lymphatic system, as detailed above.

The peripheral (secondary) lymphoid organs contributing to the immune system are lymph nodes and spleen. The organized structure of lymphoid tissues is called lymphoid follicles. Lymphoid follicle until activated by an antigen is called primary follicles comprising a network of follicular dendritic cells and small resting B cells. After an antigenic challenge, a primary follicle becomes a large secondary follicle. A secondary follicle is a ring of concentrically packed B-lymphocytes surrounding a center called germinal center (fig 2-11). A dense mantle of small lymphocytes surrounds the germinal center. Antigen activated B cells have to undergo proliferation, mutation and selection in germinal centers. Lymph nodes are specialized to trap antigen from regional tissue spaces. The structure of a lymph node (a reticular network packed with lymphocytes, macrophages and dendritic cells) enable it to effectively encounter and respond to trapped antigens. A lymph node has three concentric regions–the outermost layer (cortex), the middle layer (paracortex), and the innermost layer (medulla).

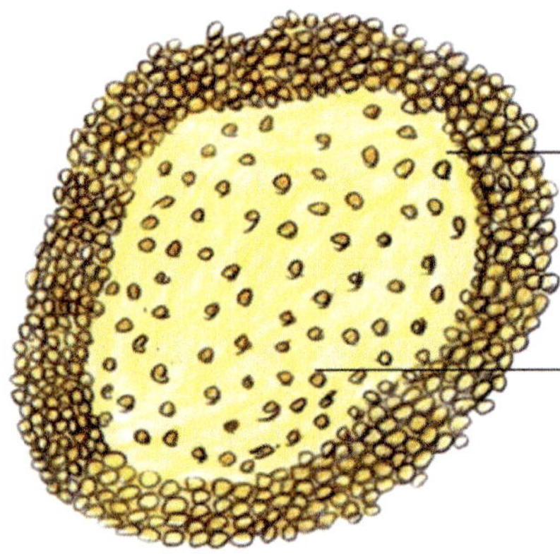

Fig-2-11. Structure of secondary lymphoid follicle consisting of the prominent germinal center surrounded by dense mental of lymphocytes.

Cortex - causes primary follicles to enlarge into secondary follicles (each containing a germinal center).

Paracortex - containing interdigitating dendritic cells present antigen to T_H cells

Medulla - secrete antibody molecules.

The interdigitating dendritic cells present in the paracortex, activate T_H cells which form small foci (consisting of proliferating B cells) at the edge of the paracortex. These foci reach maximum size within 4-6 days of infection (antigen piercing the capsule of lymph node). The medullary plasma cells have a higher concentration of lymphocytes than the afferent lymph. Due to this concentration of lymphocytes, antigens are migrated to medullary regions to be counteracted.

The spleen is specialized in filtering blood and trapping blood- borne antigens. It responds to systemic infection. Blood-borne antigens and lymphocytes are carried into the spleen through the splenic artery. The spleen has two divisions – the red pulp and the white pulp. The red pulp consisting of macrophages and numerous red blood cells (erythrocytes) is the site where old and defective red blood cells are destroyed and removed. The white pulp consists of T-lymphocytes. It also surrounds the branches of the splenic artery, forming a periarteriolar lymphoid sheath (PALS). There is a marginal zone in the periphery of PALS (fig.2-12). This zone is rich in B cells organized into primary lymphoid follicles. Upon antigenic challenge, these primary follicles develop into secondary follicles containing germinal centers (as in lymph nodes). Systemic infections – caused by *Streptococcus pneumoniae, Neisseria meningitides* and *Haemophilus influenzae* are the result of splenectomy.

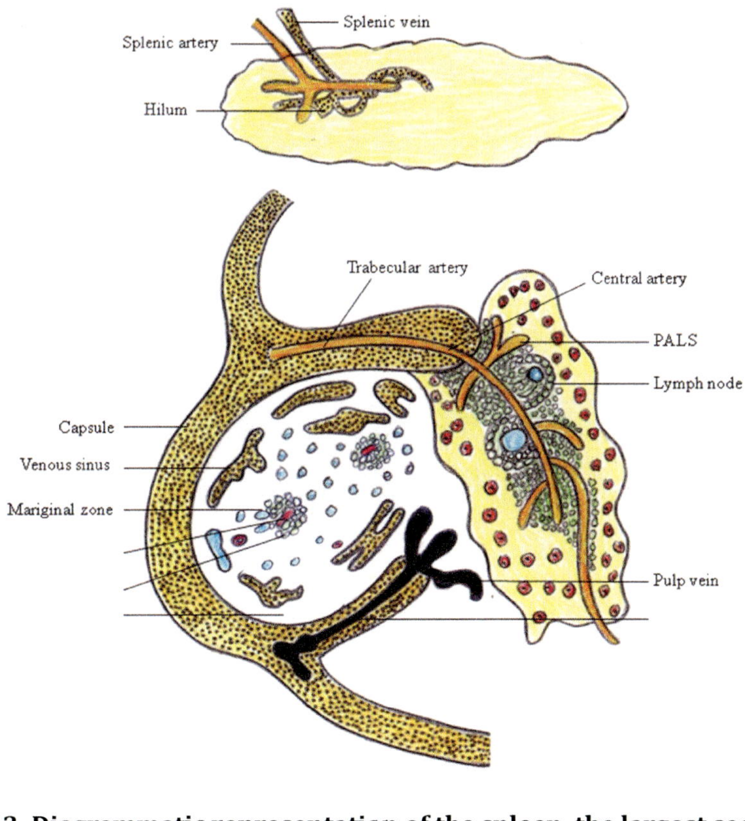

Fig.-2-12. Diagrammatic representation of the spleen, the largest secondary lymphoid organ, and its cross-section (lower panel) showing different internal structure.

MALT – A group of organized lymphoid tissues is known as MALT (mucosal-associated lymphoid tissue). Ranging from barely organized clusters of lymphoid cells in the lamina propria of intestinal villi to well organized structures such as tonsils, appendix, and Peyer's patch (a prominent focus lymphoid tissue embedded in the wall of the intestine). These tissues defend the body against infection with antibody plasma cells, exceeding the number produced in the spleen and bone marrow combined. The entry sites of pathogen being the mucous membrane lining the digestive, respiratory and urogenital system, the lymphoid cells found in these areas known and identified as mucosal-associated. Tonsils defend the body against antigen entering through the nasal and oral epithelial routes. There are three types of tonsils in oral cavity; palatine tonsils below the palate, lingual tonsils below the tongue and pharyngeal or adenoid tonsils in upper pharyngeal cavity. T cell activity is abundant in tonsils. Antigens making their way through intestinal routes are counteracted by population lymphoid cells, and antibody plasma cells. The sub mucosal layer under the lamina propria contains Peyer's patches, nodules of 30-40 lymphoid follicles, which develop into secondary follicles with germinal centers. The epithelial cells of mucous membranes deliver small samples of foreign antigen from the lumen of the respiratory, digestive and urogenital tracts to the underlying mucosal- associated lymph. The delivery is made by M-cells. M-cells are populated with a cluster of B cell, T cells and macrophages. Fig. 2-13 is amply illustrative of the situation of various potential layers in the intestinal lumen.

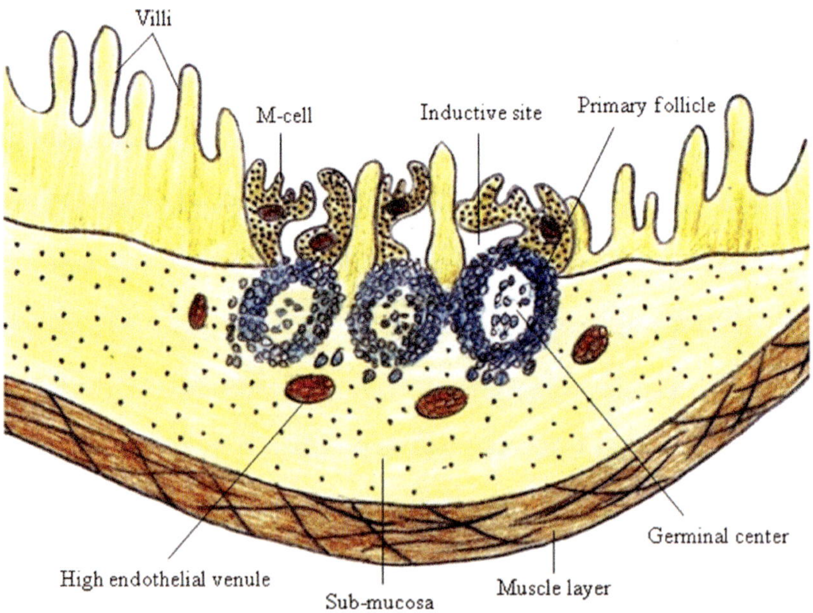

Fig.-2-13. Cross-sectional diagram of intestine through Peyer's patch showing loose clusters of lymphoid cells and diffused follicles. M-cells are the characteristic feature of mucus membrane at Peyer's patches that form an inductive site beneath.

Cutaneous Associated Lymphoid Tissues (CALT)

The skin acts as a tough barrier for the majority of potential pathogens and play are indispensible role in innate immune defense against infection. In addition to the anatomical/physical barriers to infection, the skin can also serve as an inductive site for the adaptive immune response. The outer epidermal layer of the skin is comprised of specialized epithelial cells known as keratinocytes. Keratinocytes can be induced to secrete a large number of different cytokines, and during intense inflammation can also be induced to express Class II MHC molecules and function as antigen-presenting cells.

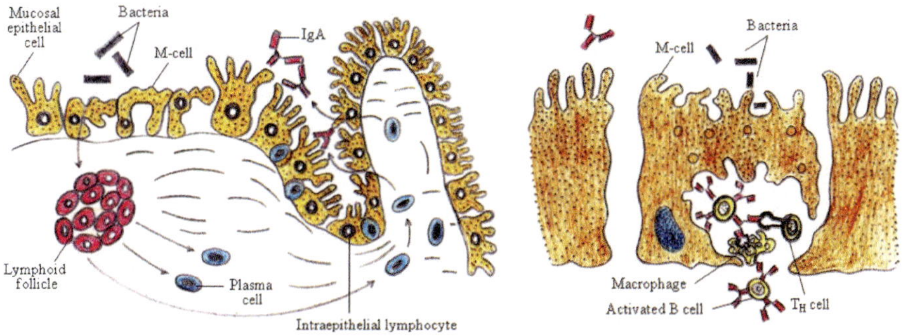

Fig.-2-14. A. Antigens are trapped by M-cells of the epithelial layer and activate B cells in lymphoid follicles at the inductive site. The activated B cells differentiate into plasma cells, which produce IgA that passes through and goes to the intestinal lumen and binds to the antigens; and B, showing close structure of M-cells with a distinct inductive site, which produces the IgA.

Fig.2-14 (a) and (b) is a sequel to the structures of routes of pathogen entry and activity of cells. Although the mucous membrane are potent barrier to the entry of pathogen, some as *Salmonella* species, *Vibrio cholerae* and Polio virus bind to the M-cell and disrupt the cell, enabling entry of pathogens. Another important barrier in the group of peripheral lymphoid tissues is those associated with epidermal layer of the skin. These tissues are known as cutaneous-associated lymphoid tissues.

The outer or epidermal layer of the skin is composed of specialized epithelial cells called Keratinocytes. These cells secrete cytokines, a potent antibody against antigens. Langerhans cells are scattered among the epithelial cells matrix of the epidermis expressing high level of class II MHC molecules, and function as activators of T cells.

The dermal layer of the skin contains scattered $CD4^+$ and $CD8^+$ T cells and macrophages, as strategy to combat pathogens. There are also numerous dendritic cells scattered throughout the skin tissue. Dendritic cells present in the skin are known as Langerhans Cells. The Langerhans cells can internalize antigens which enter into the skin. Upon activation, these cells then migrate via the efferent lymphatics and travel to the lymph nodes where they can interact with antigen-specific T lymphocytes.

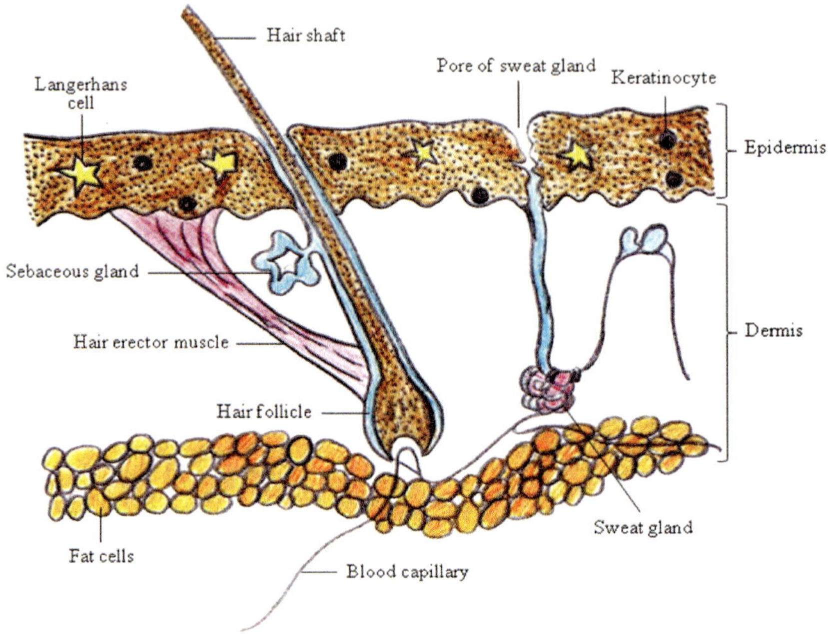

Fig-2-15. Cross sectional diagram of skin showing the distribution of Keratinocyte and Langerhans cells in the epidermal region.

There are also scattered lymphocytes in the skin (intraepidermal lymphocytes and dermal lymphocytes). The T lymphocytes present in the intra-epidermal and dermal region are mainly γδT cells and generally they are pre-activated or memory T cells.

CHAPTER 3

IMMUNOGEN AND IMMUNOGLOBULIN

A substance that induces a specific immune response is called an immunogen. An immunogen is usually called an antigen. In understanding the immune system, our main objective is to know the way in which the molecular properties of antigens give rise to the specific immune response. Immunogenicity is the ability to induce a humoral and/or cell mediated immune response. Antigenicity is the ability to combine specifically with the specific immune response. Thus, immunogenicity and antigenicity are related.

Therefore, an antigen is a molecule that is recognized by the immune system. It binds specifically to an antibody, or any molecule or molecular fragment that can be bound by a major histocompatibility complex (MHC) and presented to a T cell. In other words, antigen can also be defined as anything that can stimulate an immune response in the host body. They can be divided into "Self" and "Non-self" depending upon their origin. Antigens which are usually not identified or identified as its own are "Self" antigen where as, those which are which are identified as intruders and attacked by the immune system are called "Non-self" antigens. Similarly, an immunogen is defined as a substance that is able to provoke an adaptive immune response if injected on its own. In other words, immunogen is able to induce an immune response, while an antigen is able to combine with the products of an immune response once they are made. The overlapping concepts of immunogenicity and antigenicity are thereby subtly different.

According to new concept "Immunogenicity is the ability to induce a humoral and/or cell-mediated immune response" while, "Antigenicity is the ability to combine specifically with the final products of the immune response i.e. secreted antibodies and/or surface receptors on T cells. Although all molecules that have the property of immunogenicity also have the property of antigenicity, but the entire antigenic particle can not provoke immune response i.e they are not immunogenic.

Origin of Antigens

Exogenous Antigens

Exogenous antigens are antigens that have entered into the body from outside, for example by inhalation, ingestion, or injection. The immune system's response to exogenous antigens is often subclinical. By endocytosis or phagocytosis, exogenous antigens are taken into the antigen presenting cells (APCs), and processed into fragments. APCs then present the fragments to T helper cells ($CD4^{+}$) with the help of MHC class II molecules on their surface. Most of the T cells are MHC restricted and therefore recognize antigen only after binding with MHC molecules. After recognition, T cell become active and starts secreting various cytokines which in activate cytotoxic T lymphocytes (CTL), antibody-secreting B cells, macrophages, and other particles. Some exogenous antigens later become endogenous (for example, intracellular viruses) but upon the destruction of the infected cell, they released back into circulation and become exogenous antigens once again.

Endogenous Antigens

Endogenous antigens are the peptides or other aberated molecule generated in the normal cells during their abnormal physiology or because of viral or intracellular bacterial infection. These proteins are digested and processed into small presentable fragments which are then presented on the cell surface with the help of MHC class I molecules. The activated cytotoxic T cell recognize that protein as an antigen and start to secrete various toxins that cause the lysis or apoptosis of that infected cell. In order to keep the cytotoxic cells from killing the cells that present self-proteins, self-reactive T cells are deleted from the repertoire as a result of tolerance (also known as negative selection). Endogenous antigens include xenogenic (heterologous), autologous and idiotypic or allogenic (homologous) antigens.

Autoantigens

An autoantigen is usually a normal protein or complex of proteins that is recognized by the immune system of patients suffering from specific autoimmune diseases. These antigens should, under normal conditions, not be the target of the immune system, but, mainly due to genetic and environmental effect, the normal immunological tolerance for such antigen has been lost in these patients and their immune system recognizes as non-self antigen. Sometimes, DNA and RNA also act as autoantigen if they generated during an autoimmune diseases.

Tumor Antigens

Tumor antigen is a protein or any molecule produced during mutation or tumor development in a normal cell. These molecules or proteins which are not previously exposed to the immune system act as non-self and presented by MHC class I or MHC class II molecules on the surface of the cell to recognize by T cells. After recognition by T cell it triggers immune response in the host against that particular protein. Cytotoxic T lymphocytes that recognize antigens through MHC class I molecule may destroy the tumor cells before they proliferate or metastasize. Tumor antigens can also be present on the surface of the tumor in the form of a mutated receptor and recognized by B cells. Normal proteins in the body are not antigenic because of self-tolerance, a process in which self-reacting cytotoxic T lymphocytes (CTLs) and autoantibody-producing B lymphocytes are gather in thymus and bone marrow respectively. Sometime, tumor antigens are also called as neoantigens. These antigens are presented by tumor cells and never by the normal cells and therefore called as tumor-specific antigens (TSAs). Some important tumor antigens are alphafetoprotein (AFP), carcinoembryonic antigen (CEA), epithelial tumor antigen (ETA), Melanoma-associated antigen (MAGE).

Nativity

A *native antigen* is small antigen that is not yet processed by an antigen presenting cells (APC) into smaller fragments. A Native antigen cannot be recognized by T cells because T cell recognition is MHC restricted process and it require that these antigen must be processed into small fragment and can be bound and presented by MHC molecules. Though, native antigen could not be recognized by T cells, B cells recognize them and can be activated them.

Superantigens

Superantigens (SAgs) are a class of antigens including certain bacterial toxins which activates T cells that express a particular set of V beta T cell receptor genes resulting in polyclonal T cell activation and massive cytokine production. It is so called because of their ability to polyclonally activate large fractions (2-20%) of the T cell population at picomolar (ρM) concentration. The potent immunostimulatory properties of superantigens are a direct result of their simultaneous interaction with the V domain of the TCR and MHC class II molecules on the surface of an APC.

Kappler and Marrack and coworkers in 1989 coined the term 'superantigen' to describe massive T cell receptor (TCR) V-restricted primary response to bacterial toxin. It is broadly classified as exogenous (comprising bacterial toxins) and endogenous (comprising virally encoded) superantigens. Endogenous superantigens have not been identified so far in humans. The basic difference between a conventional antigen and a superantigen is that the former after antigen processing interacts with both Vα and Vβ regions of TCR while the latter bypasses the antigen processing mechanism and interacts only with TCR Vβ region. As a result, a specific antigen can stimulate only a small amount of T cells population (about 0.001%-0.0001%) while, a superantigen can stimulate a large clone of T cells. Therefore, the action of a specific antigen is MHC restricted and that of superantigen is not MHC restricted.

It has been postulated that it is the structure of a superantigen, which contributes to its biological properties. Superantigens are medium sized 22- 29kDa globular compact proteins. Structurally they all show conserved 2-domain architecture (N and C terminal domains and helix in the center of the molecule). The N terminal domain is characterized by presence of hydrophobic residues in solvent exposed regions and determines the lethality of the toxin. C terminal domain has four stranded sheet capped by the central helix and determines the superantigenicity of the toxin. Another common feature is the presence of disulphide loop formed by the cystinyl residues (except in TSST - I) in the N terminal domain, which is implicated in the emetic properties of the enterotoxins. It has been argued that these two fold have different ancestral origin and came together through gene recombination. It is highly resistant to proteases and to denaturation by heat.

Table 3-1: Superantigens in various diseases

Bacterial	*Viral*	*Fungal*
Pyrogenic Toxin Superantigens (PTSAgs)		
Toxic shock syndrome	Murine mammary tumor virus	Malassezia furfur (AD patients)
Staphylococcus enterotoxin	Rabbies nucleocapsid protein	
treptococcus endotoxin	Epstein Barr Virus	
Kawasaki syndrome	Cytomegalo virus superantigens	
Food poisoning	Retrovirus encoded superantigens	
Non-pyrogenic Toxin Superantigens (NPTSAgs)		
Mycoplasma arthritidis mitogen		
Yersenia enterocolitica superantigen		
Yersenia pseudotuberculosis superantigen		

Superantigens are known to play a role in allergy and autoimmune mechanisms. Moreover, superantigens results into apoptosis of vβ bearing T cells brought about by increased FAS expression. It has property of enhancing the endotoxin activity by 100,000 folds in rabbits. This enhancement causes the release of monokines like TNF, which contributes to endothelial toxicity. The endotoxin superantigen reduces the threshold of endotoxicity to an extent that circulation of endogenous endotoxins may be sufficient to contribute to shock and in addition, coexistence of gram positive and gram negative organisms results in shock in the absence of typical focus of bacterial infection. It can cause various malaise, detail of which is given in Table. However, recently its therapeutic implication has also been demonstrated. It has been found that genetically engineered Staphylococcal interotoxin A (SEA) fuse with a Fab fragment specific for tumour cell in the lung of rabbits is capable of activating about 10% of T cells, almost the strength of activation of a toxin superantigen against the tumor cells.

Factors Affecting Immunogenicity

Unless the immune system recognizes bacteria, bacterial products, fungi, parasites and viruses as immunogen, the system may not be responsive to immunogen or infectious agents. The immune system recognizes the particular macromolecules of an infectious agent, or as to the proteins or polysaccharides, lipids and nucleic acids, because those macromolecules are potent enough to serve as agent for inducing an immune response. Actually, the macromolecules are immunogen, but for this the macromolecules of an infectious agent must be combined and presented together with MHC molecules on the membrane of a cell to be recognized as immunogen. This holds good to both the immune responses i.e., humoral and cell mediated.

There are four factors that are responsible for producing immunogenicity by an immunogen, as follows;

- the degree of foreignness of an antigen into the body of an organism,
- the molecular mass of a substance to serve as an antigen/ immunogen,
- the four levels of protein organization-primary, secondary, tertiary and quaternary, of an immunogen,
- susceptibility to antigen processing and presentation with MHC molecules.

Summing up the above properties of immunogens to produce immunogenicity, it may be added that in order to elicit an immune response, a molecule must be recognized as non-self (foreign) by the biological system. If a molecule has not been exposed to immature lymphocytes, it may be later recognized as non-self or foreign. The foreignness of a molecule depends on the degree to which it has been exposed to immature lymphocytes. That is greater the phylogenic distance between two species greater the structural (and therefore the antigenic) disparity between them. For example, the BSA (Bovine Serum Albumin) would be more immunogenic into a rabbit than when injected into a cow, because of the disparity between these two species. Further, the molecular mass (size) of a macromolecule also determines its immunogenicity. Generally, substances with a molecular mass less than 5000-10000 Daltons (Da) are poor immunogens.

The foreignness and size of a molecule by themselves are not only responsible for immunogenicity, but the chemical complexity of a molecule also matters as an immunogenic factor. For example, the copolymer of sufficient size (mass) becomes highly immunogenic if they contain two or more different amino acids such as tyrosine or phenylalanine. That is to say that the immunogenicity of an immunogen is enhanced by all the four levels of protein organization – primary, secondary, tertiary and quaternary, as they contribute to the structural complexity of a protein and thus contribute to its immunogenicity (Fig 3-1). Macromolecules that cannot be degraded and presented with MHC molecules are poor immunogens, because the degradative enzymes within the antigen-presenting cells can degrade only proteins containing L-amino acids, polymer of D-amino acids cannot be processed. Larger molecules are more readily phagocytosed and processed, so they are more immunogenic than smaller molecules. However, phagocytosis may be facilitated by increasing the insolubility of macromolecules.

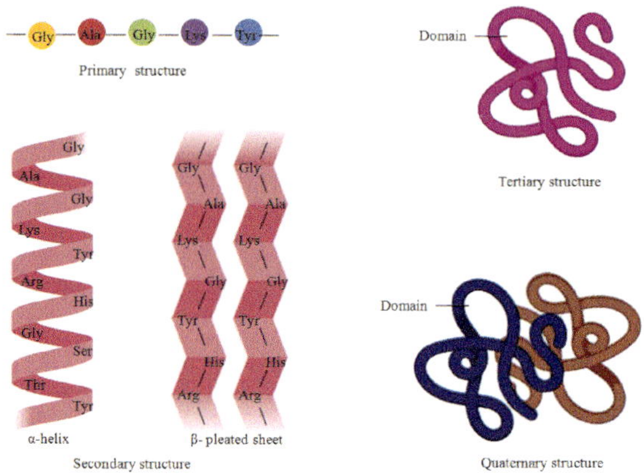

Fig.-3-1. Diagrammatic representation of four different levels of organization of protein molecules. Folding of polypeptide forms a distinct domain to bind with antigen/antibody.

Immune response depends on certain properties of the biological system, that the antigen encounters. These properties are as follows;

- enetic variability of the recipient organism,
- antigen dosage and route of immunization,
- administration of adjutants with the immunogen.

Both humoral and cell-mediated immune responses require interaction of T cells with antigens.

In this process the genetic constitution (genotype) of an immunized organism influences the type of immune response, the organism manifests as well as the degree of response. The immune response is controlled by the genetic constitution of the recipient organism. Experimental data indicate that MHC gene products, which function to present processed antigen to T cells, play a central role in determining the degree of immune responsiveness. Activation of lymphocytes depends on the dose and the route of administration of an immunogen. For an optimum immune response, the dose of the immunogen should be sufficient to activate lymphocytes, as low or as high doses cause lymphocyte to enter into a state of non-response. Experiments have also shown that more than one dose may be required to stimulate a strong immune response (known as booster). Booster increases the clonal proliferation of antigen-specific B- or T cells. In addition to dose, the route through which an immunogen is administered into the body counts for arousal of the immune response. The common routes are:

IV= Intravenous (into vein)

ID= Intradermal (into the skin)

IM= Intramuscular (into the muscle)

SC=Subcutaneous (beneath the skin)

IP=Intraperitoneal (into the peritoneal cavity), and now

IN = Intranasal

Out of the above, since an antigen carried through the virus reaches the spleen, the IV type of administration is more popular than other types. The administration of antigens by any of the above route is called parenteral (para = around, enteric = gut). Administration of adjuvant (helper) with the immunogens boosts the immune response in case of low immunogenicity of an immunogen, or when only small amount of an antigen are available. Adjutants are substances and they exert one or more of the following effects;

- prolong antigen persistence,
- enhance co-stimulatory signals,
- induce granuloma formation, and
- stimulate lymphocyte proliferation nonspecifically.

The common adjutants and their actions are as given below (refer Table 3-2)

- Alum (aluminum potassium sulfate) – prolongs the persistence of antigen (sustained release)
- Freund's water-in-oil adjutants –

(a) Freund's incomplete adjuvant – release antigen very slowly (contains antigens in aqueous solution of mineral oil as an demulsifying agent).

(b) Freund's complete adjuvant – slow release (contains heat killed mycobacterium).

The mode of action of antigen presentation with an adjuvant is to increase immunogenicity by sustained release.

- Synthetic polyribonucleotides and bacterial lipopolysaccharides – stimulate the non-specific proliferation of lymphocytes.

Table 3-2: Some commonly used adjuvants to enhance immune response.

Adjuvant	*Immune Response Observed*
Freund's incomplete adjuvant	+
Freund's complete adjuvant	++
Aluminum potassium Sulfate (alum)	+
Mycobacterium tuberculosis	+
Bordetella pertusis	+
Bacterial lypopolysccharides	++

Antigenic Determinants

In the process of enacting immunogenic activity, there are certain sites or region on an antigen that bind to antigen-specific membrane receptors on lymphocytes. These regions or sites on the immunogen molecule are called antigenic determinants or epitopes. Lymphocytes and a complex antigen may interact on several levels of antigen structures. All four elements of a protein may be involved in interaction with an epitope on a protein structure of an antigen whereas epitopes are faced by glucosidic bonds with polysaccharide structure of an antigen. It has also been revealed in the studies with small antigens that B- and T cells recognize different antigenic determinants on the same molecule. In Mice immunized with glucagons (a small human hormone of 29 amino acids), reaction to epitopes takes place in the amino- terminal portion of an antigen, but T cells react to epitopes in the carboxyl- terminal portion. However, B- and T cells cannot be considered to be functioning independently, since recognition of antigens involves combining of antigen with MHC. Fig. 3-1 and 3-2 are amply illustrative of this aspect of antigen binding versus recognition by lymphocytes.

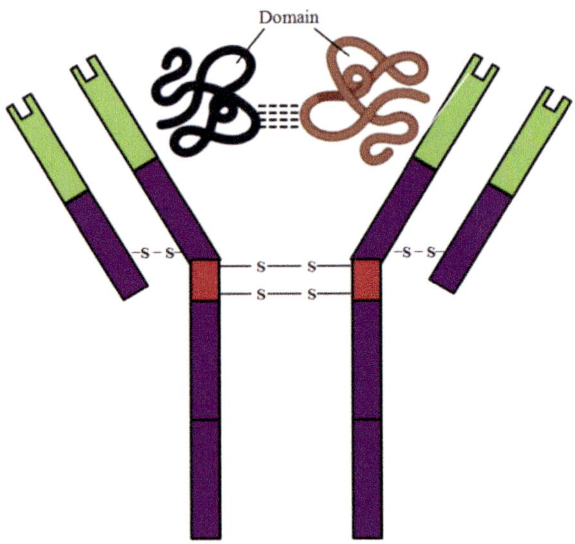

Fig.-3-2. Schematic representation of the interaction of antibody with the distinct binding domains (D) of antigen.

Nature of Antigenic Determinants or Epitopes

- ***Size*** – In general antigenic determinants are small and are limited to approximately 4-8 residues. Approximately, antigen with 4-8 amino acid residues is best suited for an antibody to bind and neutralize.
- ***Number*** – Although, each 4-8 residues can constitute a separate antigenic determinant, it is found that number of antigenic determinant is limited to those portions of the antigen that are accessible to antibodies.

Characterization of Epitopes Recognized by B-Cells

A comparison of antigen recognition by T cells and B cells has been shown in Table 3-3. B cells recognize soluble antigen, when it binds to their membrane bound antibody. T cells recognize only peptides combined with MHC molecules on the surface of antigen-presenting cells and altered self cells. This is the fundamental difference in the nature of recognition of antigen by B-cells and T cells.

Studies have further revealed the confirming properties of epitopes as they are recognized by B and T cells. The properties of epitopes to be recognized by B-cells are as follows;

- the antibody's binding site and the epitope must have complementary shapes, i.e., the size of the epitope recognized by a B cell cannot be larger than the size of the antibody's binding sites. This is essential because the shape (size) of the antigen is assumed by the sequences of chemical properties of an antigen in the binding sites and the environment that they produce. All eight amino acid residues of the octapeptide are in van der Waals contact with 14 residues of the antibody's groove. Protein antigens and peptide antigens interact with antibody in different ways. In the process of recognition, all antibody-binding sites are assembled from the same regions of the antibody molecule.

Table 3-3: Comparison of Antigen Recognition by T cells and B cells

Characteristic	*B cells*	*T cells*
Interaction with antigen	Involves membrane bound Ig and Ag	Involves T cell receptor, Ag, and MHC molecule
Binding of soluble antigen	Yes	No
Involvement of MHC Molecules	No	Required to display processed antigen
Chemical nature of Antigen	Protein, polysacch arides, lipid	Mostly proteins, but some lipids and glycolipids presented on MHC-like Molecules
Epitope properties	Accessible, hydrophilic,Mobile peptides containing Sequential or nonsequential Amino acids	Internal linear peptides produced by processing of antigen and bound to MHC molecules

- A B cell cannot recognize an antigenic determinant (epitope) unless the epitope is accessible to bind to an antibody. This is called topographical characteristic of an antigen–antibody complex analyzed to date. The interface between antibody and antigen has been shown to have numerous complementary protrusions and depressions (Fig. 3-3).

- protein B cell epitopes tends to be amino acid sequences that are hydrophilic and mobile. Epitopes may be composed of sequential contiguous residues along the polypeptides chain or non-sequential residues from segments of the chain brought together by the folded conformation of an antigen. Non-sequential epitopes are also called conformational determinants. Fig. 3-4 shows the residues widely spaced in the amino acid sequence but are brought into nearness by folding of the protein. Sequential and non-sequential epitopes generally behave differently when a protein is denatured, fragmented or reduced.

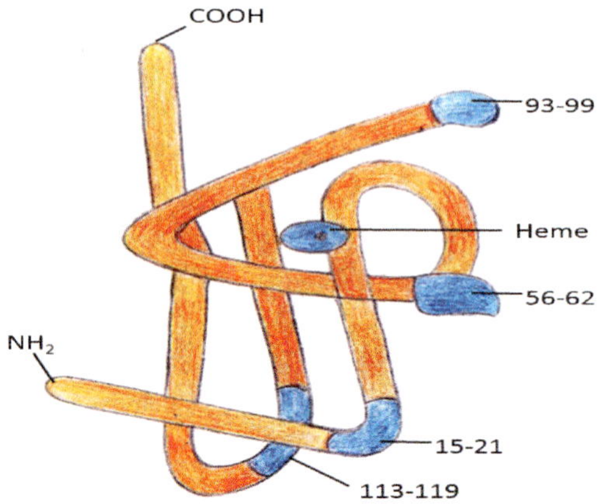

Fig-3-3. Binding domain or epitopes formed by several protrusion and depression of polypeptides chain as in myoglobin (Adapted from J. Klein. Immunology: The science of self-nonself discrimination. New York, John Wiley & Sons, 1982, p356) .

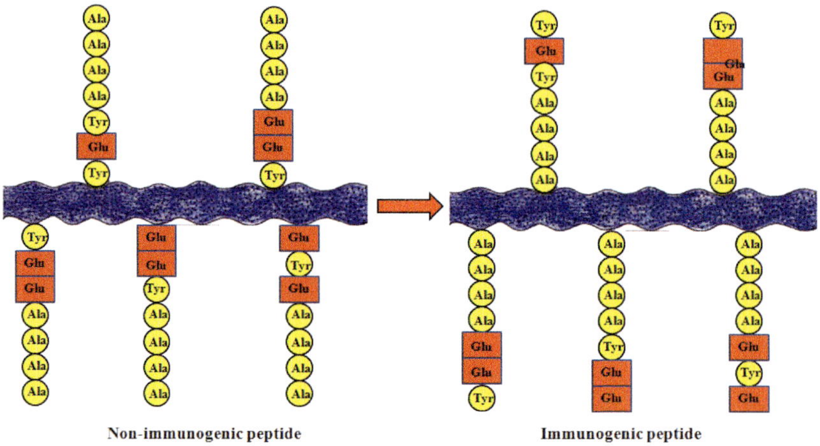

Fig.-3-4. Conversion of non-immunogenic polypeptide to immunogenic polypeptide by inversion of amino-acid sequences during protein folding.

- Antigenic determinants are located in the most mobile regions, permitting an antibody to react with it. However, because of the loss of entropy due to binding to a flexible site, the binding of antibody to a flexible epitope is generally of lower affinity than the binding of antibody to a rigid epitope.
- By comparing the antigen binding profiles of different monoclonal antibodies to various globular proteins, it has been observed that most of the surface of a globular protein is potentially antigenic. This surface is known as immunodominant, because they induce a more pronounced immune response than other epitopes in an organism.

Hapten

Haptens are small molecules which could never induce an immune response when administered by themselves, but when coupled with a carrier molecule such as protein, it become immunogenic. Once the body has generated antibodies against a hapten-carrier conjugate, hapten may also be able to bind to the antibody, but it will usually not initiate an immune response. Sometimes, hapten can block immune response to the hapten-carrier conjugate by preventing their binding to the antibody. A well-known example of a hapten is urushiol, which is a toxin found in poison ivy. It is absorbed by poison ivy plant through the skin and undergoes oxidation within the skin cells to generate the actual hapten, called a quinone. This hapten then reacts with skin proteins to form hapten-carrier conjugate. Usually, the first exposure of this conjugate causes sensitization and proliferation of T cells, but their second exposure activate them.

Some haptens can induce autoimmune disease due to the lack of immunogenicity. An example is hydralazine, a blood pressure-lowering drug that occasionally can produce drug-induced lupus erythematosus in certain individuals. Similarly, Halothane can cause a life-threatening hepatitis, whereas, penicillin-class drugs cause autoimmune hemolytic anemia.

Characterization of Epitopes to be Recognized by T cells

In a study held in 1959 by P.G.H. Gell and Baruj Benacerrat, it was found that a qualitative difference exists in the nature of response of T cells and B-cells to protein antigens. They found that a secondary response mediated by T cells was induced by denatured protein antigens even when the primary immunization had been with native protein. It was later established that T cells do not recognize soluble native antigen but rather recognize antigen that has been processed into antigenic peptides, presented in combination with MHC molecules. Proliferation follows in only those peptides that contain epitopes recognized by T cell receptors of responding T cell populations. Further studies have revealed the following properties of T cell epitopes;

- there are two distinct interaction sites for the antigens to be recognized by T cells; the epitope and the agretope. Epitope interacts with an MHC molecule (Fig. 3-5).
- Agretope shares certain structural features when binding to MHC molecules.
- Peptides interact specifically with MHC molecules. Different intracellular pathways process endogenous and exogenous antigens. This processing of antigens yields antigenic peptides that combine with class I or class II MHC molecules resulting in peptide-MHC complexes. The peptide-MHC complexes when presented on the cell surface are recognized by T cells. This process will be discussed later in detail.

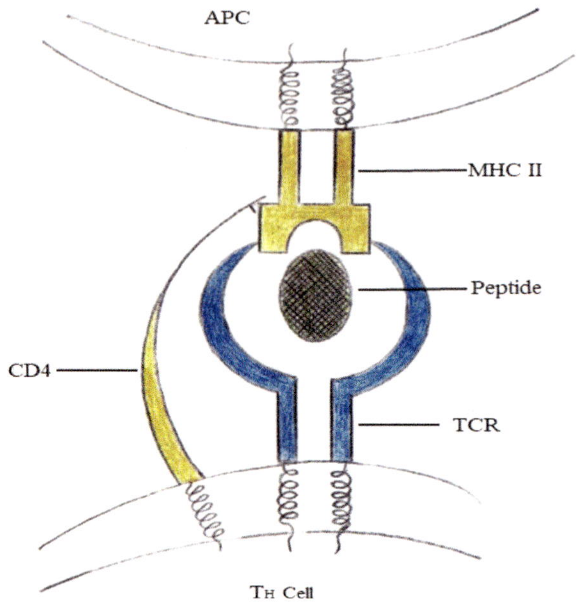

Fig.-3-5. Formation of complex between T cell receptor (TCR) on a T-helper cell and an antigen presented by MHC class II molecule on antigen presenting cells (APCs).

- T cell epitopes are found on the "inside" of the protein molecule. Epitopes are internal peptides.
- The T cell epitope that are immunodominant within an antigen are determined in part by the particular MHC molecules expressed by an individual.

Structural basis of antibody specificity

Haptens are small molecules that can bind to antibodies but cannot by themselves induce an immune response. To study the binding of an individual antibody to a unique epitope on a complex protein antigen, Karl Landsteiner (in 1920s and 1930s) employed various haptens in combination with a large protein (as carrier) and found that this combination works as determinant to the immune system. This experiment has led to the concept of structural basis of antibody specificity. Landsteiner in his studies with immunized rabbits found that an antihapten antibody could initiate a reaction with other haptens having a slightly different chemical structure. He called it as cross-reaction (antigen-antibody interactions). Hapten specific antibodies may be produced by the combination of hapten with large protein carriers and these specific antibodies may be employed to measure the presence of various substances in the body. The HCG (human chorionic gonadotropin) test of urine to determine pregnancy of a woman is an example of such an antihapten antibodies interaction. Likely the drug -protein combination in the body can be helpful to understand drug allergies. Antihapten antibodies can be used in assay for non-immunogenic biological substances, including drugs, peptides hormones and steroid hormones.

Immunoglobulins

It is well know that antigen binding proteins are present on the B cell membrane and are also secreted by plasma cells. Antibody circulating in the blood functions in the immune system to search out and neutralizes antigens or recognizing them for elimination. Immunoglobulins function as antibodies. The immune system usually responding by producing antibodies is adapted to antigenic determinants. For this purpose several clones of B cells are bound to single antigenic determinants. These monoclonal antibodies make up the polyclonal and heterogeneous serum antibody response to an immunizing antigen. These and various other factors of immunoglobulins functioning as antibodies are detailed in this chapter.

Basic Structure of Immunoglobulins

A. Tiselius and E. A. Kabat in 1939 demonstrated that γ globulin fraction are present in serum antibodies. They immunized rabbit with the albumin of egg white and then divided the immunized serum of rabbit into two aliquots. Electrophoresis of serum aliquot revealed four peaks that corresponds to albumin and alpha (α), beta (β) and gamma (γ) globulins. The other serum aliquot was reacted with ovaalbumin (albumin of egg whites) and the precipitate that formed was removed; the remaining serum proteins, which did not react with the antigen, were then electrophoresed. A comparison of these two electrophoresed serum aliquots shows that there was a noticeable drop in the γ globulin peak in the aliquot, which had been reacted antigen (Fig. 3-6). This γ-globulin fraction was identified as containing serum antibodies, which were called immunoglobulins. Further studies have confirmed that besides the main class of antibody molecules found in the γ-globulin fraction, considerable amount of antibody molecules are also present in the β and α fractions of serum, as other important classes of antibody molecules

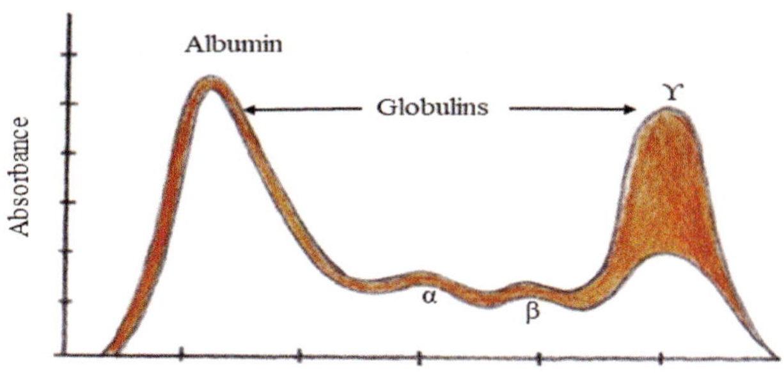

Fig. 3-6. Peaks of albumin, α, β, γ globulins of electrophoresed samples as obtained by A. Tiselius and E. A. Kabat in 1939 (with permission).

The structure of an antibody molecule is a mass of four common peptide chain. The basic structure of molecules consists of two identical chains of each light (L)

and heavy (H) polypeptides of molecular weight 25000 and so on respectively, linked by a disulphide bond. This polypeptide is further linked by non-covalent linkage as salt linkages, hydrogen bonds, and hydrophobic bonds. The interaction between the covalent and non-covalent linkage and the disulphide bond link the antibody molecules to form the basic four chain immunoglobulin structure, the dimer (H-L)2.

There are various amino-terminal regions of amino acids among antibodies of different specificity. The segments of highly variable sequence are called V-regions found in light and heavy chain both. The segments found in light chains are called VL and those found in heavy chains are called VH (Fig. 3-7). The specificity of these segments depends on the different content of amino acids sequences, and accordingly antibodies display different specificity. The antigen-binding site of the immunoglobulin molecule formed by both light chains and heavy chains correspond to the differences in regions. The area of these regions (V) is called complementarity-determining regions (CDRs). However, within the same immunoglobulin class for fewer differences are found, when sequences are compared with the rest of the molecule. Immunoglobulins are site for attachment of carbohydrates which affects the rate at which immunoglobulins are cleared from the serum by the liver. By this, the solubility of immunoglobulins is also increased.

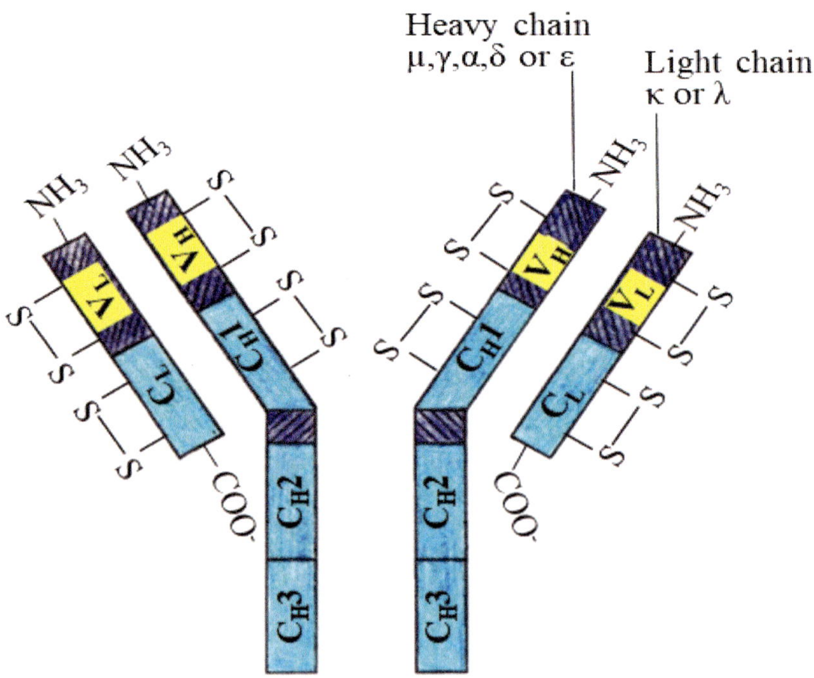

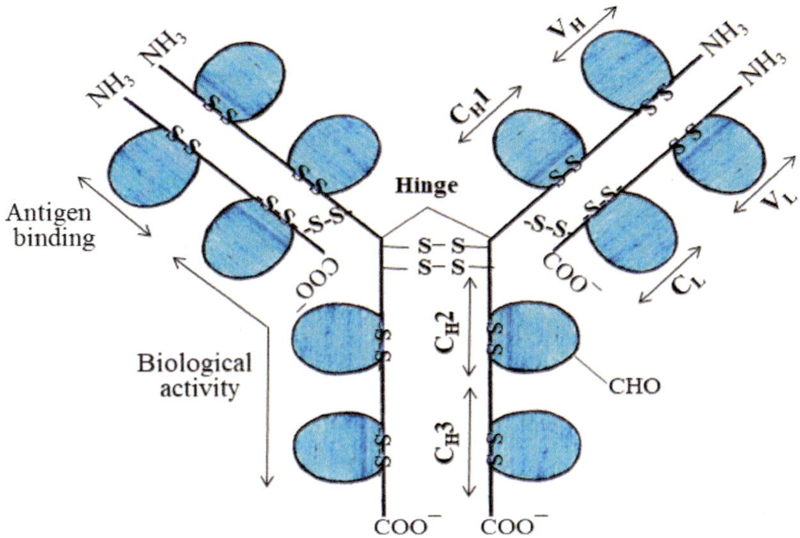

(Fig.-3-7. Structure of immunoglobulin. Diagram A depicts the two light chains and two heavy chains with constant and variable regions. Heavy chains represents some segments γ, δ, α and ζ, whereas light chain contains κ and λ. In diagram B, immunoglobulin molecule is divided into two segments. The anterior segments of immunoglobulin molecule are coupled with light and heavy chains, which are responsible for antigen binding, whereas posterior end contains only the heavy chains, which are responsible for biological activity. Looping in immunoglobulin molecule is formed by disulfide bonds.

Structure of Antibody

The basic structure of the immunoglobulin molecule was established in experimental observations made by Rodney Porter and Gerald Edelman who shared a Nobel Prize in 1972. The experimental approaches though were quite different, yet these complemented to each other. Porter cleaved the immunoglobulin molecule with enzymes to obtain fragments, where as Edelman dissociated the molecule by the interchain disulphide bond. They both studied the low molecular weight fraction, which contained a 150,000 Da molecular weight globulin designated ad immunoglobulin G (IgG). Papain digestion of IgG (brief fragment) produced two identical fragments (each with MW of 45000) called Fab fragments (fragment antigen binding) because they possessed antigen binding activity and one fragment (MW of 50,000) called the Fc fragment (fragment crystallisable) because of its ability of crystallizing during cold storage producing a single 100,000 MW fragments of two Fab.

Pepsin digestion (Alfred Nicozoff) precipitated antigen but Fc fragment were not recovered under this experiment. The IgG of MW of 150,000in the experiments carried out by Porter and Edelman, revealed that it was composed of two polypeptide chains, designated as Light (L) and Heavy (H) (Fig. 3-8). In his experiment, Porter using antisera from goats immunized with either the Fab fragment or Fc fragments of rabbit IgG found that antibody to the Fab fragment could reach with both the H and L chains, where as antibody to the Fc fragment reacted only with the H chain (see fig.4-3). Thus, it was experimentally established that the IgG molecule consist of two identical H and L chains linked by disulphide bonds. Separation of the individual heavy and light chain is allowed by mercaptoethanol reduction and alkylation.

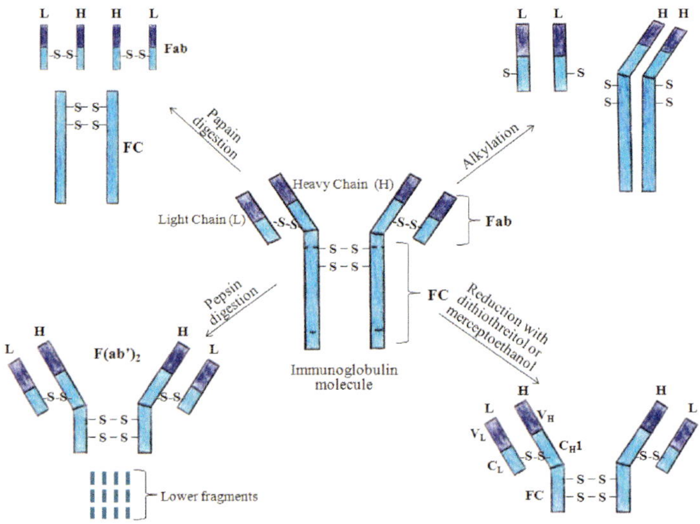

Fig.-3-8. Fragmentation of rabbit immunoglobulin by the enzyme papain and pepsin. It depicts that papain cuts the molecule into two parts, *Fab* and *Fc*, whereas pepsin cuts the lower fragment of immunoglobulin and produces *F(ab')*^{2} fragment.

Multiple Myeloma and Bence-Jones Proteins

The plasma cells in an individual are end-stage cells as they secrete specific antibodies for a few days and then die (end-stage). Cancerous plasma cells are not an end stage cell, since the cell divides over and over in an unregulated way without requiring any activation by antigen to induce clonal proliferation. Since cancerous plasma cell is called a Myeloma cell. Its protein synthesizing machinery and secretory functions go unabated and the cell continues to secrete specific antibody. The specific antibody is called Myeloma protein. Myeloma cells also secrete excessive amounts of light chains. Bence & Jones identified these chains in the urine of myeloma patients, and after them the chains are called Bence-Jones proteins.

Multiple myeloma develops of its own in an animal and humans or induced by injecting mineral oil in the peritoneal cavity. The clones of cancerous plasma cells are called plasmacytoma or MOPCs (mineral-oil induction of plasmacytoma cells).

Immunoglobulin Domains

An immunoglobulin molecule is rendered with organization of protein, so that it can specifically bind antigen and perform several biological effector functions. The organization may be termed as primary, secondary, tertiary and quaternary. Primarily the amino acids sequence accounts for the variable and constant region of the light and heavy chains. The secondary structure is formed by folding of the extended polypeptide chain back and fold upon itself into an antiparellel-β pleated sheet. In the tertiary structure, the chains are folded into compact globular domains, connected to neighboring domains by continuation of peptide chain lying outside the β-pleated sheets. An interaction of light chains puts the molecule to specific actions. The carboxyl terminal half of the molecule has two basic amino acid sequences, called the constant region with two light chain types, κ and λ. A single antibody molecule contains only one light chain type either κ or λ. On the basis of differences in amino acid sequences, λ light chains are classified into λ1, λ2 and λ3 in mice and λ1, λ2, λ3 and λ4 in humans. Similarly, the amino-terminal part of the heavy chain (consisting of 100-110 amino acids) showed great sequence variation among myeloma heavy chains and therefore called the variable (V) region. The remaining part of the protein revealed fine sequence patterns called constant (C) regions (μ, δ, γ, ε and α), each called an isotype. The length of the constant region is about 330 amino acids for δ, γ and α, and 440 amino acids for μ and ε. Table-3-4 details chain composition of the five immunoglobulin classes in humans.

Table 3-4: Types of chain contained in different immunoglobulin molecule

Class	*Heavy Chain*	*Sub class*	*Light chain*	*Molecular Configuration*
IgA	α	α1, α2	κ or λ	$(\alpha_2\kappa_2)_n$□□ $(\alpha_2\lambda_2)_n$
IgD	δ	None	κ or Λ	$\delta_2\kappa_2/\delta_2\lambda_2$
IgE	ε	None	κ or λ	$\varepsilon_2\kappa_2/\varepsilon_2\lambda_2$
IgG	γ	γ1, γ2, γ3, γ4	κ or λ	$\gamma_2\kappa_{2/}\gamma_2\lambda_2$
IgM	μ	None	κ or λ	(μ2κ2)n□ (μ2λ2)n

*where, n is 1 or 5.

**where, n is 1, 2, 3, or 4

Both the light and heavy chains contain several homologous units of about 110 amino acids residues. Each of the unit is called a domain. In each unit a loop of about 60 amino acids is formed by an intra chain disulphide bond. Light chains contain one variable domain (V_L) and one constant domain (C_L). Heavy chain contains one variable domain (V_H) and either three or four constant domain (C_H1), (C_H2), (C_H3) and (C_H4). The contents of heavy chain depend on the class of antibody (fig.2 b). Immunoglobulin domains are folded into a characteristics structure called immunoglobulin fold.

Amino acids sequences of V_L and V_H have been compared. This comparison shows that sequence variability is connected in some regions of V_L and V_H domains.

These regions are called hypervariable regions (HV). There are three such hypervariable regions in each heavy and light chain consisting 15% to 20% of the variable domain. Fewer variations are seen on the remainder of the V_L and V_H domains, called Framework regions (FRs). The antigen binding site of the antibody molecule is hypervariable region because of concentration of sequence variability of amino acids. The antigen binding site is complementary to the structure of epitope, so the hypervariable regions are also called complementarity determining factors (CDRs). CDR regions are located on the loop that connects the β strands of V_H and V_L. There are six hypervariable loops in each Fab fragment, scaffolded by framework region. For an overall antibody function a rigid framework is required. This is provided by the three dimensional structure of the immunoglobulin variable region. The antigen binding specificity is achieved by the three heavy chains and the three light chain CDRs. The analysis confirm that the V_H domain often contribute more to antigen binding than the V_L domain. This was further established in a study in which a single heavy chain specific for a glycoprotein antigen of the human immunodeficiency virus (HIV) was combined with various light chain of various antigenic specificities. It was found that all of the hybrid antibodies bound to the HIV glycoprotein antigen bound to heavy chain alone was capable to contribute specificity. However, with all these results, it should not be concluded that the light chain are irrelevant, rather in some antibody-antigen reaction the light chains render more specificity than the heavy chains.

Conformational Changes Induced by Antigen Binding

Successive x-ray crystallographic analysis of Fab fragments have revealed many facts making it clear that in some cases binding of antigen induced corresponding or conformational changes in the antibody, antigen both. Some such changes are listed below;

- A close fit between the epitopes and the antibody site results the formation of complex of antigen and antibody accompanied by a change in the epitope and the antigen binding site as likely in the complex formation between neuraminidase and anti-neuraminidase.
- Comparison of anti-hemagglutinin Fab fragment before and after binding to a hemagglutinin peptide antigen shows remarkable changes in the heavy chain CDRs loop and in the accessible surface of the binding site.
- Significant changes in the Fab upon binding of the antigen are visible in the light chain CDR1 (L1) and the heavy chain CDR3 (H3). In fact, upon antigen binding, the CDR1 region of the light chain moves as much as 1Ao and heavy chain CDR3 moves 2.7Ao. Thus, besides variability in the length and the amino acid composition of the CDR, loops enable antibodies to resume a shape more effectively complementary to that of their epitopes.

Hinge Region

There is another region in the γ, δ, and α heavy chains of extended peptide sequence between the CH1 and CH2 domains. This region has no homology with other domains and rich in proline residues being flexible. The flexibility of this region enables the two Fab arms to assume various angles to reach other to bind the antigen. This region is called the hinge region (Fig. 3-7). It also serves as a tether allowing the Fab components and the Fc to move relative to each other. As a result Fab arms move and twist to align the CDRs with epitope displayed on cell surfaces. Proline and cysteine are the two prominent amino acids found in the hinge region give it an extended polypeptide conformation. Enzymatic digestion of antibody occurs due to the presence of proline in the hinge region. The two heavy chains are held together by the disulphide bonds made from cysteine residues.

The hinge region of IgG, IgD and IgA gives segmental flexibility to the heavy chains having three constant domain regions. The heavy chains in IgE and IgM lack hinge region, but have four constant region domains- C_H1/C_H1, C_H2/C_H2, C_H3/C_H3 and C_H4/C_H4. The two domains of IgE and IgM are much more accessible than the others to the aqueous environment. Complements activated by IgG and IgM are the result of this accessibility.

The J-chain

The joining (J) chain is a small 137 amino acid long polypeptide secreted by mucosal and glandular plasma cells. It was first discovered as a vital component of IgA in human and rabbit colostrums, and later as a component of IgM. The J-chain shows large degree of structural and functional homology between avian and human species. It is made up of two β-pleated sheets of four ribbons folded against each other. The cystine residues present in the j-chain effectively serving as the "glue" between two Fc regions of the antibody. The J Chain present in IgA and IgM is required for secretion in to mucosa. Initially, it was hypothesized that IgM and IgA are the only two types of antibody that polymerize due to the presence of J chain. Subsequently, it was found that IgA and IgM polymerized in the absence of J chain, however, organisms lacking J chains showed lower number of IgA and IgM. The J-chain present in IgA and IgM adds several salient features like increases the number of antigen binding site which make them suitable for agglutinating bacteria and viruses, and provide flexibility to operate in a non-inflammatory fashion due to little or no complement-activating potential. Furthermore, present of J-chain increases the affinity of IgA and IgM against polymeric Ig receptor (pIgR), also known as transmembrane secretory component (SC) to form secretory complex by interacting with the receptor protein. The secreted antibodies form the 'first line' of defense against foreign particles and block any harmful substances that favor the entry of pathogen through mucosa.

The Carboxyl-Terminal Domains

B cells express different classes of immunoglobulin at different stages and stimulatory conditions. Five classes of immunoglobulins (IgG, IgD, IgA, IgM and IgE) and their sub-classes can be expressed as either membrane bound or secretory immunoglobulins. The structure and function of carboxyl-terminal differ in secreted immunoglobulin (sIg) and membrane bound immunoglobulin (mIg)

(Fig). The carboxyl terminus of C_H domain in secreted immunoglobulins have a hydrophilic amino acid sequences of different length, whereas in membrane bound immunoglobulins, carboxyl-terminal domain have three different regions:

- A short cytoplasmic cell
- An extracellular hydrophilic "spacer" sequence consisting 26 amino acid residues, and
- A hydrophobic transmembrane sequence.

A detailed discussion about the functions of five immunoglobulins secreted by B cell follows in a separate chapter. It may however be noted here that even when different classes of immunoglobulins are expressed sequentially on a single cell, the antigenic specificity of all the membrane bound antibodies remain same and binds to the same epitope.

Effector Functions of Immunoglobulins

Various effector functions of antibodies (role of antibodies to invoke responses to remove the antigen and kill pathogens) are a result of interactions between heavy chain-constant regions and other serum proteins or cell membrane receptors. In this all the classes of immunoglobulins have different effects on the antigen in so far as the binding and its elimination is concerned. All the effector functions are mediated by the domains of the constant region. There are three immunoglobulin mediated effector functions:

- Opsonisation
- Complement activation, and
- Transcytosis (ADCC)

Opsonisation – promotes antigen phagocytosis by macrophages and neutrophils. IgG molecules are present on the surface of macrophages and neutrophils. The interaction between a single protein molecule (Fc receptor-FcR) and the IgG molecule, if at the same time are bound by several antibody molecules to the same target, resulting in the phagocytosis of the antigen-antibody complex. The pathogen inside the phagocyte turns a target to the various damaging processes like enzyme digestion, oxidative damage, and the membrane-damaging effects of antimicrobial peptides. Thus, in opsonization, all these phagocytic process leads to the damage of the pathogen because of recognition.

Complement Activation – In it, protein fragment binds non-specifically to cell-and antigen-antibody complexes near the site of activation. The activation is that of serum glycoproteins, so the process is called the complement system or complement activation. Complement induces a collection of proteins that can perforate cell membranes. In complement activation, the protein fragment called C3b is an important pathway. Red blood cells and macrophages have receptors for C3b and so bind cells or complexes to which C3b has adhered. Binding of adherent C3b by macrophages leads to phagocytosis of the cells or molecular complexes attached to C3b. Binding of antigen-antibody complexes by the C3b receptors of a red blood cell allows the erythrocyte to deliver the complexes to liver or spleen, where resident macrophages remove them without destroying the red cells. This collaboratory function between the antibody and the complement system causes inactivation or removal of antigens and killing of pathogens.

ADCC (Transcytosis) – Where a cell-mediated phagocytosis is expressed by movement or transport of antibody to mucosal surface, it is called transcytosis. Antibody-dependent cell-mediated cytotoxicity (ADCC) can cause NK-cell-mediated death of target cells when antibody bound to the target cells association with the Fc receptor of the NK cells. The immunoglobulins- IgM, IgA, and IgG are some important antibodies that undergo such transport and thereby causing transcytosis. The transport of IgG class immunoglobulins is responsible for immunization of fetus across the placenta tissue that separates mother and fetus. The fetus receives a sample of mother's repertoire of antibody as protective endowment against pathogens. This transfer is known as passive immunization (acquisition of immunity by receipt of performed antibodies rather than by active production of antibodies after exposure to antigens). These days, passive immunization is in vogue in treating immunodeficiency diseases and as a protection against the anticipated exposure to infectious agents against which one lacks immunity.

Classes of Immunoglobulins

There are five classes of immunoglobulin; immunoglobulin G (IgG), immunoglobulin M (IgM), immunoglobulin A (IgA), immunoglobulin E (IgE), and immunoglobulin D (IgD). All these differ in their ability to carry out various effector functions in their average serum concentration and in half life (Table 3-5). A brief description of the above antibody classes are as follows;

Table 3-5: Properties of different classes of human serum immunoglobulin molecule

Property/Activity	*IgA1*	*IgA2*	*IgD*	*IgE*	*IgG1*	*IgG2*	*IgG3*	*IgG4*	*IgM*
Molecular weight	150,000-600,000	150,000-600,000	150,000	190,000	150,000	150,000	150,000	150,000	900,000
Normal Serum level (mg/ml)	α1	α2	δ	ε	γ1	γ2	γ3	γ4	μ
Intervention to complemnt system	3.0	0.5	0.03	0.0003	9	3	1	0.5	1.5
Crosses to placenta	-	-	-	-	+	+/-	++	-	+++
Fc binding on Phagocytes	-	-	-	-	+	+/-	+	-	-
Mucosal transport	++	++	-	-	-	-	-	-	+
Effect on mast Cells	-	-	-	+	-	-	-	-	-

Immunoglobulin G (IgG)

Consisting of two γ heavy chains and two κ or λ light chains (Fig. 3-9 a) .There are four subclass of this antibody depending upon their differences in γ-chain sequence and average serum concentration- IgG1, IgG2, IgG3, IgG4 (Table 3-3). Owing to the size of the hinge region and the number and position of the inter-chain disulfide bonds between the heavy chains these subclasses are of different structures. The biological activity of the molecule differs in the sub-classes, according to the suitable amino acid differences. The biological activities of different subclasses are;

- Developing fetus is protected by IgG1, IgG3 and IgG4.
- Complement activation is mediated by IgG3 and to a little extent by IgG1 and IgG2
- Opsonization is mediated by IgG1 and IgG3. IgG4 has an intermediate affinity in binding on phagocytic cells.

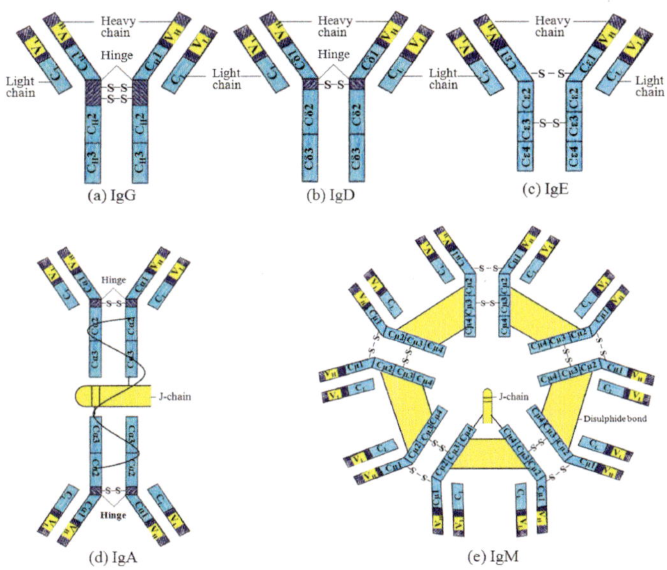

Fig-3-9. General structures of the five major classes of secreted antibody. IgA secreted in dimmer form attached with J-chain and IgM is secreted in pentameric form attached with di-sulfide bonds and a J-chain.

Immunoglobulin M (IgM)

The total serum concentration of this subclass antibody is 1.5 mg/ml. This is a membrane bound antibody on B cells. IgM is secreted by plasma cells as a pentamer. In the pentamer, five monomer units are held together by disulfide bonds arranged with their Fc region in the center of the pentamer and the ten antigen-binding sites on the periphery of the molecule. Prior to secretion of the pentamer IgM, the monomeric units get subjected to polymerization.

IgM is specific immunoglobulin class with 10 antigen-binding sites which provides it a higher valency molecule than other isotopes. It is the first antibody synthesized by the neonate. The higher valency enables the IgM to bind with multidimensional antigens such as viral particles and RBCs. Because of the pentameric structure, a single molecule of IgM is capable of activating the Fc regions unlike immunoglobulin.

Immunoglobulin A (IgA)

This is a prominent immunoglobulin found in external secretion such as saliva, tears, breast milk and mucous of the other bronchial, genitourinary and digestive tracts. Polymeric forms are also seen. The secretory component is a 70,000 D MW polypeptide produced by epithelial cell of mucous membranes. IgA is a secreted up to 5-15g daily in humans. Fig.3-10 shows the body system on which IgA exerts effector activity. On mucosal epithelia (e.g. the lining of the digestive, respiratory and genital tracts) and on glandular epithelia (e.g. mammary, saliva and lachrymal glands), this IgA immunoglobulin is profusely expressed. It prevents the attachment of the pathogens to the mucosal pathogens. Newborn are protected against infections during the first few months of life because of presence of this antibody (IgA) in breast milk and that is why the breast feeding is highly recommended to keep the newborn healthy in early months of life.

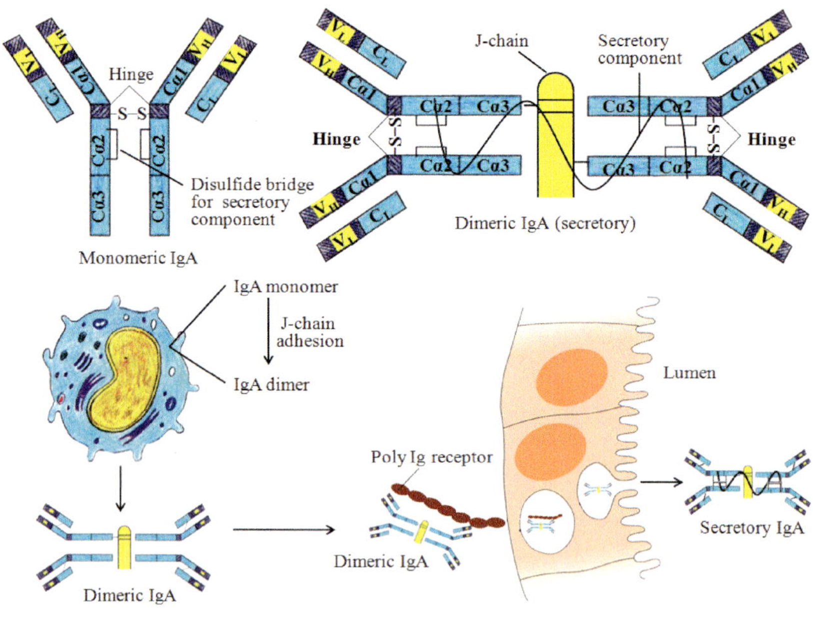

Fig.-3-10. (a) Structure of the secretory IgA with two monomers attached with J-chain and a secretory component. (b) IgA is secreted from the plasma cell and attached to poly-Ig receptor on mucosal membrane epithelial cells. After internalization of dimeric IgA with poly-Ig receptors, enzymatic cleavage takes place, and then secretory IgA is released into the lumen.

The J-chain

The joining (J) chain is a small 137 amino acid long polypeptide secreted by mucosal and glandular plasma cells. It was first discovered as a vital component of IgA in human and rabbit colostrums, and later as a component of IgM. The J-chain shows large degree of structural and functional homology between avian and human species. It is made up of two β-pleated sheets of four ribbons folded against each other. The cystine residues present in the j-chain effectively serving as the "glue" between two Fc regions of the antibody. The J Chain present in IgA and IgM is required for secretion in to mucosa. Initially, it was hypothesized that IgM and IgA are the only two types of antibody that polymerize due to the presence of J chain. Subsequently, it was found that IgA and IgM polymerized in the absence of J chain, however, organisms lacking J chains showed lower number of IgA and IgM. The J-chain present in IgA and IgM adds several salient features like increases the number of antigen binding site which make them suitable for agglutinating bacteria and viruses, and provide flexibility to operate in a non-inflammatory fashion due to little or no complement-activating potential. Furthermore, present of J-chain increases the affinity of IgA and IgM against polymeric Ig receptor (pIgR), also known as transmembrane secretory component (SC) to form secretory complex by interacting with the receptor protein. The secreted antibodies form the 'first line' of defense against foreign particles and block any harmful substances that favor the entry of pathogen through mucosa.

Immunoglobulin E (IgE)

IgE is secreted in extremely low concentrations (0.3μg/ml) and mediates allergic reactions. The first biological assay for the activity of this antibody was demonstrated by K. Prausnitz and H. Kustner who injected serum from an allergic person into a non-allergic person intradermally. On injection of an appropriate antigen at the same site of skin, a wheal and flare reaction developed. The reaction was known as P-K reaction (after Prausnitz and Kustner); late established by K. and T. Ishisaka. Anti-isotype serum antiserum was prepared by immunizing rabbits from serum obtained from an allergic person. The rabbit antiserum was then allowed to react with each class of human antibody (i.e., IgG, IgA, IgM, IgD) known then. Each or the above antibody was precipitated and removed from the rabbit antiserum. What remained was a specific unidentified class of anti-isotype antibody, which block the P-K reaction. This new antibody was in reference to the E antigen of ragweed pollen, called IgE. This antibody is a potent antiparasitic agent.

Immunoglobulin D (IgD)

IgD is secreted at low level in normal human serum. Failure of myeloma protein to react with anti-isotype antisera, the then known isotypes; IgA, IgM, and IgG led to the discovery of this class of antibody. It is the major membrane bound immunoglobulin expressed by mature B cells. The Biological effector function of this class of immunoglobulin is yet to be established (Fig. 3-11).

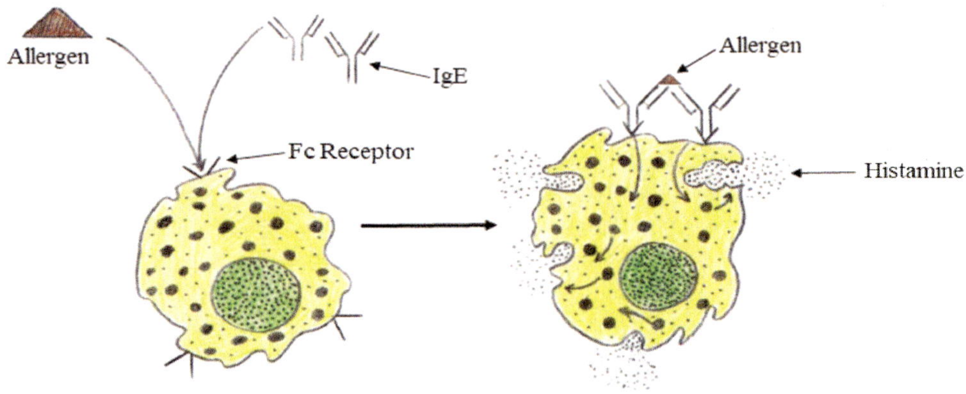

Fig.-3-11. Allergen attaches to Fc receptor, which is present in the mast cell. After binding, mast cells get activated and degranulation takes place. Degranulation leads to the release of histamine that mediates the allergic reaction.

Antigenic Determinants on Immunoglobulin

There are three types of antigenic determinants or epitope located in characteristic position of the immunoglobulin molecules, isotypic, allotypic, and idiotypic.

Isotypic determinants are constant region determinants passing through all members of a species. Thus, within a species each normal person will express all isotopes in the serum. That is why the different species expres different constant-region genes (inheritance), and the infusion of an antibody into different species is recognized as foreign.

Allotypic determinants are multiple alleles. Multiple alleles are encoded subtle amino acid deficiencies which occur in some not all the members of a species. In humans, allotypes have been characterized for all four IgG subclasses, for one IgA subclass and for the κ-light chain. The γ-chain allotypes are reported to as Gm markers. By the number of alleles Gm marker are designated into subclasses G1m (1), G2m (23), G3m (11), G4m (4a). Antibodies to allotypic determinants can be developed during pregnancy and blood transfusion.

Idiotypic determinants are produced by this confirmation of the heavy and light chain variable region. Each individual antigenic determinant of the variable region is called an idiotype (fig.4-17c). There are multiple idiotypes in each antibody. The idiotype of the antibody is the sum total of the individual idiotypes. An idiotype may be the antigen binding site or may consist of variable region sequences outside the antigen binding site. Antibodies having minimal variation in their isotype and allotype help to determine the idiotypic difference, and anti-idiotype antibody may be produced by injecting such antibodies into a syngenic recipient.

B Cell Receptor

How does a membrane-bound immunoglobulin mediates an activating signal after contact with an antigen? This feature of immunoglobulin has ever been a mystery for immunologists. The state of affair is that antigen binding receptor is complexes of membrane bound immunoglobulin (mIg) and disulfide linked heterodimer called Ig-α/Ig-β. The tail in both Ig-α/Ig-β are long enough to interact with intercellular signaling molecules. B cell activation has been discussed in detail in a subsequent chapter (Fig. 3-12).

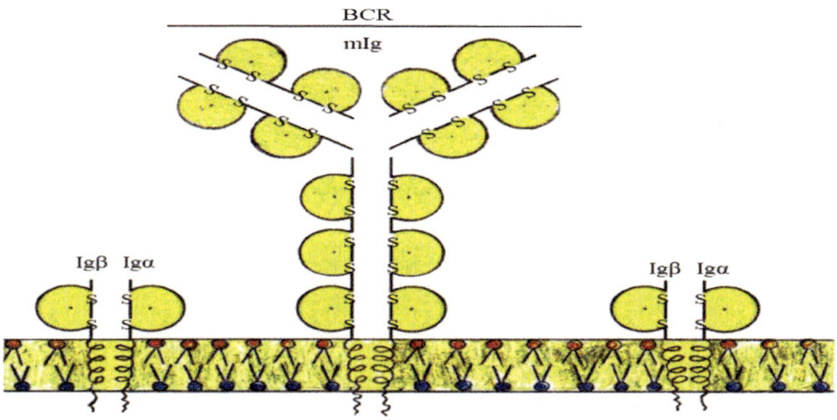

Fig.-3-12. Generalized structure of B cell receptor (BCR). BCR is composed of membrane bound immunoglobulin (mIg) and two heterodimers Igα/Igβ.

The Immunoglobulin Super Super Family

Various immunoglobulin heavy chain light chains have a common evolutionary ancestry. The genes encoding immunoglobulin heavy and light chains arise from a common gene existing from an earlier stage. Later gene duplication and divergence is a function of the ancestral heavy and light chain genes (Fig. 3-13). The domain structure of immunoglobulin is suggestion of it. Super family is a term which is used to denote proteins whose corresponding genes are derived from a common pre-existing gene encoding the domain structure. Immunoglobulin super family has a number of proteins viz.,

- T cell receptor,
- Class-I and class-II MHC molecules,
- T cell accessory proteins, including CD2, CD4, CD8, CD28 and the γ, δ and ε chains of CD3,
- Various cell adhesion molecules, encoding VCAM-1, ICAM-1, ICAM-2 and LFA-3,
- Platelet-derived growth factor,
- Poly-Ig receptor which contributes the secretory component to secretory IgA and IgM,

- Part of the B cell receptor, Ig-α/Ig-β,
- β-micro globulin, and
- Other proteins

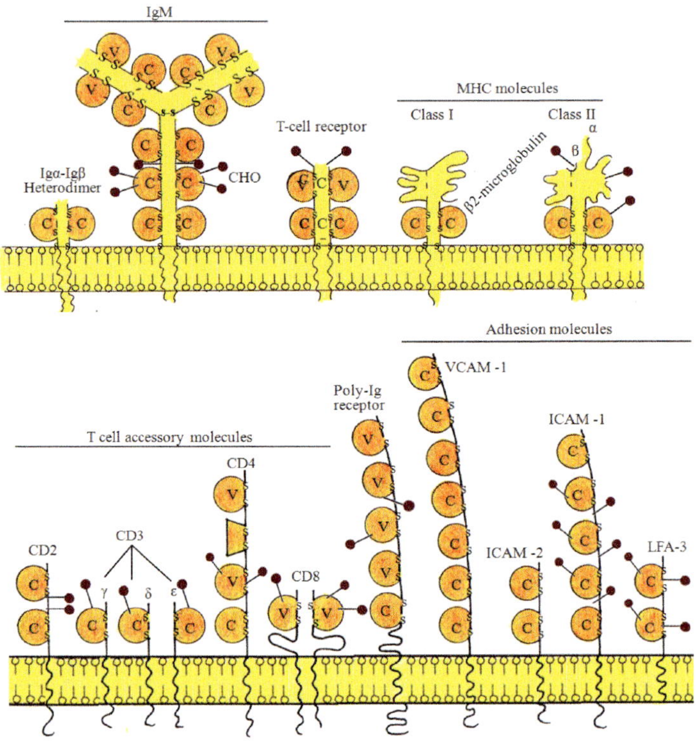

Fig.-3-13. Schematic representation of members of Ig super family, which are a group of structurally related membrane bound glycoproteins.

The primary amino acid sequence of these proteins suggest that all contain immunoglobulin fold domains containing 110 amino acids arranged in pleated sheet of anti-parallel β-strands and invariant interaction disulfide bond spanning 50-70 residues. The folds keep interactions between proteins.

Monoclonal Antibodies

The B cell clones are produced by proliferation and differentiation (induces by multiple epitopes) being each specific for one epitope. This is called polyclonal antibody response, causing the location, phagocytosis and complement mediated lysis of antigen. The efficacy of an antiserum for various *in vitro* uses is reduced for diagnostic and therapeutic purposes. Monoclonal antibody derived from a single clone specific for a single epitope is preferable. The method to prepare monoclonal antibody was devised by George Kohler and Ceaser Milstein in 1975. They fused a normal activated, antibody producing B cell with a myeloma cell (a cancerous plasma cell) and generated a hybrid cell called hybridoma. The hybrid cell possessed the immortal growth properties of the mother cell (myeloma cell) and secreted the antibody produced by the B cell **(Fig. 3-14).**

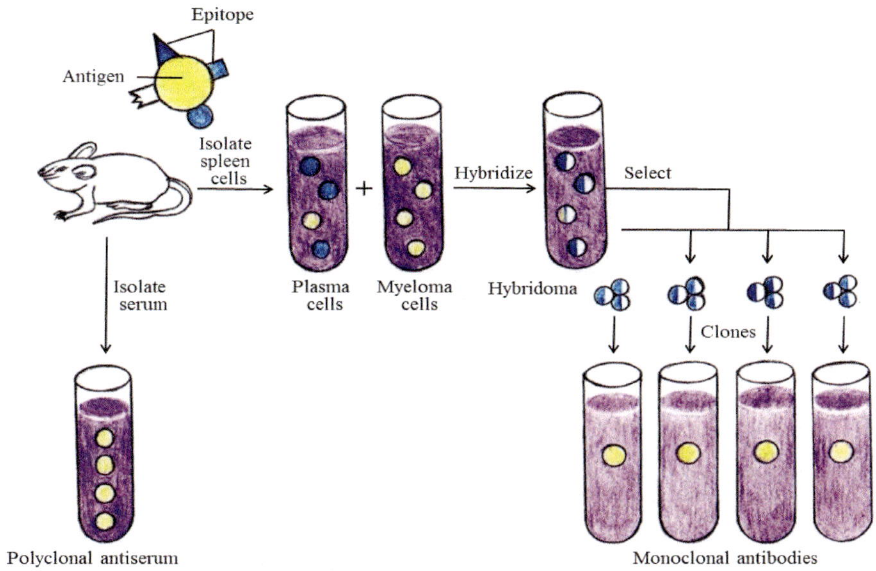

Fig.-3-14. The method to prepare and produce monoclonal antibodies was demonstrated by George Kohler and Ceaser Milstein in 1975.

Production of Monoclonal Antibodies

Antigenic specificity matters for antibodies secreting hybridoma. For this, two most common screening techniques are in practice; ELISA and RIA. The techniques will be detailed in the last chapters. Monoclonal antibodies are available when hybridoma cells secrete at very high densities (>107/ml). The production of human monoclonal antibodies presents technical difficulties, because of the lack of human myeloma cells that exhibit immortal growth and are susceptible to HAT selection. Alternatively, B-lymphocytes can be transformed with Epstein-Barr Virus (EBV) to secrete antibodies. Cloning of such prime, transformed cells has allowed the production of human monoclonal antibodies, though the amount secreted is generally small. Further, antigen activated B cells are not readily obtainable from the human lymphoid tissue equal to the mouse spleen. To overcome this problem B cells are transplanted from humans into SCID mice (because they lack functional immune system) to produce human monoclonal antibodies, by isolation of B- and T cells of the mice.

Hybrid Cell and HAT medium

There are two methods of B cell formation using polyethylene glycol (PEG). For fusion, myeloma cells are taken from animals that have been immunized with the antigen, against which an antibody is desired. Since, in this method not all of the cells fused and the procedure gives a complex mixture of unfused myeloma and spleen cells (parental cells) as well as the fused cells, the survival and growth of the B cells into hybridoma is not insured. If the mixture of hybridoma and unfused parental cells is placed in the HAT medium (i.e. three components-hypoxanthine, aminopterin and thymidine in the culture medium), the parental myeloma cells die off. Another common method employs the use of deficient (because of previously selected mutation), for one of the nucleotides salvage pathways, making them unable to grow in HAT medium. Actually, the HAT selection is very important in formation and selection of hybridoma. The enzymes catalyzing the Salvage pathways include hypoxanthine-guanine phosphoribosyl transferase (HGPRT) and thymidine kinase (TK). Cells containing HGPRT for the capacity to utilize the Salvage pathways survive in HAT medium. Fig. 3-14 & Fig. 3-15 is amply illustration of the mechanism and function of the two pathways which determines the selection of hybrid cells.

Basic Hybridoma Technology

The term Hybridoma is used for a cell which is formed due to fusion of an antibody producing B cell of mouse and a myeloma cancer cells which are capable of multiplying indefinitely. The resulting hybrid cells or hybridoma cell have the antibody producing capability inherited from lymphocytes and the ability to grow continuously from cancer cells.

A mouse is immunized with a specific antigen for the production of specific antibody. Simultaneously, tumor is produced in a mouse by injecting tumor causing antigens. After that spleen cells from immunized mouse and myeloma cells from tumor bearing mouse are isolated and cultured separately. The cultured myeloma cell is unusual in two ways, first it has lost their antibody producing ability and second its lacks Hypoxanthine Guanine Phosphoribosyl transferase (HGPRT) enzyme responsible for the use of salvage pathway. Now the fusion of the spleen cells to myeloma cells is induced by using polyethylene glycol (PEG). The hybrid cells thus formed are grown in selective Hypoxanthine aminopterin thymidine (HAT) medium. HAT medium contain a drug Aminopterin which blocks the pathway for nucleotide synthesis making the cell to depend upon an other pathway that needs HGPRT enzyme.

Therefore, myeloma cells which do not fuse with the B cells will die, since they are $HGPRT^-$ Similarly, B cells which do not fuse with the myeloma cells will die, because they lack tumorgenic property of immortal growth. Therefore HAT medium allows selection of Hybridoma cells which inherit HGPRT gene from B cells and tumorigenic property from myeloma cells. Now the selection of hybridoma cells is done for cloning and antibody productions. This is facilitated by preparing single cell colonies that produces single type of antibodies called monoclonal antibodies. The hybridoma cells are kept stored at low temperature for future use in the production of monoclonal antibodies.

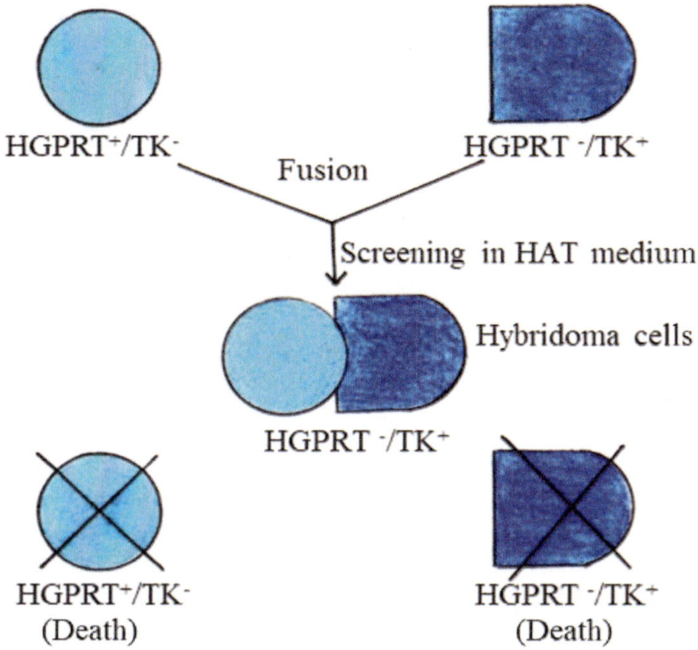

Fig.-3-15. HAT selection of hybridoma cells for the production of monoclonal antibodies.

Clinical Relevance of Monoclonal Antibodies

Monoclonal antibodies reagents are now used for the diagnosis of numerous pathogenic micro-organisms, detecting pregnancy, measuring the blood level of various drugs, matching histocompatibility antigens and antigen shed by certain tumors. Immunotoxins are specially inactivated by therapeutic agents, in reference to monoclonal antibody. Immunotoxins; ricin *Shigella* toxin and *Diphtheria* toxin, all inhibit protein synthesis, each of these toxins with two types of functionally different polypeptide components present inhibitory chain and binding chain. The binding polypeptide chain is identified with the help of monoclonal antibody especially for tumor, cell and attended therapeutically.

Applications of Monoclonal Antibodies

The applications of Monoclonal antibodies are as follows:

- The first possible therapeutic application of monoclonal antibodies was selective elimination of undesired cells, such as tumor cells or activated T lymphocytes in transplantation patients.
- It is used in radiological scanning for tumor localization
- Counting and differentiation between different lymphocyte subsets at different stages of differentiation.

- Depletion of a particular type of T cell subsets from a mixed population of bone marrow cells to prevent graft verses host reaction.
- Analysis of complex antigen mixtures.
- Analysis of embryological relationship.
- Treatment of drug overdose.

Antigen-Antibody Interaction

The antigen-antibody interaction unlike enzyme-substrate interaction, does not lead to an irreversible chemical alteration in either the antibody or the antigen, so it is reversible. The antigen-antibody (Ag-Ab) interaction depends on noncovalent interactions including hydrogen bonds, ionic bonds, hydrophobic interactions and van der Waals interaction (Fig. 3-15). For Ag-Ab interaction a high degree of complementarity is required between antigen-antibody, operating in a close fit with each other.

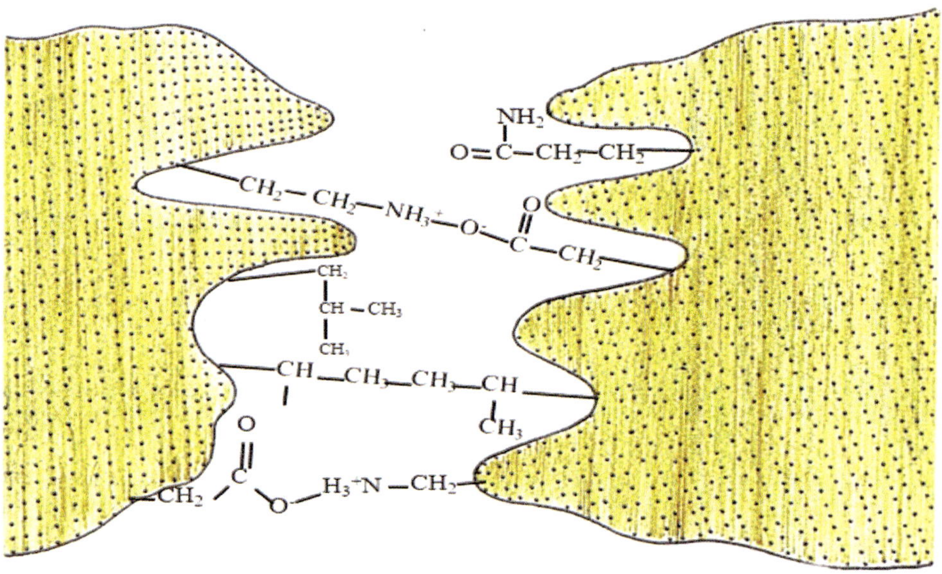

Fig.-3-15. Figure shows the four possible types of non-covalent bond formation between antibodies and antigens. These are hydrogen bonds, ionic bonds, hydrophobic bonds and van der Waals interaction.

Strength of Antigen-Antibody Interaction

Antibody interacts with an antigen in the similar pattern as the enzyme-substrate interaction takes place, involving non-covalent interaction (similar to "lock and key" fit of enzyme-substrate). Though antibody interacts with antigen corresponding to the enzyme-substrate interaction, yet it does not lead to irreversible alteration of either antibody or antigen. This interaction can be enhanced by induced fit (~1 Å at the peptide backbone level, and more at the side chain level) due to their mutually deformable conformations. The antibody-antigen interaction involves hydrogen bonds, ionic bonds, hydrophobic interactions and Van der Waals forces and hence, some points of interaction between antibody and antigen are weak and some are strong depending upon the level of complementarity. The antibody-antigen interaction is referred to by many terms as shown in Table 3-5.

Table 3-6: Summary of some of the terms used to describe antigen-antibody interaction.

Antigen	*Antibody*
Epitope	Paratope
Ligand	Receptor
Antigenic Determinant	Combining Site
Protein	Binding Site

Types of bonds involved in antibody-antigen interaction (Table 3-6.)

Hydrogen-bonding

- reversible hydrogen bridges between hydrophilic (hydroxyl, amino and carboxyl) groups.
- relatively weak, essentially electrostatic interactions that are enhanced by displacement of water molecules.

Electrostatic interactions

- oppositely charged ionic groups, such as carboxyl and amino groups, attract each other.
- force of these interactions is inversely proportional to the square of the distance between the charges.
- mutual attractiveness between them increases exponentially as the charged groups come closer together.
- displacement of water molecules, with their high dielectric constant, increases the force of these interactions.

Van der Waal's forces

- depend upon interaction between external electron clouds that lead to induced dipole interactions.
- force of these interactions is inversely proportional to the 7th power of the distance between the electron clouds.

Hydrophobic interactions

- nonpolar groups associate with one another in an aqueous environment.
- association of water molecules with one another as they are displaced from the non-polar groups provides the driving force for these interactions.

Conformational complexity is crucial for close interaction of epitope and paratope. Here, conformation allows the molecules to be very close together resulting into multiple weak interactions to form antibody-antigen complexes.

However, the specificity to which antibody (paratope) interacts with antigen (epitope or antigenic determinant), is not absolute and therefore certain degree of cross-reactivity may occur.

Table 3-7: Summary of the chemical and physical forces involved in antigen-antibody binding

Force	*Summary*
Hydrogen Bonds	Bonding with hydrogen atoms
Ionic (coulombic)	Oppositely charged groups
Van der Walls	Electron clouds interact
Apolar (hydrophobic)	Seeking minimum energy state in the absence of water
Steric factor/steric repulsive	Basis of discrimination, non-complementary electron cloud resulting in repulsive force

Nature of interactions

Antibodies may interact with antigen at single epitope or may be at multiple epitopes and therefore, interactions can be characterized into affinity and avidity.

Affinity

Antibody affinity is the strength of the reaction between a single antigenic determinant and a single combining site on the antibody. It is the sum of the attractive and repulsive forces operating between the antigenic determinant and the combining site of the antibody as illustrated

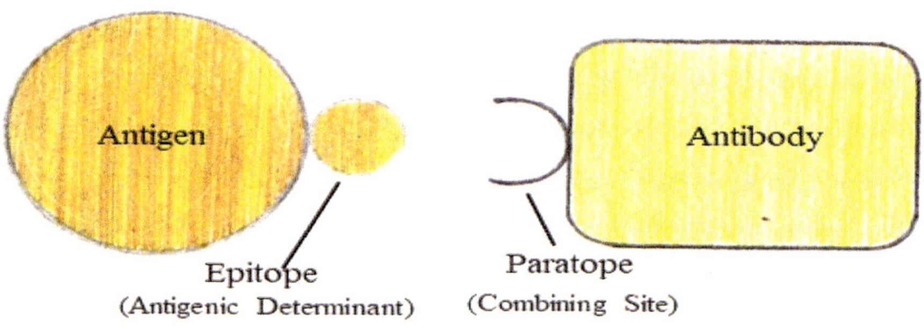

Fig.-3-16. Antigen-antibody interaction.

In Fig.-3-16 Affinity is the equilibrium constant that describes the antigen-antibody reaction as illustrated below. Most antibodies have a high affinity for their antigens.

$Ag + Ab \leftrightarrow Ag - Ab$

Applying the Mass of Action,

$$\text{Keq} = \frac{[\text{Ag - Ab}]}{[\text{Ag}]}$$

Avidity

In proteins, 'avidity' is a term used to describe the combined strength of multiple bond interactions between antigen and antibody. Avidity is different from affinity, which is a term used to describe the strength of a single bond. As such, avidity is the combined synergistic strength of bond affinities rather than the sum of bonds. Therefore, the bonus effect is that avidity is a multiple of the affinities of the individual interaction (Fig.-3-17). It is the strength of interaction of antigen molecules having multiple epitopes with antibodies having more than one paratope. Therefore, it is also termed as the functional affinity of antibodies. This type of interaction can be represented by the following formula;

n Ab + m Ag = Abn Agn

Where ***n*** is the number of paratopes and ***m*** is the number of epitopes.

Keq = 10^4 Affinity; 10^6 Avidity; 10^{10} Avidity

Fig.3-17. Avidity is a multiple of the affinities of the individual interaction.

Due to involvement of multiple epitope-paratope interaction, factors contributing to avidity are complicated, but this multivalency leads to the probability that all antibody-antigen interaction will dissociate simultaneously is exceedingly small, and if one interaction is dissociated, the others will remain associated, thus enhancing the probability that those dissociated will reassociate.

It is commonly applied to antibody interactions in which multiple antigen binding sites simultaneously interact with a target. Individually, each binding interaction may be readily broken. However, when many binding interactions are present at the same time, transient unbinding of a single site does not allow the molecule to diffuse away, and binding of that site is likely to be reinstated. The overall effect is synergistic, strong binding of antigen to antibody (e.g. IgM is said to have low affinity but high avidity because it has 10 weak binding sites as opposed to the 2 strong binding sites of IgG, IgE and IgD). The more antigen-binding sites an antibody has, the higher will be its avidity. Some antigens have many copies of the same epitopes on their surface; as a result two different Ig molecules bind to the same antigen. This type of interaction is termed as cross reactivity, and is crucial for activation of both complement system and B cells.

Equilibrium of Ag-Ab and antibody avidity

The affinity of an antibody for an antigen corresponds to the strength of the noncovalent interaction between Ag-Ab. The interaction between a single antigen binding site on an antibody and a single epitope is the affinity of the antibody for that epitope. The affinity of antibody can be measured in terms of low and high-affinity antibodies, which account for association or dissociation to an antigen. To begin with the following equation is essential-

$$Ag + Ab \underset{K_{-1}}{\overset{K_1}{\rightleftharpoons}} Ag - Ab$$

Where, K_1 indicates association and K_{-1} indicate dissociation. The ratio of K_1/K_{-1} is the association constant K_a

Now

$$K_a = \frac{(Ag - Ab)}{[Ab][Ag]}$$

Thus K_a is dependent upon both the Ag-Ab complexes as also vary for different Ag-Ab complexes. Table 6-1 is illustrative of the interaction between an antibody and antigen, as also the role of K_{-1} in different associations. The point at which Ag-Ab $\rightleftarrows$ Ab + Ag dissociate, an equilibrium constant K_d is established. It is reciprocal to K_a and a quantitative indicator of the stability of an Ag – Ab complex; very stable complexes have very low values of K_d and less stable ones have higher values.

The association constant K_a can be determined by a method known as equilibrium dialysis. Fig.-3-18 shows this method, where in a dialysis chamber containing two equal compartments separated by a semipermeable membrane antibody is placed in one compartment and a radioactively labeled ligand (heptanes, oligosaccharides, oligopeptides composing the epitope of polysaccharides or protein antigens) small enough to pass through the semipermeable membrane is placed in the other compartment.

In the absence of antibody, ligand added to compartment B will equilibrate on both sides of the membrane. In the presence of antibody, ligand will bound to the antibody at equilibrium and the unbound ligand will be equally divided in both the compartments. Thus the total concentration of ligand will be greater in the compartment containing antibody. The higher the affinity of the antibody, the more ligand is bound. Thus the concentration of antibody is related to the concentration of ligand.

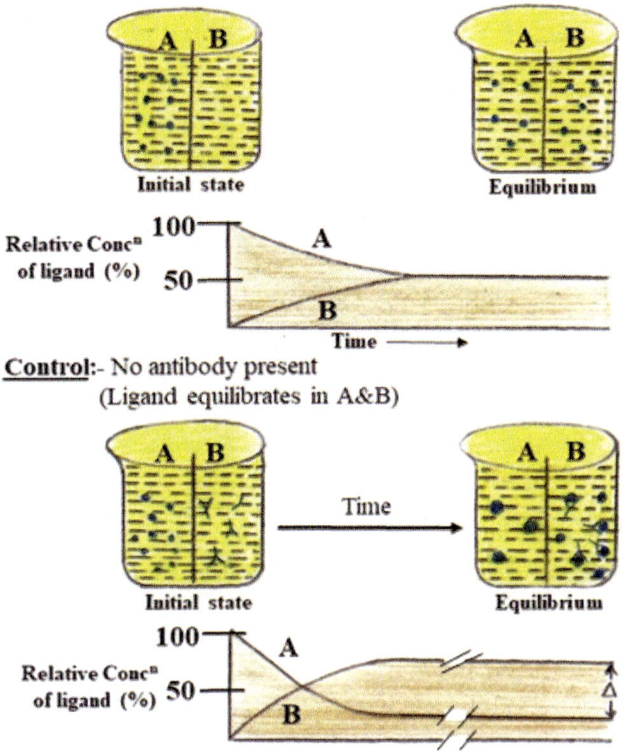

Fig.-3-18. Determination of antibody affinity by equilibrium dialysis.

Antibody Aridity refers to the reaction of an antibody with antigenic determinants on a virus or bacterial cell. Thus, antibody avidity is a better measure of its binding capacity within biological systems being the overall strength of the interactions between a multiple antibody molecule and a multivalent antigen molecule at multiple sites. This antigen-antibody interaction is the basis of several immunologic assay viz, Immunodiffusion, RIA, ELISA, western blotting, flow cytometry etc. which will be discussed in last chapter.

Cross Reactivity

Cross-reactivity can occur when two (or more) antigens share similar structural features. Antibody produced in response to Ag1 is very specific and would, therefore, have a large affinity constant (K) when combining with Ag1. However, Ag2 is similar in shape to Ag1 and is capable of interacting with anti-Ag1 antibody via two of three sites. The interaction between Ab and Ag2 is not as strong as the interaction between Ab and Ag1 (i.e. K is much smaller) but is still strong enough to allow binding. Hence, Ag1 and Ag2 are said to cross-react. Ag3, in contrast, cannot interact very well with anti-Ag1 antibody and would have a K value so low that significant binding would not occur. Ag3, therefore, would not cross-react with Ag1 (Fig.-3-19).

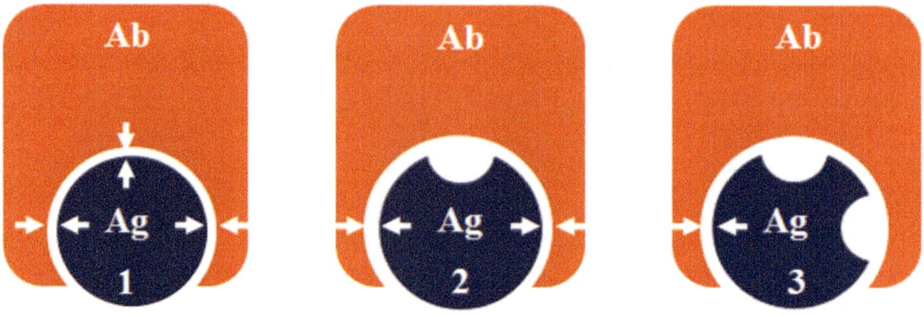

Fig.-3-19. Cross-reactivity of antibody molecules.

Crossreactivity also forms the basis for several diagnostic tests. For example, infection with *Treponema pallidum* (syphilis) causes the production of antibodies that cross-react with a substance found in cardiac muscle, cardiolipin. Since it is much easier to obtain pure cardiolipin than pure *Treponemal* antigens, this cross-reaction is used to test for syphilis (Wassermann test). Likewise, antibodies produced against certain *Rickettsia* cross-react with antigens from *Proteus*. Since the latter are much easier to obtain, they can be used to test for the former.

Microbial antigens induce the formation of antibodies in individuals lacking these antigens on their red blood cells. It was on this very study that vaccinia virus to induce immunity was developed to prevent cowpox (caused by varida virus). In reference to blood groups an individual with O blood group has Anti-A and Anti-B antibodies (ABO blood group). Table 3-7 shows the different types of blood group and their cross-reactivity.

Table 3-8: Different types of blood groups

Blood type	*Antigens on RBCs*	*Antibodies in serum*
A	A	Anti-B
B	B	Anti-A
AB	A and B	Neither
0	Neither	Anti-A and Anti-B

The blood-group antibodies although elected by microbial agents, will cross-react with similar oligosaccharides on red blood cells, or hosT cell components resulting in a tissue damaging autoimmune reaction.

Special Form of Antibodies

There are antibodies that have some specialized characteristics very distinct from the conventional types. They are found in the nature in a very restricted manner but have immense diagnostic and therapeutic implication. However, some of them are biochemically engineered owing to the fact that antibodies may have immunotherapeutic relevance as mentioned above. Examples of such antibodies of peculiar characteristics are abzymes and nanobodies.

Abzymes (Catalytic Antibodies)

Classically, antibodies have been characterized as proteins produced by the immune system that have the sole function of binding other molecules, called antigens, to elicit an immune response. Most of the natural antigens are proteins, and their interactions with antibodies are very strong (*K*d values varies between 0.01 and 10 nM (Karush 1978). In this classic conception, antibody function is similar to that of enzymes in specifically binding other molecules. However, antibodies are different from enzymes in that they do not have the ability to catalyze chemical conversions of their bound partners. In 1946, consideration of the catalytic function of enzymes, Linus Pauling first hypothesized that the active center of an enzyme is closely juxtaposed to a "strained configuration" of its substrate (that is, targeted against the structure of the transition state) rather than to the native conformation of the substrate molecule. On the basis of this hypothesis, it is logically followed that the binding energy between enzyme and substrate could play a significant role in lowering activation energy, thereby catalyzing a chemical change. This idea led Jencks in 1969 to propose that antibodies generated in an anti-hapten immune response against chemically stable analogs of the transition state of a reaction. This prediction was confirmed in 1986 when the first anti-hapten monoclonal catalytic antibodies were obtained and termed "abzymes". Therefore, an abzyme (antibody and enzyme), also called catmab (catalytic monoclonal antibody), a monoclonal antibody with catalytic activity.

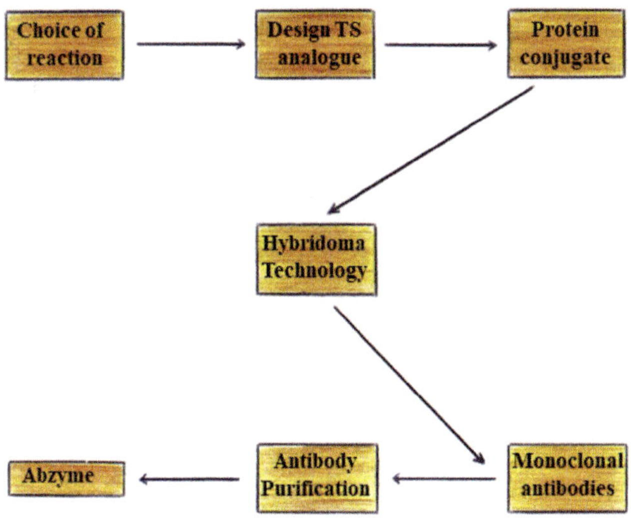

Fig.-3-20. Stages in the synthetic production of abzymes

Abzymes are usually artificial constructs, but are also found in normal humans (anti-vasoactive intestinal peptide autoantibodies) and in patients with autoimmune diseases such as systemic lupus erythematosus. They are generated by immunizing an animal against a transition state analog (TSA) of the desired reaction, and therefore they are specific for a particular reaction. It can also be produced for any non-lethal antigen and very easy to humanize. Detail about the synthesis of abzymes is given in Fig.-3-20.

Through the use of protein engineering, catalytic property of abzyme can be improved and even to surpass the activity of natural enzymes. Molecular biologists have developed methods to clone the array of genes that encode IgG molecules. The millions of gene products from an immunized animal are screened for the production of antibodies with desirable catalytic activities. Once a desired product is isolated, can be produced in large amount by using recombinant technology, and antibodies so produced are called recombinant antibodies. However, it is found that sIgA present in the milk of healthy parturient women, catalyzes protein phosphorylation, and casein (and more than 10 other milk proteins) and ATP were acts as substrates for this reaction. At present, this is the only example of natural abzymes catalyzing a bi-substrate reaction. Abzyme levels are especially high in milk of women who were infected with viral infections or allergic diseases during pregnancy.

Catalytic antibodies have great potential in the pharmaceutical industries. Abzymes have been implicated for use in the detoxification of cocaine. Catalytic antibodies have been generated that cleave the cocaine molecule at specific bonds, thereby eliminating the toxic effect of the drug. Therefore, as a pharmaceutical reagent, anti-cocaine abzymes could treat patients who are addicted to cocaine, or reverse the lethal effects of a cocaine overdose. Perhaps the most exciting application of abyzme technology is the specific targeting of cancer cells for destruction.

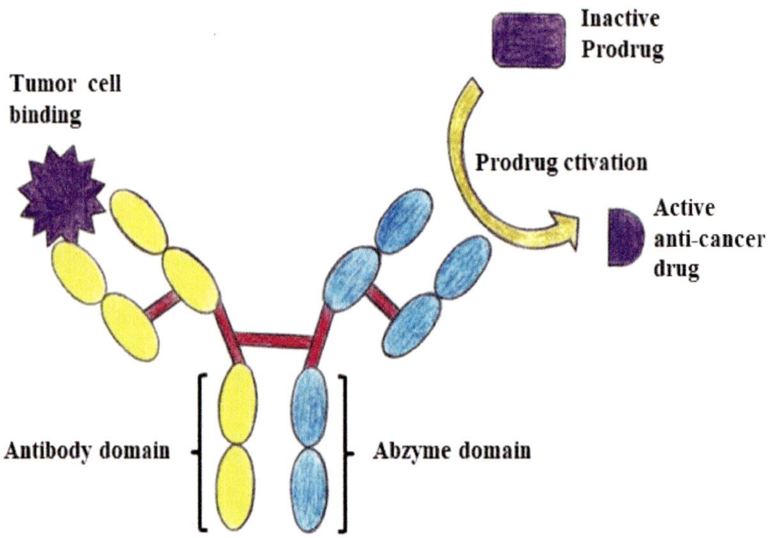

Fig.-3-21. Tumor drug specific abzymes

Cancer cells contain unique determinants, called tumor cell antigens, on their surface that are lacking in normal cells. By utilizing antibodies that specifically bind these tumor cell antigens, cancer drugs can be delivered directly to the tumor. In this case, abzymes may have two distinct antigen binding sites: one site binds with high affinity to a tumor cell antigen, while the second site catalyzes the cleavage of a prodrug (Fig.-3-21) which is a non-toxic precursor of a cytotoxic drug. In this case antigen specific antibody is administered to patients which binds the tumor cells with high affinity and then prodrug is introduced into the bloodstream which become active only in the vicinity of the targeted antibody. By this technique, tumors are selectively destroyed while healthy cells are spared from the toxic effect of cancer drugs. Researchers have also developed abzymes that catalyze the specific destruction of viral genes. Perhaps in the future, we will have the tool to treat a wide variety of diseases through the use of catalytic antibody technology.

Nanobodies

Nanobodies are single-domain antigen-binding fragments derived from heavy-chain antibodies that are devoid of light chains, which constitute an alternative to monoclonal antibodies. Nanobodies are relatively simple proteins about a tenth the size of antibodies and just a few nanometers in length. They have retained the ability of an antibody to bind to very specific targets. Nanobodies have huge advantage over monoclonal antibody in their small size.

The great size of the antibody imposes practical and medical limitations. High temperatures or extremes of pH make MAbs unravel. They typically expire in weeks unless stored near freezing temperatures. Antibodies are digested quickly in the gut, blocked from entering the brain and held to the periphery of solid tumors. Many diseases are thus unreachable by monoclonals, and patients who can use MAb therapies must receive them by injection at a clinic.

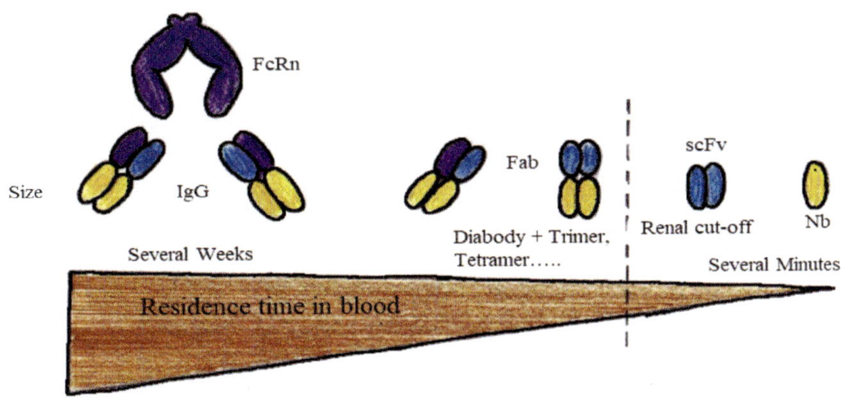

Fig.-3-22. The residence time of antibodies in the peripheral circulation

In certain conditions in which MAbs do not work well, smaller proteins might perform better. In 1980s, protein engineers began experimenting with antibody fragments created by chopping off the stem of the Y-shaped antibody, or sometimes the stem and an arm, leaving just one "hand" to do the chemical duty of the antibody. Like full-size MAbs, these antibody fragments (Fabs) can treat diseases by binding to toxins, pathogens or aberrant cell signals or alternatively to the cell receptors to which those undesirable molecules dock. Unlike a complete antibody, fragments of antibody cannot recruit other components of the immune system, such as killer T cells, because they lack the protein stem that performs that task. On the other hand, Fabs tend to fall apart or filter out of the bloodstream quickly, and so their active half-life decreases to few hours rather than the weeks that full-size antibodies have within the body as shown in Fig.-3-22.

In 1989, Muyldermans and his group investigated an odd observation on how dromedary camels (the one-humped, Arabian variety) and water buffalo fight off parasites. One of the tests for antibodies in the dromedary blood seemed to show an error and they found that unlike normal four-chain antibodies, it indicated the presence of simpler antibodies composed solely of a pair of heavy chains (Fig.-3-23). Equally surprising, they found, these "incomplete" antibodies are able to grasp their targets just as firmly as normal antibodies do, despite having only half as many CDRs. And unlike Fabs, the heavy-chain-only antibodies do not stick to one another.

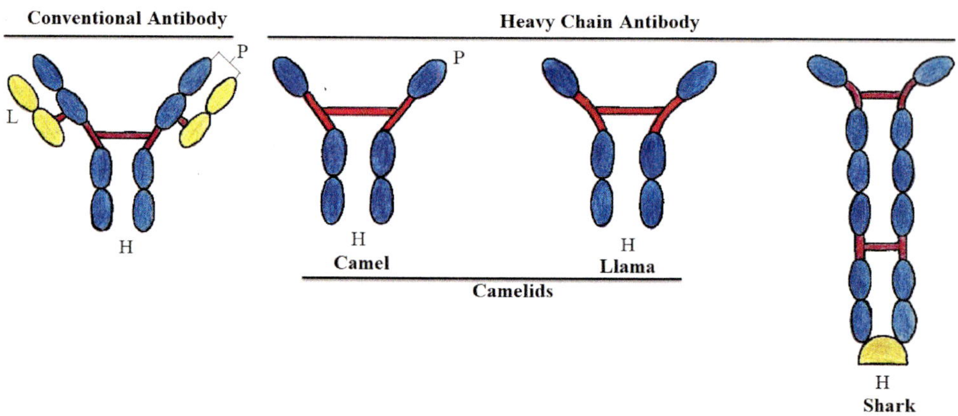

Fig.-3-23. The presence of simple a pair of heavy chains is the characteristic of camelids and sharks.

When compared these novel molecules down to just their distinctive, variable segments, the segments amazingly retained strong affinity for their targets, virtually equal to a full antibody 10 times their size. These shortened proteins were also more chemically agile, able to engage targets-including the active sites of enzymes and clefts in cell membranes-too small to admit an antibody, and thus nanobodies were born. Because nanobodies are so much simpler than antibodies in their chemical composition and shape, they can be encoded by a single gene and are easier for microbes to synthesize (Fig.-3-243).

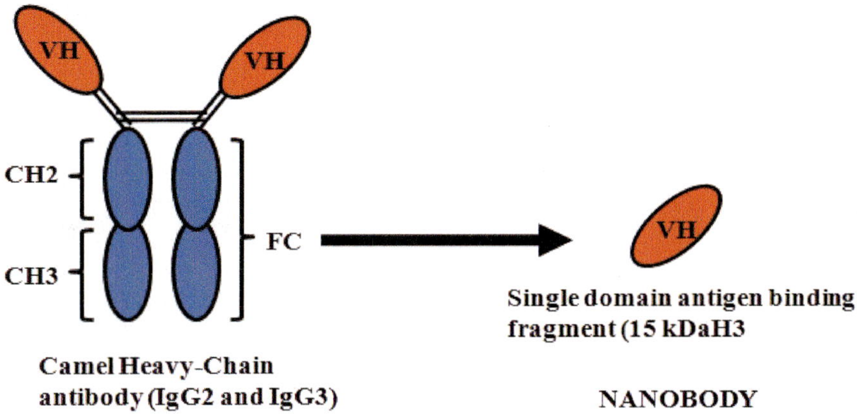

Fig.-3-24. Nanobodies are very small in size and have low toxicity and high specificity equivalent to conventional antibodies and even more. But, unlike conventional antibodies, nanobodies have a short half life and can diffuse easily to their target. Due to the small and much simpler nature of nanobodies, they can be encoded by a single gene and are easier for microbes to synthesize. Synthetically produced nanobodies are very stable at room temperature and have a long shelf life without refrigeration.

Nanobodies thus produced have many medical implications. Nanobodies against ion channel have been produced, which proved to be highly relevant therapeutic targets especially in areas such as neuropathic pain, neurological disorders, inflammation and cardiovascular disorders. The antagonistic nanobodies are now being further evaluated for their ability to mobilize hematopoietic stem cells in certain types of cancers. Moreover, nanobodies are now being employed in *in vivo* imaging, which can certainly prove to be a powerful diagnostic tool in various pathological conditions in the future.

CHAPTER – 4

IMMUNOGLOBULIN GENE ORGANISATION AND EXPRESSION

The magnitude of the studies in molecular biology of immunoglobulin genes lies in the fact that by the knowledge of molecular biology of immunoglobulin genes engineering of antibody molecules for therapeutic application can be vastly developed to combat most of the life threatening ailments, which otherwise would be difficult to be handled. Today antibody engineering has assumed a phase conducive to the therapy of malignancies. When an engineered antibody is used in combination with a chemotherapeutic drug, efficacy against metastatic malignancies is highly multiplied than with single chemotherapeutic drugs. The detailed organization of the immunoglobulin genes, the process of Ig gene rearrangement and the mechanisms by which the dynamic immunoglobulin genetic system generates more than 1010 different antigenic specifications, are some of the concepts developed in the chapter.

Genetic Model Compatible with Ig Structure

Immunoglobulin genes account for the following properties of antibodies:

- The existence of isotype with the same antigenic specificity (as is found from the association of a given variable region with different heavy-chain constant regions).
- Vast diversity of antibody specifications.
- Presence of a variable region at the amino-terminal and a constant region at the carboxyl-terminal end of Ig heavy and light chains.

Germ-Line and Somatic-Variation Theories

A vast number of potential antigens can be responded by a genetic system in the mammals comprising antibodies (proteins). This shows that genetic mechanism in the mammals may develop diverse responses to combat antigens of vast strata. What is this genetic mechanism that might generate such diversity has been a matter of imagination for immunologists and two theories have been construed- (i) the germ-line theories and the somatic-variation theories. Genetic mechanisms to account for antibody diversity has not been perfectly postulated by the germ-line theories, as the theory hold that the genome contributed by the germ cells, egg and sperm, contains a large repertoire of immunoglobulin genes. Thus, the exact mechanism to account for antibody diversity has not been forwarded by this theory. The somatic-variation theory holds that the genome contains a small number of immunoglobulin genes, from which a large number of antibody specificities are generated in the somatic cells by mutation or recombination.

With the determination of more and more amino acid sequences in the immunoglobulin it became clear that there must be mechanisms for generating antibody diversity as also for maintaining constancy. This proposition contrasts with the central idea of immunoglobulin structure, as how could stability be maintained in the constant (C) region while some kind of diversifying mechanism generated the variable (V) region. Thus, more of the theories- the germ-line or somatic-variation could fit in the riddle of diversity of antibody mechanism.

The Two-Gene Model of Dreyer and Bennett

With regard to the structure of Immunoglobulins, W. Dreyer and J. Bennett suggested (1965) that two separate genes encode a single immunoglobulin heavy or light chain, one gene for the V region and the other for C region. These two genes come together at the DNA level to form a continuous signal to excite a single Ig heavy or light chain. The Dreyer-Bennett theory combined elements of the earlier germ-line and somatic-variation theories, as it could account for those Immunoglobulins in which a single V region was combined with various C regions. This theory could also account for the conservation of necessary biological effective functions while allowing for the evolutionary diversification of V region genes. As the theory postulated that two genes encode a single polypeptide, it was contradicted with the existing one gene-one polypeptide theory.

S. Tonegawa and N. Hozumi found the first direct evidence that two separate genes encode a single immunoglobulin heavy or light chain, one gene for the V region and the other for the C region. The genes are rearranged in the course of B cell differentiation. Fig. 4.1 refers to the experimental works of Tonegawa and Hozumi affirming the two-gene theory of Dreyer and Bennett, that one gene encoding the V region and one gene encoding the C region apply to both heavy and light chain genes. Working on the theory of two genes Tonegawa and Hozumi using various restriction endonucleases, cleaved DNA from embryonic cells and from adult myeloma cells with fragments. They suggested that during differentiation of lymphocytes from the embryonic state to the fully differentiated plasma cell stage the V and C regions genes undergo rearrangement. In the embryo, the V and C genes are separated by a large DNA segment that contains a restriction endonuclease site. During differentiation, the V and C genes are brought closer together and the intervening amino acid sequence in the immunoglobulin DNA sequence is eliminated.

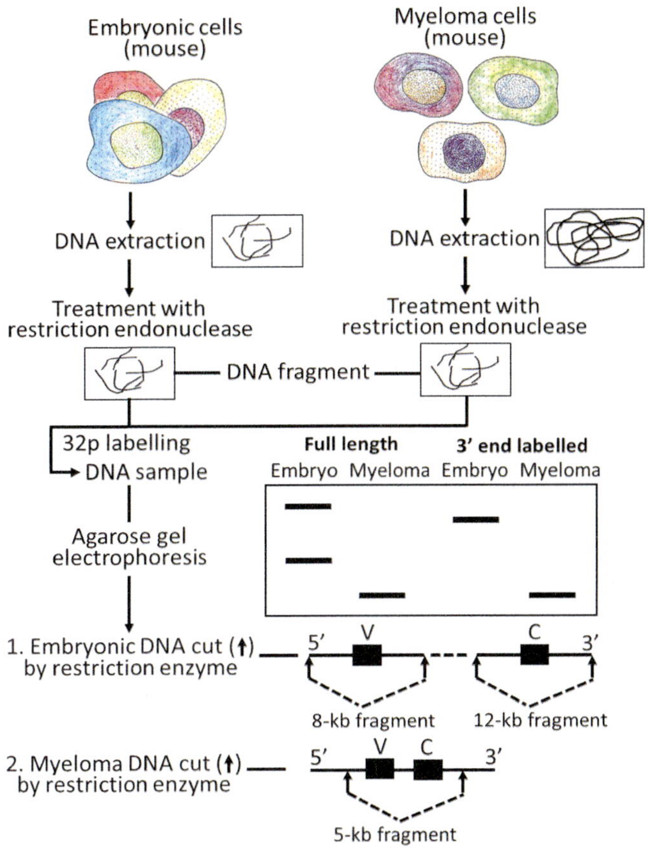

Fig.-4-1. Experimental demonstration of an arrangement of light chains during the B cell development. After digestion by restriction enzymes, the DNA bands appear to depict the differences between the myeloma DNA and embryonic DNA. The Lower panel shows the fragments of DNA after the action of restriction enzymes on both myeloma and embryonic DNA.

Multigene Organization of Ig Genes

Immunoglobulin k and l light chains and heavy chains are encoded by three separate multigene families (containing several coding sequences) called gene segments. The k and l light chain families contain V, J and C gene segments. The k and l heavy chain families contain V, D, J and C gene segments. The rearranged VJ segments encode the variable region of the light chains and VDJ gene segments encode the variable region of the heavy chains. In each gene family C gene segments encode the constant regions. Each V gene segment is preceded at its 5' end by a small exon that encodes a short signal or leader (L) peptide, which guides the heavy or light chain through the endoplasmic reticulum. The signal peptide is cleaved from the nascent light and heavy chains prior to assembly of the finished immunoglobulin molecule. This system prevents amino acids encoded by this leader sequence from appearing in immunoglobulin molecule. Table 4.1 shows k and l light and heavy chains encoded by separate multigene families situated on different chromosomes. Multigenes of Ig genes are organized into various segments viz. l-chain multigene family, k-chain multigene family, and heavy-chain multigene family.

Table 4-1: Chromosomal location of immunoglobulin genes in human and mouse

	Chromosome	
Gene	**Human**	**Mouse**
λ light chain	22	16
κ light chain	2	6
Heavy chain	14	12

Lambda (l)-chain Multigene Family - cloning of the germ-line DNA confirmed that the light-chain variable region is actually encoded by two gene segments. This determined its complete nucleotide sequence. However, when the nucleotide sequence was compared with the known amino acid sequence of the l-chain V region, it was observed that though the first 97 amino acids of the l-chain V region corresponded to the nucleotide sequence, the remaining 13 carboxyl-terminal amino acids of the protein's V region failed to correspond to the nucleotide sequence. So, it turned out that many base pairs away a separate, 39 bp gene segment called J (joining region) encoded the remaining 13 amino acids of the l-chain variable region. Thus, a functional l variable-region gene contains two coding segments – a 5' V segment and a 3' J segment (fig. 4. 2 a). In un-rearranged germ-line DNA these two segments are separated by a non coding DNA sequence.

The mouse germ line contains two Vl gene segments, four Jl gene segments and four Cl gene segments besides pseudogenes- Jlj and Cl4 gene segments, which are incapable of encoding protein. The Vl and the three functional Jl gene segments encode the variable region of the light chain and each of the three functional Cl gene segments encode the constant region of one of the three l-chain subtypes (l1, l2 and l3). In humans, there are 30 Vl gene segments, 4Jl segments and 4Cl segments.

Kappa (k)-chain Multigene Family - Figure 4. 2 b is amply illustrative of k-chain multigene family in the mouse and humans. It comes out that the arrangement of the gene segments is quite different in k and l gene families. The k-chain multigene family is the same in the mouse and the humans i.e. 40 Vk gene segments, 5Jk segments and a single Ck segment.

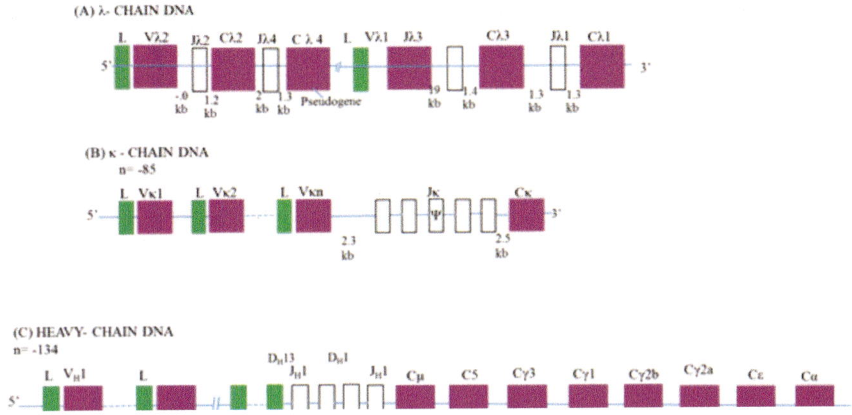

Fig.-4-2. Organization of germ line immunoglobulin gene segments for λ, κ and heavy chain DNA in the mouse.

Heavy-chain Multigene Family - In it, the VH gene segment encode amino acids 1 to 94, and J_H gene segment encodes amino acids 98 to 113 (amino acids 95 to 97 are not encoded) in additional nucleotide sequence between V_H and J_H gene segments correspond to amino acids 95 to 97. A third germ-line gene segment, which encodes amino acids within the third complementary-determining region (CDR3), is called D (diversity) for its contribution to the generation of antibody diversity. D gene segments lies between the V_H and J_H gene segments. Upstream from a cluster of 27 functional D_H gene segments there are 51 V_H gene segments. Downstream from the D_H gene segments there are six functional J_H gene segments followed by a series of C_H gene segments. The C_H gene segments consist of coding exons and non-coding introns. Heavy chain constant region is encoded by exon. At first encounter of an antigen by the B cell a sequential arrangement of C_H gene segments in the order - Cμ, Cδ, Cγ, Cε, Cα (fig. 4. 2 c) is expressed in both human and mouse.

Variable Region Gene Rearrangements

The variable region gene arrangements occur in an ordered sequence during B cell maturation in the bone marrow. The heavy and light chain variable-gene rearrangement occurs in series with a single functional variable-region DNA for each chain. By this rearrangement mature, immunocompetent B cells are produced. Each such cell is committed to produce antibody with a binding site encoded by the particular sequence of its rearranged V genes. The variable-region gene arrangement though ordered in sequence is intervened by some random events, which actually determine the order, mechanism and results/consequences of these rearrangements. The particular random events may be clear through the following discussion of -

- V-J rearrangements in light-chain DNA
- V-D-J rearrangements in heavy-chain DNA

V-J rearrangements in Light-Chain DNA: In human or mouse k light-chain DNA, any of the Vk gene segments can be joined with any of the functional Jk gene segments. Figure 4.3 is amply illustrative of the light chain gene rearrangement and further joining. The rearranged light-chain sequence is transcribed by RNA polymerase from the L (leader) exon through the C segment to the generation of a light-chain primary RNA transcript. The leader sequence is, however, not present in the finished light-chain protein product.

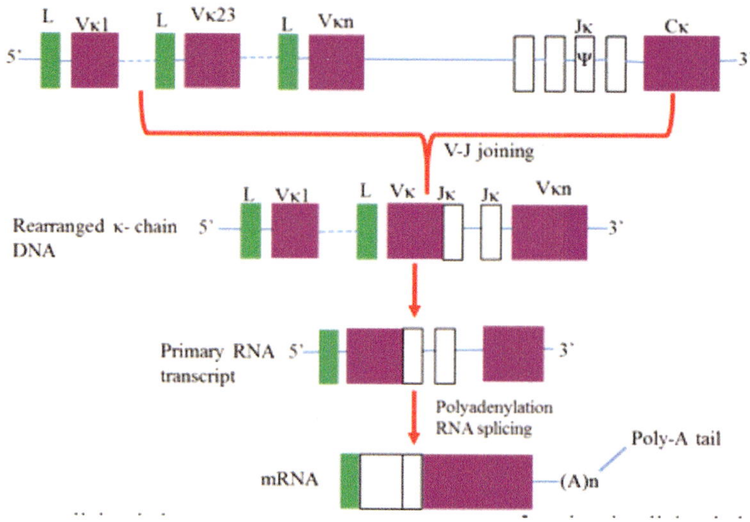

Fig.-4-3. Kappa light chain gene rearrangement to generate functional κ light chain protein by joining Vκ23 and Jκ4 regions of germ line DNA.

V-D-J rearrangements in Heavy-Chain DNA: In heavy-chain DNA variable region rearrangements, a DH gene segment first combines with a JH segment (DH JH) and then attaches with a VH segment (VH DH JH). This rearrangement encodes the entire V region. Following this rearrangement, RNA polymerase combined to the promoter sequence and transcribes the entire heavy-chain gene including the introns. Later, differential poly adenylation and RNA splicing remove the introns and process the primary transcript to generate mRNA including either the Cm or the Cδ transcript. The B cell on its surface expresses both IgM and IgD antigenic specificity owing to the very presence of two different heavy-chain mRNA as above.

Mechanism of Variable-Region DNA Rearrangements

Variable-region gene rearrangement during the maturation of B cell is an interesting and worth examining phenomenon. There are two closely related conserved sequences in variable-region germ-line DNA-one-turn recombination signal sequences and two-term signal sequences- for the sequences serve as signal for the recombination processes that rearrange the genes. One recombination signal sequence (RSS) is located 3' to each V gene segment, and 5' to each J gene segment and on both sides of each D gene segment. In different sequences, different spacers are employed. Signal sequences having a one-turn spacer can join only with sequences having a two-turn spacer. This is known as one-turn/two-turn joining rule. Applying this rule, a V_L segment joins only to a J_L segment and not to another V_L segment. Likewise, V_H, D_H and J_H segments join another type of segments and not V_L segments.

V (D) J Recombination

Two recombination-activating genes RAG1 and RAG2 were first identified by David Schatz, Marjorie Oettinger and David Baltimore. It was reported that RAG's encoded proteins act synergistically to mediate V(D)J combination. The RAG1 and RAG2 proteins and the enzyme terminal deoxynucleotidyl transferase (TdT) are the only lymphoid-specific gene products that are required for V(D)J rearrangement.

A multi-step process, as given below, catalyses the recombination of variable region gene segments:

- Normal double-strand break repair (DSBR) enzymes catalyze the coding sequences to join the signal sequences.
- Sites for the addition of P-region nucleotides are germinated by the cutting of the hairpin, followed by the trimming of a few nucleotides from the coding sequence by single-strand endonuclease.
- At the junctures of signal sequences and coding sequences, one strand of DNA is cleavage by RAG1 and RAG2.
- TdT is optionally added up to 15 nucleotides (N-region nucleotides) at the cut ends of the V, D, and J coding sequences of the heavy chain.

When, the two gene segments are in the same transcriptional orientation, joining results deletion of the signal joint and intervening DNA as a circular excision product (Fig. 4-4). In case the two gene segments have opposite orientations, joining occurs by inversion of the DNA, whereby both the coding joint and the signal joint are retained. In the human k locus half of the VK gene segments are inverted with respect to JK and a number of mechanisms are involved through the diversity of the coding joints, formed by the recombination of two gene segments.

RAG deficiency versus SCID

Ig-gene rearrangement is responsible for immunity. Several defects in Ig-gene rearrangements contribute to immunodeficiency. Deficiency of RAG1 and RAG2 cannot initiate the recombination process because they cannot introduce double strand DNA breaks between the recombination signal sequences (RSSs) and coding sequences in germ-line Ig DNA. Due to this defect, the V, D and J gene segments remain unarranged. As the B and T cells utilize the same recombination system, the RAG-1 and RAG-2 knockout mice lack mature B and T cells. This is why they show a severe combined immunodeficiency (SCID). Another recombination defect is found in CB-17 SCID mice, because they lack mature functional B and T cells. However, both the above types of mice exhibit SCID.

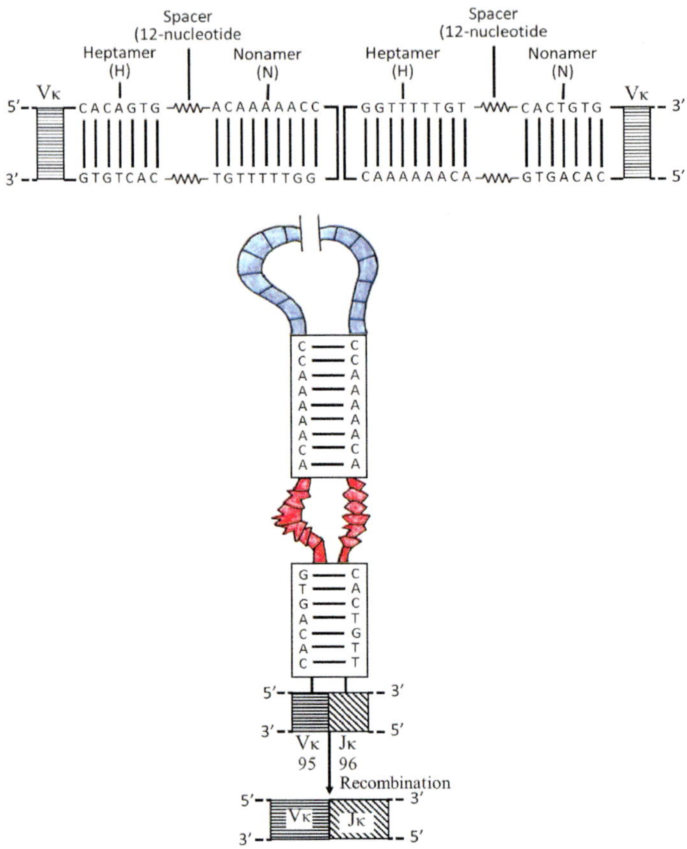

Fig.-4-4. General process of recombination of Ig-gene segments at 5'-end, where loop forms, V_k and J_k region come closer to each other, and the rest of the part is deleted.

Imprecise Joining of Signal and Coding Sequences

Antigen binding site is closely related to the joining of signal and coding sequences. V-(D)-J rearrangements are introduced precisely at the junctions of signal sequences and coding sequences, the subsequent joining of the coding sequences is imprecise. This imprecision in the joining process helps generate antibody diversity by contributing to the hyper variability of the antigen binding site.

Gene segments that join out of phase, is also an expression of imprecise joining because in such a state the triplet reading frame for translation is not preserved. This is called non-productive rearrangement, for the resulting VJ and VDJ unit will contain numerous stop codons, interrupting translation (Fig. 4-5). Gene segments that join in phase are called a productive rearrangement, for here reading frame is maintained, yielding a complete antibody. It is estimated that only one in three attempts at V_L-J_L, D_H-J_H and V_H-D_HJ_H joining are productive. Mature and immunocompetent B cells are the results of this productive joining.

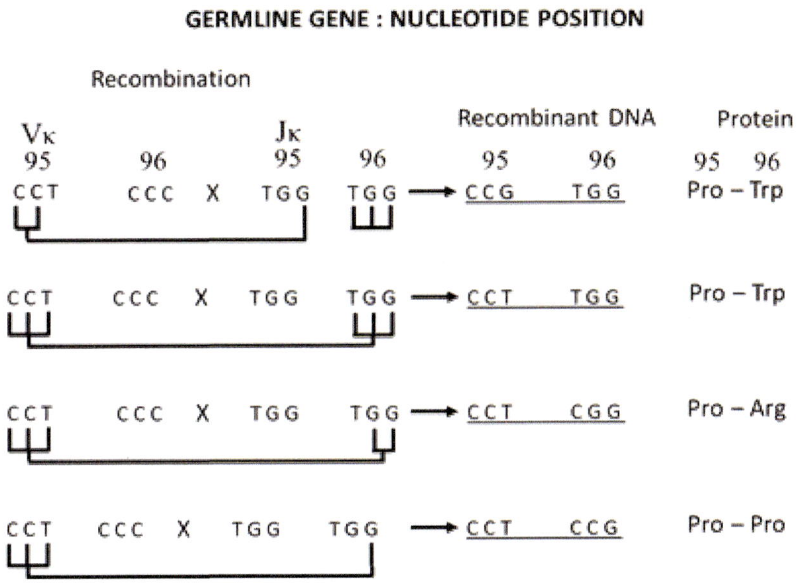

Fig.-4-5. Increased antibody diversity by joining flexibility at V_k and J_k recombinant products after joining flexibility of V_k and J_k segments.

Allelic Exclusion

B cells are diploid like other somatic cells and contain rearranged light and heavy chain genes from only one chromosome of parents. This is known as allelic exclusion. It suggests that once a productive VH-DH-JH rearrangement and a productive VL-JL rearrangement have occurred the recombination system is turned off, so that the heavy and light chain genes on the homologous chromosomes are expressed.

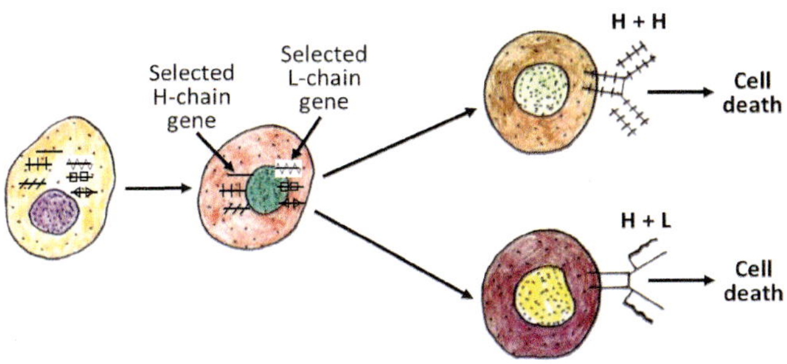

Fig.-4-6. The diagram depicts the allelic exclusion of the B cells, which die by apoptosis due to improper gene re-arrangements of light and heavy chains.

A model proposed by G.D. Yancopoulos and F.W. Alt may be referred to Fig. 4-6. As per this model encoded protein of a productive rearrangement acts as a signal to prevent further gene rearrangement and m heavy chains signal the maturing B cell to turn off rearrangement of other heavy chain alleles and turn on rearrangement of k light chain genes. If neither l allele rearranges productively the B cell dies by apoptosis. Studies have suggested that the remaining alleles are regulated by protein encoded heavy and light chain genes. The m-transgene product rearrange homologous chain genes. This is how allelic exclusion prevents gene rearrangement.

Antibody Diversification

As per germ-line doctrine, the entire variable region gene is hereditary (egg and sperm). As per somatic variation theory during the development of the immune system, the variable region genes are diversified in the somatic cells by mutational or recombinational events. So far seven means of antibody diversification have been identified:

- P-region nucleotide addition (P-addition)
- N-region nucleotide addition (N-addition)
- Somatic hyper mutation
- Junctional flexibility
- Multiple germ-line gene segments
- Combinational V-(D)-J joining
- Combinatorial association of light and heavy chains

P-addition: At the junction of a variable-region gene segment and attached signal sequence a single-strand DNA is cleaved. After it, the nucleotide at the end of the coding sequence turns back to form a hairpin structure (Fig. 4-4). This structure is then cleaved by an endonuclease at a position that leaves a short single-strand at the end of the coding sequence. The addition of complimentary nucleotides to this strand (P-addition) by repair enzymes generates a palindromic sequence in the coding joint. These nucleotides are called P-nucleotides (Fig. 4-7). Any variation at the site of the cut in the hairpin may cause variation to the sequence of the coding joint.

N-addition: Amino acids in the variable-region coding joints in rearranged heavy-chain genes are encoded by nucleotides (N-nucleotides) added during the D-J and V to D-J joining process by a terminal deoxynucleotidyl transferase (TdT), catalyzed reaction (Fig. 4-7). Thus TdT is responsible for N-nucleotide addition. This has been proved in transfection studies in fibroblasts, since fibroblasts when transfected with the gene encoding TdT were accompanied by N-nucleotides at the coding joints. It has been shown that 15 N-nucleotides can be added to both the D_H-J_H and V_H-D_HJ_H joints. This implies that N-region nucleotides addition is quite large because N-region consists of wholly random sequences. Owing to this N-region nucleotides also generate, by subsequent addition, heavy-chain diversity at V-D-J coding joints.

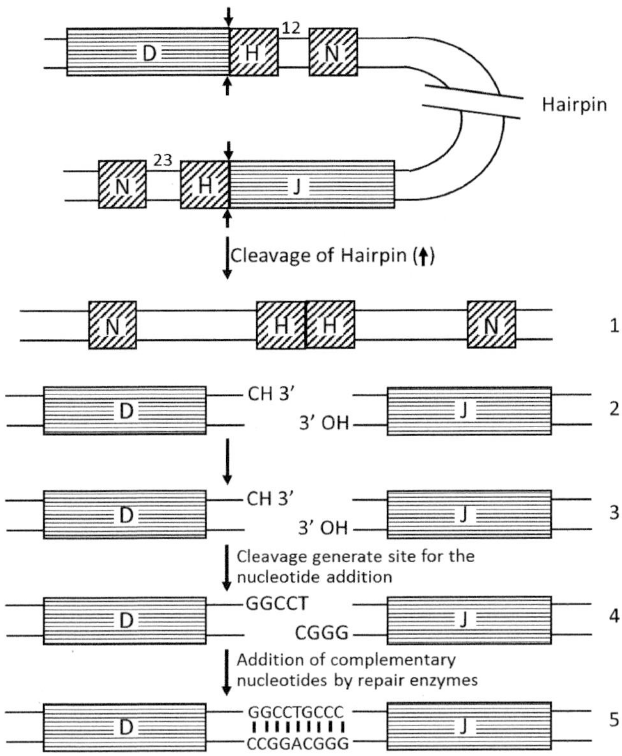

Fig.-4-7. Addition of p-nucleotides and N-nucleotides during joining of D_k and Jk-region of Ig-gene segments by recombination.

Somatic hypermutation: Antibody diversity arises due to a mechanism that operates during formation of specific variable region by gene arrangement, and the assumption that once functional copy is formed, it is not further altered, is false, instead, antibody diversity arises at this stage too by a process called somatic hypermutation. Somatic hypermutation occurs only within germinal centre structures that form in secondary lymphoid organs within a week or so of immunization with an antigen that activates a T cell dependent B cell response. The frequency of mutation is very dynamic approaching 10-3 per base per generation, so it is called hypermutation (unlike 10-8/bp/generation) in other genes. Studies suggest that certain nucleotide motifs and palindromic sequences within V_H and V_L cause this hypermutation. Although somatic hypermutations occur throughout the V_J or VDJ segment in mature B cells they are clustered within the CDRs of the VH and VL sequences. Since CDRs are directly involved in antigen recognition mutation within the CDRs is more likely to influence the affinity for antigen, a process called affinity maturation. Somatic hypermutation was efficiently demonstrated by Claudia Berek and Cesar Milstein.

Junctional Flexibility: Joining of the coding sequences is often imprecise (Fig. 4-8) while that of the signal sequences is always precise. This diversity is caused by a phenomenon called junctional flexibility. The sequence constancy in the signal joints contrasts with the sequence variability in the coding joints. Antibody diversity is increased by several productive combinations that encode alternative amino acids at each coding joint (Fig. 4-5). Junctional flexibility generates changes in amino acids, thereby expressing antibody diversity.

Multiple germ-line gene segments: There are variations in gene segments in different individuals. Despite these variations multiple germ-line V, D and J gene segments contribute to the diversity of the antigen-binding sites in antibodies.

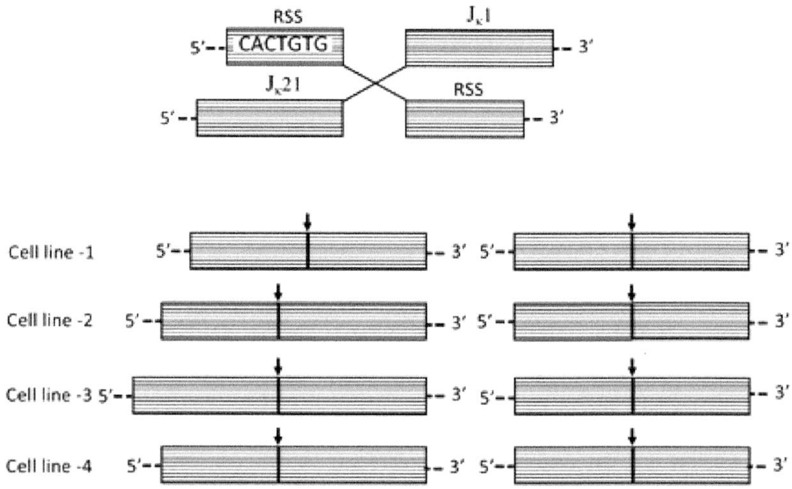

Fig.-4-8. Experimental evidence for junctional flexibility in immunoglobulin genes in B cells. Recombination results in the formation of Vk21/Jk-1 and RSS/RSS complex in four different cell-lines.

Combinatorial V(D)J Joining: Random rearrangement by multiple germ-line segments in somatic cells is known as the combinatorial joining of VDJ and V-J gene segments. This contributes to the diversity of antibody. A calculation of these gene segments may be seen in Table 4-2. In humans, these are minimal calculations of diversity (the antibody diversity in mice is also expressed in the above table). However, exact calculation of the contribution of gene segments is estimated as 1010 or even 1011 different antibody combining sites, being the set of possible variations carried by an individual. The variations may be of much smaller subset.

Combinatorial Association of Light and Heavy Chains: The specificity of an antibody's antigen-binding site is determined by the variable regions in both heavy and light chains, as the final source of antibody diversity. The potential number of heavy and light chain combination is 2,644,240. This number may be higher than the actual number of combinatorial diversity generated in an individual, because all V_H and V_L do not pair with each other. The recombination process is since random; all the $V_{H,}$ D or V_L gene segments are not used at the same frequency. The structure of the antibody binding site is influenced by the new sequences created by other means, viz. junctional flexibility, P and N-nucleotides addition etc. of antibody diversity. This is because of these segments that potential antibody diversity is boosted to 1011 and above.

Table 4-2: Antibody diversity produced in diversity in humans and mice by Ig gene rearrangement

		Light chain	
Multiple germ-line segment	**Heavy chain**	**λ**	**κ**
Estimated number of segment in humans			
V	51	40	30
D	27	0	0
J	6	5	4
V-D-J and V-J joining	51x 27 x 6 = 8262	40 x 5 = 200	30 x 4 = 120
Possible number of combinations		8262 x (200 x 120) = 2.64 x 106	
Estimated number of segment in mice			
V	134	85	2
D	13	0	0
J	4	4	3
V-D-J and V-J joining	134 x 13 x 4 = 6968	85 x 4 = 340	2 x 3 = 6
Possible number of combinations		6969 x (340 x 6) = 2.41 x 106	

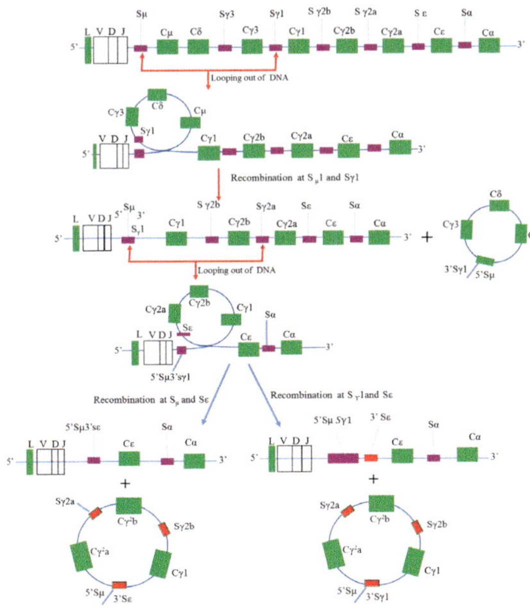

Fig. 4-9. Proposed model for Ig class switching induced by IL-4 and IL-13. It occurs through different stages involving recombination, circular excision, etc.

Class Switching Among Constant-Region Genes

V_H, D_H, J_H gene segment can combine with any C_H gene segment during stimulation of B cell. This process is called class (isotype) switching. The DNA flanking sequences are called switch regions composed of multiple copies of short sequences. An immunoglobulin class is a series of class specific recombinase proteins, which bind to switch regions and facilitate DNA recombination. Thus recombinase is essential for cell development. Various cytokines induce B cells to class-switch to a particular isotype.

Expression of Ig Genes

The heavy chain C-gene segments are organized as a series of coding exons and non-coding introns. Primary transcripts are required to produce functional mRNAs (produced from rearranged heavy and light chain genes containing intervening DNA sequences). By primary transcripts, intervening DNA sequences are removed and their flanking exons are connected by a process called RNA splicing. This splicing occurs at sites located at the intron-exon boundaries within a primary transcript. The mRNAs thus go to polyribosomes to become heavy or light chains.

Membrane-bound Forms of Immunoglobulin and concomitant expression of IgM and IgD

A particular immunoglobulin can exist in either membrane-bound or secreted form. The secreted form has a hydrophilic sequence of about 20 amino acids and the membrane-bound form has a sequence of about 40 amino acids containing a hydrophilic segment that is outside the cell (Fig. 4-10).

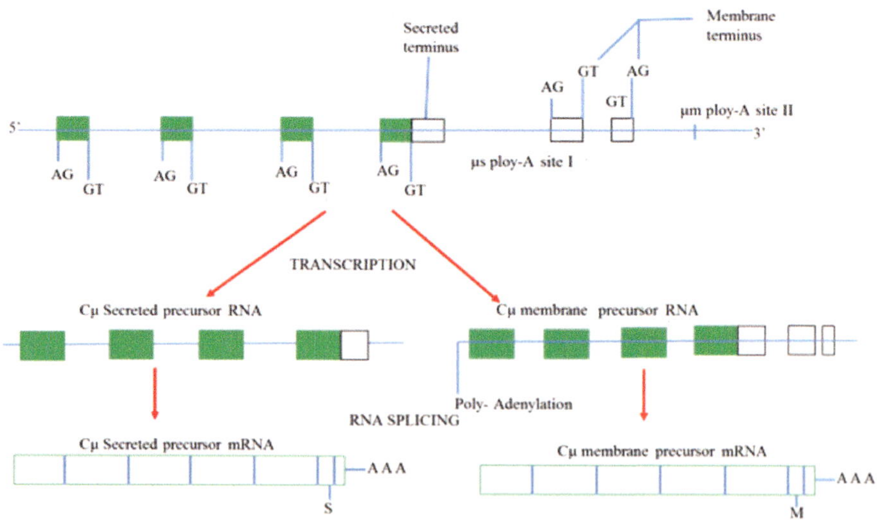

Fig.-4-10. Expression of secreted and membrane bound antibody primary transcript of the re-arranged heavy chain genes showing the Cμ exons at 5'-end and poly-A at 3'-end.

In membrane-bound IgM the M1 exon (Cm gene segment consists of four exons C_μ^1, C_μ^2, C_μ^3 and C_μ^4) encodes the cytoplasmic segment of the C_{11}^4 domain and M2 exons encode the transmembrane and the cytoplasmic segments. The processing of a common primary transcript determines the production of secreted or membrane-bound form of immunoglobulin. A rearranged m heavy-chain gene contains two poly adenylation signal sequences or poly-A sites. Different patterns of splicing result in the process of cleavage of the primary transcript. In case of cleavage occurs at site 1 (located at the 3' end of the C_μ^4 exon) the M1 and M2 exons are lost. In case cleavage occurs at site 2 (located at the 3' end of the M2 axon, mRNA is produced for the membrane form of the heavy chain. However, there is some mechanism for the production of mRNA in either form of immunoglobulin in naive B cells or in plasma cells.

As mentioned above, primary transcripts contain both the C_μ and C_δ gene segments located at about 5 kbp apart and sites 1 and 2 are associated therewith. Sites 3 and 4 are located in the C_δ gene segments. Expression of IgM and IgD is through mature B cell by two pathways occurring simultaneously or concomitantly i.e. if the heavy chain transcript is cleaved and poly adenylated at site 2 after the C_μ exons, the mRNA will encode the membrane form of the m heavy chain, and if the cleavage and poly-adenylation is at site 4 after the C_δ exons the Cm exons are removed and mRNA will encode the membrane form of the δ heavy chain. Cleavage and poly-adenylation of the primary heavy-chain transcript at site 1 and 3 or at site 1 or 3 will cause secretion of m or δ heavy chain. Figure 4-10 is amply illustrative of these phenomena of concomitant expression of IgM and IgD.

Synthesis, Assembly and Secretion of Immunoglobulins

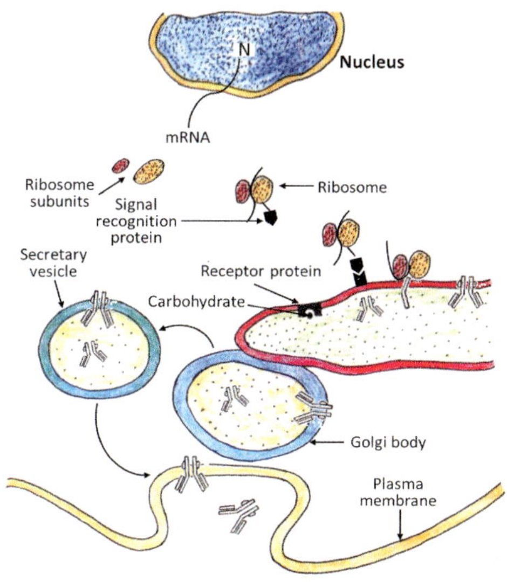

Fig.-4-11. Diagrammatic representation of synthesis, assembly and secretion of Ig-molecules in B cells.

The heavy and light chains are synthesized on separate polyribosomes (polysomes) containing an amino terminal leader sequence, which serves to guide the chains into the lumen of the rough endoplasmic reticulum (RER). The chains are then cleaved and pass through the cisternae of the RER into the Golgi apparatus and then into secretory vesicles. Fig. 4-11 may be reported to for the illustration of this pathway of synthesis of L and H chains.

Chain assembly differs by the classes of immunoglobulins. In the case of IgM assembly of chain takes place within the RER and two half molecules form one complete molecule. In the case of IgG, first two it chains assemble then an H_2L intermediate is assembled to develop into one complete molecule H_2L_2.

Transcription Regulation of Ig genes

At different development stages, immunoglobulin genes are expressed in B cells at different rates. Immunoglobulin genes are regulated by three types of DNA regulatory sequences- promoter, enhancers and silencers. A promoter is located just upstream of the leader sequence associated with each V_H and V_L gene segment whereas enhancers and silencers are located at considerable distances from the starting of the transcription site (Fig. 4-12). The immunoglobulin promoters contain a highly conserved AT-rich sequence called the TATA box wherein the actual process of transcriptiontakes place.

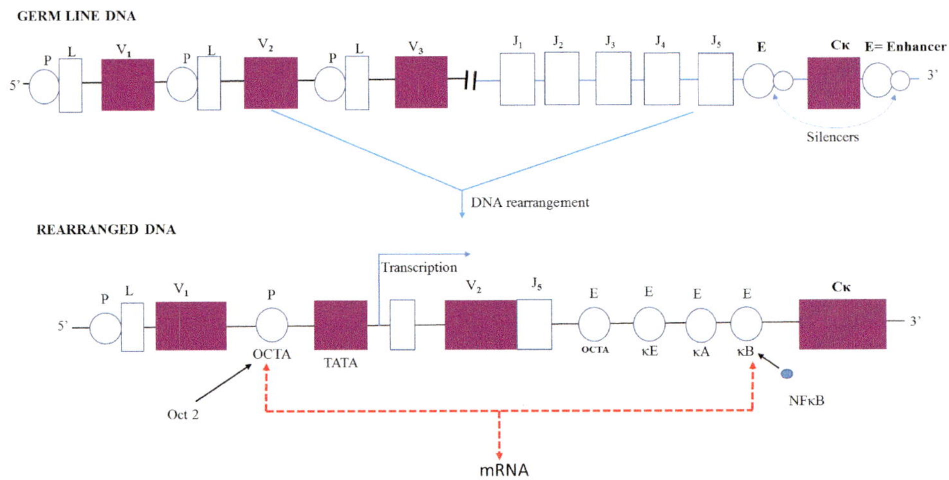

Fig.-4-12. Generalized location of promoter, enhancer and silencer on Ig-genes. During the synthesis, the silencers come closer in gene re-arrangement by recombination.

Transcription is performed by RNA polymerase II, which starts transcribing DNA from the initiation site. How enhancers and silencers modulate transcription has so far not been fully understood. It is thought that at the promoter site DNA binding factors bound to an enhancer changing the chromatin structure, which facilitates the formation of a stable transcription. Chromatin structures are rendered inaccessible for transcription when silencer binding proteins interact directly with histones. In germ-line DNA transcription of V_H and V_L gene segments is negligible, gene rearrangement brings enhancers close to the promoter they affect greatly increasing the transcription rate of rearranged VLJL and VHDHJH units.

Effect of DNA Rearrangement on Transcription

In unarranged germ-line DNA, the rate of transcription of V_H and V_L coding regions is almost negligible, because the promoters associated with the immunoglobulin V gene segments bind RNA polymerase II very weakly. The variable region gene enhancers affect transcription from the promoter to cause rearrangement of V_LJ_L or $V_HD_HJ_H$ unit to rise to 10^4 times the rate of transcription of unarranged V_L or V_H segments. Enhancers transcribe the transfected genes at a rate higher than those wherein enhancers are deleted. Thus the role of enhancers in transcription of immunoglobulin genes is confirmed. Enhancers also activate genes that promote cellular proliferation or prohibit cell death (cellular oncogenes). Translocations of the *C-myc* and *bcl-2* oncogenes have each been associated with malignant B cell lymphoma. The transcription of *C-myc* leads to constitutive expression of *C-myc* and translocation of *bcl-2* leads to suspension of programmed cell death in B cells.

Inhibition of Ig-Gene Expression in T Cells

Ig gene rearrangement of H and L chains occurs only in B cells and complete T cell receptor (TCR) gene rearrangement is limited to T cells. Studies by Hitoshi Sakano and coworkers have revealed that when a sequence known as the PU.1 binding site within the 3 k-chain enhancer was mutated, Vk-Jk joining occurred in both T cells and B cells. That is the binding of a protein expressed by T cells (not B cells) to the un-mutated k-chain enhancer usually inhibits the Vk-Jk joining in T cells.

Recombinant DNA Technology for Antibody or Antibody Engineering

The recombinant DNA technology is the potential to genetically grafting of antigen-binding proportions of an antibody molecule made in one species onto the constant regions of the antibodies of a different species (Fig. 4-13, 4-14). The introduction of mouse monoclonal antibodies in humans (being foreign) is cleared from the human blood stream by human anti-mouse antibodies (HAMA) besides causing allergic reactions due to circulating complexes of mouse and human antibodies. To use human monoclonal antibodies for grafting user poses technical problems. For this reason, studies have set in, in the field of recombinant DNA technology. The work of Cesar Milstein has paved the way to design and construct genes that encode Ig molecules in which the variable region genes come from one species and the constant region genes come from another. By it molecular hybrids or chimeras can be made to deliver a powerful toxin to a particular antigenic target such as a tumor cell. Transplanted Ig genes have endowed the receiver to respond to immunization by producing antibodies encoded by that of the donor. The entire antibody repertoire of individuals may be reconstructed by the selected L and H chain variable region genes.

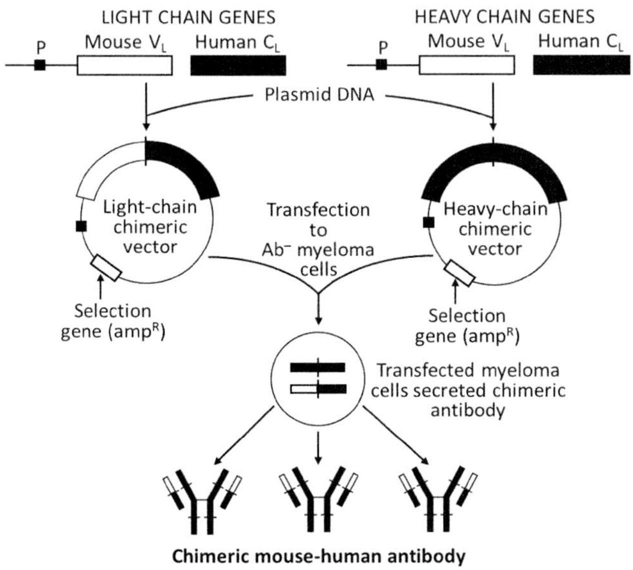

Fig.-4-13. Experimental demonstration of production of monoclonal antibody chimeric for both human and mice with heavy and light chains.

Additional approaches to recombinant DNA technology are humanized chimeric antibodies, bispecific antibodies (heteroconjugates), derivation and use of Ig-gene combinatorial libraries (H and L chain libraries). and Cloning the recombinant DNA from mouse containing the promoter leader and variable region sequences from a mouse antibody gene and the constant-region exons from a human antibody gene is called humanized chimeric antibodies. The humanized chimeric antibodies are potential to trigger human competent activation or FC receptor binding.

This technique has been used to treat patients of Hodgkin's lymphoma (cancers of lymphatic tissues). Unlike the major therapies used in treatment of lymphomas i.e. radiation or chemotherapy, wherein a broad spectrum of normal cells as well as tumor cells are killed or severely damaged, the humanized chimeric antibodies have proved to be a novel therapeutic agent. This grafting of entire B cell populations including B lymphoma cells were cleared from the body in a way that other cell populations are spared. The therapeutic agent used in the humanized chimeric antibodies, is known as CD20 a membrane-bound phospho-protein. The important issues with such a treatment- efficacy, safety and immunogenicity are however under vigorous and intensive studies. Fig. 4-14 is illustrative of the application of CD20.

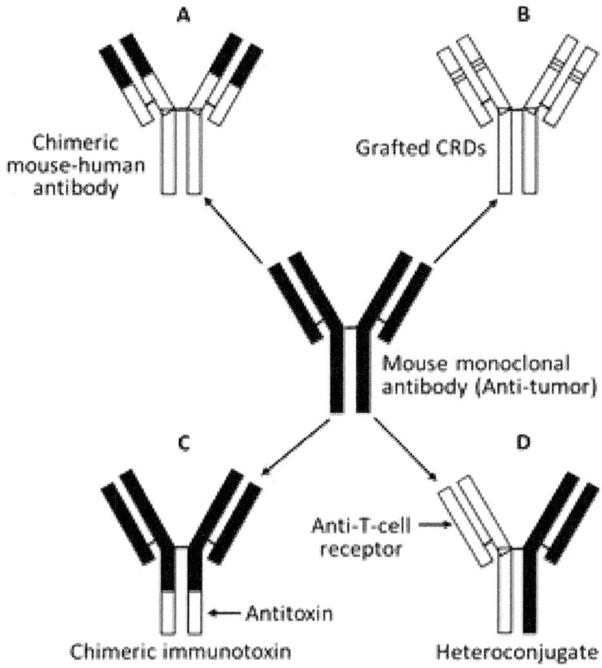

Fig.-4-14. Diagrammatic representation of chimeric hybridoma monoclonal antibodies engineered by recombinant DNA technology.

Bispecific antibodies are hybrids of two different antibody molecules, constructed by chemically cross-linking two different antibodies or synthesized monoclonal antibody producing cells (completely fused). The destruction of tumor cells begin when the effector cell is cross linked to the tumor cells.

Producing gene libraries encoding Fab fragments is another approach for generating monoclonal antibodies. In it, the PCR (Polymerase Chain Reaction) is employed to amplify the DNA that encodes antibody H and L chain Fab fragments from hybridoma cells or plasma cells. Amplified sequences are endowed with a promoter region and EcoRI restriction site and the resultant construct is then inserted into bacteriophage l yielding separate H and L chain gene libraries. Gene libraries have yielded and impressive amount of diversify >10^{10} unique member. Specificities against a wide variety of antigens can be achieved from these gene libraries. The specific antibodies may certainly delete the need for immunization (vis-a-vis in vivo immunization). However, the immune response is important and highly potential in comparison to gene libraries, since they have different effective functions and hence greater biological versatility, and higher and higher affinity. The potentialities of the immune response are being tried to be induced in gene libraries.

CHAPTER – 5

MAJOR HISTOCOMPATIBILITY COMPLEX AND ANTIGEN PRESENTING CELLS

Genesis of Autoimmune Response

Autoimmunity or susceptibility to disease, of an individual is hereditary. The hereditary factor which accounts for this immune response of an individual to antigens of infected organisms is the major histocompatibility complex (MHC) molecules, which recognize antigens by T cells (T_H cells and T_C cells). In case the antigen is not accepted or responded as foreign, it is recognized as outsider and so it is histocompatible. The immune system rejects it by developing antibodies.

Organization of MHC Genes

The MHC (major histocompatibility complex) genes are tightly linked with DNA on chromosome 6 in humans is called HLA complex. However, the MHC genes linked with DNA on chromosome 17 in mice is called H-2 complex. The genes encode proteins associated with intracellular recognition and antigen presentation to T lymphocytes. The MHC called HLA and H-2 complex in humans and mice are organized into three regions based on the type of molecules encoded (Fig. 5-1).

Class-I MHC Genes: Encode glycoproteins expressed on nearly all nucleated cells and their function is presentation of peptide antigens to T cells. Class-1 MHC molecules are expressed in the K and D regions in mice and A, B, and C loci in humans. The function of these genes is expressed in developing cytotrophoblasts at the fetal-natural interface so that the fetus is not recognized as foreign and not rejected by maternal TC cells. Thus, it protects the fetus and enables to develop it in the material body.

Mouse H-2 Complex

Complex	H-2						
MHC Class	I	II		III		I	
Region	K	IA	IE	S		D	
Gene product	H-2K	IAαβ	IEαβ	C′-protein	TNF-α TNF-β	H-2D	H-2L

Mouse H-2 Complex

Complex	HLA							
MHC Class	II			III		I		
Region	DP	DQ	DR	C4, C2, BF		B	C	A
Gene product	DPαβ	IAαβ	IEαβ	C′-protein	TNF-α TNF-β	HLA-B	HLA-C	HLA-A

Fig.-5-1. General organization of the Major Histocompatibility Complex (MHC) in the mouse and human.

Class-II MHC genes: Encode glycoproteins expressed on antigen presenting cells (macrophages, dendritic cells and B cells) and their function is to present processed antigenic peptides to T_H cells, Class-II MHC molecules are expressed in the IA and IE regions in mice and DP, DQ, and DR regions in humans with specialized functions in the immune process.

Class–III MHC genes: Encode various secreted proteins that have immune functions, where class-I and class-II MHC genes are concerned with antigen processing, the product of class-III MHC genes, in contrast, includes various complement molecules, pro-inflammatory cytokines including tumor necrosis factor (TNF), and heat shock proteins (HSPs) and inducing an inflammatory response.

Thus one of the characteristic features of MHC molecules is their regional distribution in the immune frame, encoding glycoproteins and proteins, and presenting antigens to T cells.

MHC Haplotypes

Another feature of MHC molecules is the inheritance of MHC forming distinct haplotypes in the organisms. An individual inherits the alleles one each from the mother and the father. The alleles are expressed in the same cells, and therefore express both parental sets of alleles. The human HLA complex is highly polymorphic and multiple alleles of each class I and class II gene exist. In humans MHC loci are closely linked and usually inherited as a haplotype. When the father and mother have different haplotypes (Fig. 5-2) there is a one in four chance that siblings will inherit the same paternal and maternal haplotypes and therefore be histocompatible with each other, none of the offspring is histocompatible with the parents. This factor is responsible for diversity of the HLA loci in the human population. This very feature also accounts for any two individuals, being unrelated, to have identical HLA genes.

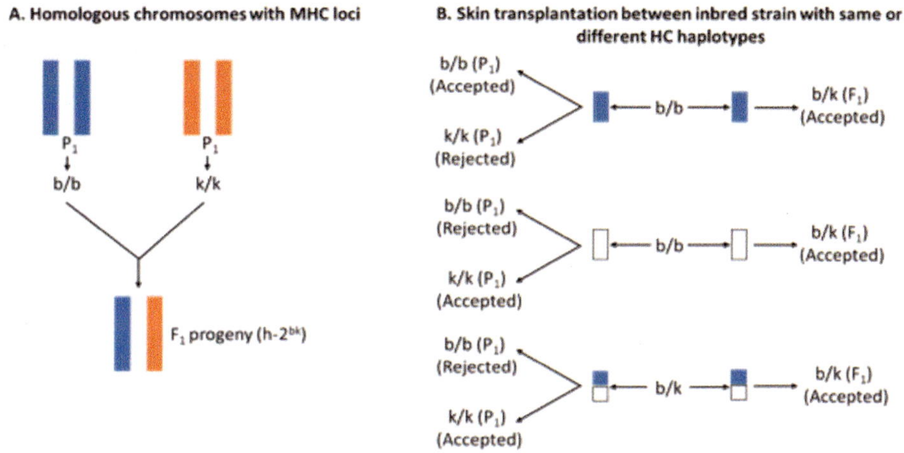

Fig.-5-2. Illustration of inheritance of MHC haplotype in inbreed mouse strain, (a) mating of inbreed mouse strains with different MHC haplotypes, and (b) skin grafting between inbreed mouse strains with same or different haplotypes.

Structure and Function of Class I and Class II Molecules

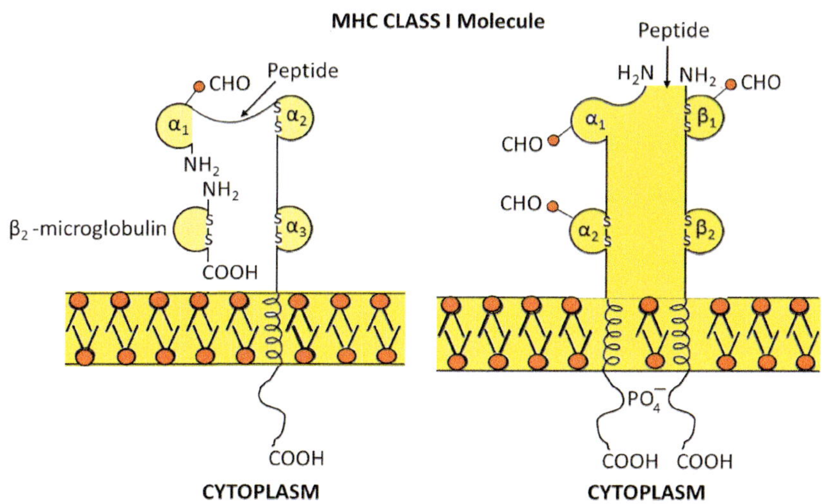

Fig.-5-3. Schematic diagrams of class I and class II MHC molecules, with a distinct α-subunits (α_1, α_2 and α_3) and β_2-microglogulin in class I MHC and α-subunits (α_1, α_2) β-sub units (β_1 and β_2) in class II MHC molecule.

Class I and class II MHC molecules as already stated function as highly specialized antigen-presenting molecules that form unusually stable complexes with antigenic peptides, displaying them on the cell surface for recognition by T cells, whereas class III MHC molecules (group of unrelated proteins) are neither common in structure nor in function to the class I and class II MHC molecules. Fig. 5-3 is schematic diagram of class I and class II MHC molecules. Isolation, purification and crystallization of the extra-cellular portion of class I and class II MHC molecules have revealed the following specificities in their structures and function;

Class I MHC molecules- (Structure and Function)

- A large a chain of the polymorphic transmembrane glycoprotein of about 45 kDa encoded by genes within A, B and C regions of the human HLA complex is associated noncovalently with the much smaller b2 microglobulin (protein of 12 kDa) molecule (Fig. 5-4). The a chain is anchored in the plasma membrane by its hydrophobic transmembrane segment and hydrophilic cytoplasmic cell. Expression of class I molecules on the cell membrane is subject to the association of the α chain with the β_2 microglobulin.
- There are three external domains of a chain of class I MHC molecules- α_1, α_2 and α_3 each containing about 90 amino acids, a transmembrane domain of about 25 hydrophobic amino acids followed by a short stretch of charged (hydrophilic) amino acids and a cytoplasmic anchor segment of 30 amino acids. There has been found striking similarity between the α_3 domain and α_2 microglobulin and the constant-region domains in immunoglobulins. Further there is continual interaction between the α_1 and α_2 domains and a membrane-proximal pair consisting of α_3 domain and α_2 microglobulin. The interaction of α_1 and α_2 domains results in the formation of a platform of antiparallel b-strands spanned by two long a-helical regions. A peptide binding cleft is located on the top surface of the class I MHC molecule bound to the α_1 and α_2 domains.

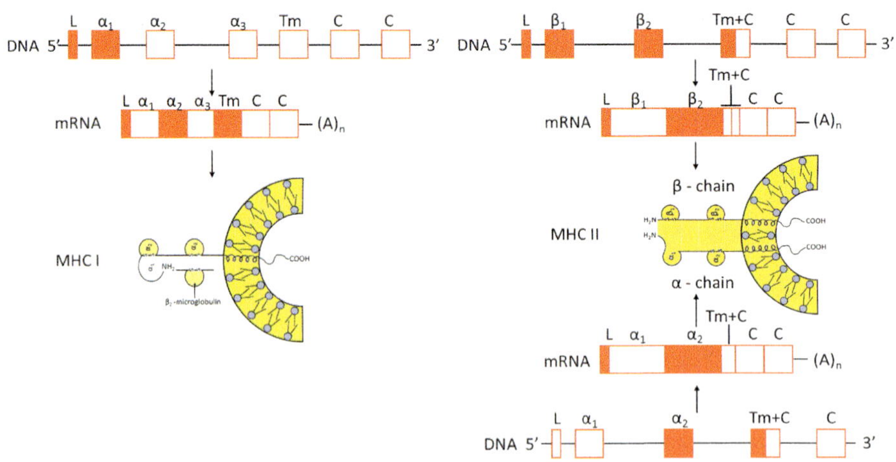

Fig.-5-4. Schematic diagram of 5'3'-end of DNA representing different regions for mRNA for MHC class I and class II molecules.

Class II MHC molecules - (Structure and Function)

- Class II MHC molecules are composed of two different polypeptide chains, a 33 KDa a chain and a 28 KDa b-chain associated by noncovalent interaction (Fig. 5-4). There are two external domains- α_1 and α_2 in one chain and β_1 and β_2 domains in the other. The α_1 and β_1 domains form the antigen binding grove (cleft) for processed antigen. Peptide binding clefts can be superimposed.
- For human DR1 the class II molecules occur as a ab heterodimer, which causes the location of peptide binding clefts in opposite direction.

Class I and Class II MHC genes, by changing and sequencing have been found to have a series of exon and introns, wherein the exons encode the extracellular α_1, α_2 and α_3 domains and one or two 3'-terminal exons encoding the cytoplasmic domains (C).

Nature of peptide binding by MHC molecules

Four wide varieties of antigenic challenges are responded by class I and class II MHC molecules. This is made possible by peptide binding by MHC molecules. A given MHC molecule can bind numerous different peptides and some peptides can bind to several different MHC molecules. Thus, the binding between a peptide and an MHC molecule is often referred to as "promiscuous". Most of the MHC molecules expressed on the membrane of a cell get associated with a peptide of self or nonself origin. This binding affinity of various peptides for MHC molecules has been experimentally established by equilibrium dialysis. In the dialysis experiment a labeled peptide is added to a solubilized class I or class II MHC molecule in a dialysis bag and the amount of labeled peptide bound to the MHC molecule is determined at equilibrium. By Seatchard plots, it comes out that the association constant K_D of the peptide-MHC molecules complex is about 10^{-6}, the rate of association is slow, but the rate of dissociation is even slower. This means that the peptide-MHC molecule association is very stable under physiologic condition.

Peptides are derived from endogenous intracellular proteins, which are digested into peptides within the cytosol. From the cytosol these peptides are transferred into the cisternae of the endoplasmic reticulum, where they interact with class I MHC molecules (which present the peptides to $CD8^+$ T cells). Like class I MHC molecules, as above, class II MHC molecules bind peptides and present them to $CD4^+$ T cells. Peptides are derived from exogenous proteins (either self or nonself), membrane bound proteins or proteins associated with the vesicles of the endocytic processing pathway.

The bound peptides isolated from different class I molecules are eight to ten amino acids in length (commonly nine) containing specific amino acid residues to bind to a particular MHC molecule. This nonamer (nine residues) is compatible to the size of peptide binding cleft in class I molecules (closed at both ends). Due to this structural feature of peptides (nonameric), they bind to class I molecules with a 100 to 1000 fold higher affinity than the peptides being longer or shorter. The amino acid residues anchor the peptide into the groove of the MHC molecules, so they are called anchor residues (Fig. 5-5). Any peptide of correct length that contains the same or similar anchor residues binds to the same class I MHC molecule. The anchor residues at both ends of the peptide are buried within the binding cleft, and this keeps the peptide firmly in place. The bound peptide, which assures an extended structure, interacts with the MHC cleft at both ends but arenas away from the floor of the cleft in middle (Fig. 5-6). This is why it is thought that the amino acids that arch away from the MHC molecules are more exposed and interact directly with the T cell receptor.

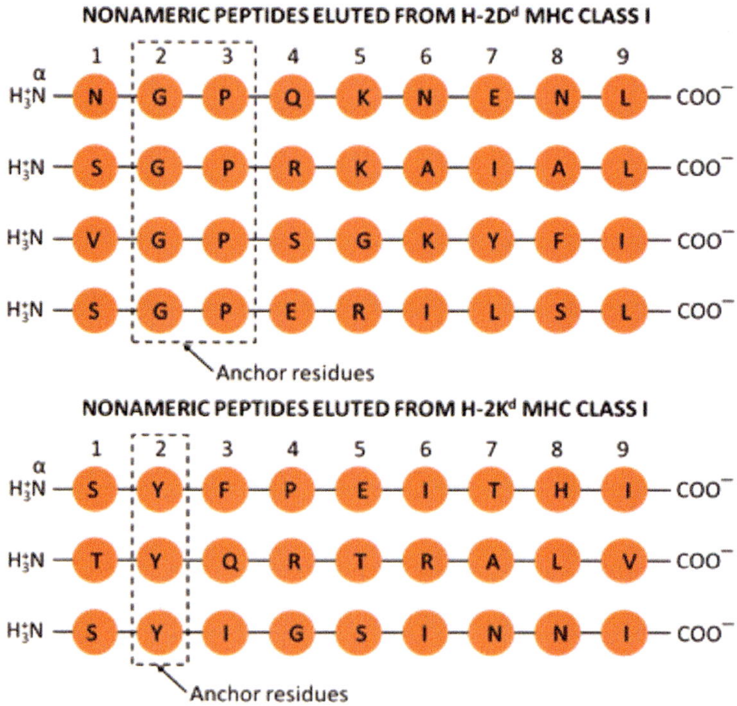

Fig.-5-5. Examples of anchor residue in nonameric peptides eluted from two MHC class I molecules. Alphabet represents the name of amino-acids (Adapted from V. H. Engelhard. 1994. Curr. Opin. Immunol. 6:13; with permission)

The bound peptides isolated from different class II molecules are slightly longer than the nonameric peptides of class I molecules and contain 13-18 amino acid residues. A central core of 13 amino acids determines the ability of a peptide to bind class II molecules. Peptides bound to class II MHC molecules maintain a roughly constant elevation on the floor of the binding cleft. This is the point of difference in peptide binding of class II from class I molecules. Peptides binding to class II molecules lack anchor residues, instead hydrogen bonds between the backbone of the peptide and the class II molecules are distributed throughout the binding site, unlike clustered at the ends of the site (as in class I molecule).

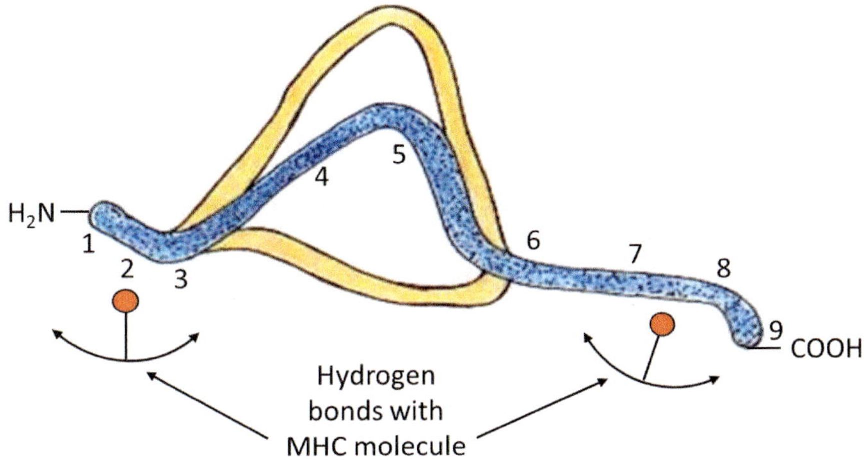

Fig.-5-6. Conformational changes take place at 4 and 5 number to form the bulge, which is responsible for the firm binding with the MHC class I molecule. Adapted from P. Parham. 1992. ***Nature 360:367*****; with permission.**

The Diversity of MHC Molecules

The MHC molecules in one individual may differ dominantly from those expressed by another individual of the same species. This diversity within a species is referred to as caused by polymorphic (the presence of multiple alleles at a given genetic locus within the species). This diversity with a species may either stem from the presence of duplicated genes with similar or overlapping functions, not unlike the isotype of immunoglobulins. Thus, the phenomenon of diversity is also referred to be polygenic. The difference between the frequency observed for a particular combination of alleles and that expected from the frequencies of the individual alleles is known as linkage disequilibrium.

X-ray crystallographic analyses of several class I molecules have determined the location of the polymorphic residues within the structure of the membrane-distal domains in class I and class II MHC molecules, as also, the relevance of different alleles to their functional differences. The analysis has shown that out of 17 amino acids 15 are involved in the peptide binding cleft of class I molecule. It has been determined that allelic differences contribute to the differences in the ability of MHC molecules to interact with a given antigenic peptide and that allelic differences in the class II molecules expressed by antigen presenting cells may influence the cells ability to recognize a particular peptide.

Class I and II molecules have a role in antigen presentation, but class III molecules have a role in immune responses. Certain regulatory disorders of genes within the class III region of the MHC may be caused by genetic association of particular MHC alleles. These class III molecules include several complement components (C2, C4a, C4b, and factor β) two steroid, 21 hydroxylase enzymes, cytokines, tumor necrosis factors α and β (TNF-a and TNF-β) and two heat shock proteins (an unusual group of highly conserved proteins that are produced by cells in response to various stresses, nutrient deprivation, oxygen radicals and viral infection).

Detailed Genomic Map of MHC Genes

Cloning of both mouse and human MHC genes has permitted restriction-enzyme mapping of the MHC and sequence analysis of clones derived from the MHC have refined and expanded the earlier serological assays used for the purpose of mapping. As a result, nonclassical MHC genes have been identified recently. Figure 5-7 shows a detailed genomic map of the mouse and human MHC including classical and nonclassical MHC molecules.

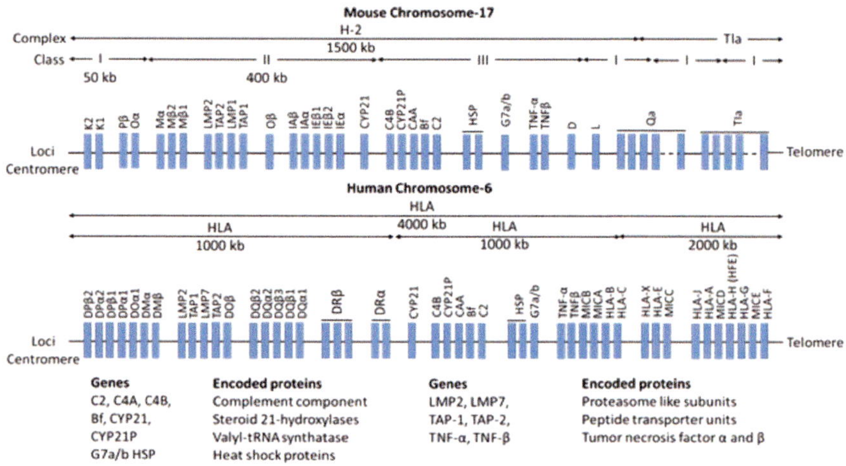

Fig.-5-7. General organization of genomic map of mouse and human MHC including gene encoding classical and non-classical MHC molecule.

Salient Features of Class I MHC Map

- Class I MHC region is about 2000 Kb long and contains approximately 20 genes encoding the well characterized classical class I MHC molecules- HLA-A, HLA-B, and HLA-C in humans and H-2K, H-2D, and H-2L in mice.
- **The nonclassical class I MHC genes are-** HLA-E, HLA-F, HLA-6, HLA-H, HLA-J and HLA-X loci, as well as MIC which includes MICA through MICE. MIC is a recently identified family of genes. These genes do not encode a protein product, but products characterized as class 1b molecules having a chain which associate with b2-microglobulin. The MIC gene products are expressed at low levels in epithelial cells, where heat shock proteins are influenced by heat or other stimuli. They present peptide to T cells.
- The class 1b molecule encoded by the H-2M locus is able to bind a self peptide derived from a subunit of NADH dehydrogenase. This H-2M encoded class I molecule may present peptides from prokaryotic organisms like

Mycobacterium tuberculosis, Listeria monocytogenes, Brucella abortus and Salmonella typhimurium

Salient Features of Class II MHC Map

- The Classical class II MHC molecules- HLA-DR, HLA-DP, and HLA-DQ in humans and H-21A and H-1E in mice, encode the a and b chains. The a and b-chain products are expressed together, increasing the number of different antigen-presenting molecules in the cell.
- The nonclassical class II MHC molecules- DM, DN, and DO have been identified. The functions of these genes differ;
    a. the DM gene facilitates the loading of antigen peptides onto the class II MHC molecules.
    b. the DO serves as a regulators of class II antigen processing.
- Besides, the above two genes LMP2 and LMP7 encode proteasome subunits and two genes TAP1 and TAP2 that encode peptide transporter subunits, have also been identified on class II MHC region.

Class III MHC Map

These genes encode several complement components, two steroid 21 hydroxylases, two heat shock proteins and two cytokines (TNF-α and TNF-β). Congenital adrenal hyperplasia like diseases is caused by these genes.

Differential Expression of the MHC Genes

The classical class I MHC molecules are expressed on most somatic cells with diverse expression on different cell types. The highest levels of class I molecules are expressed by lymphocytes, where they constitute approximately 1% of the total plasma-membrane proteins or some 5×10^5 molecules per cell. The low levels are expressed by muscle cells, liver hepatocytes, fibroblasts and neural cells.

The expression of class MHC I molecules allows each cell to display a large number of peptides. If a cell is normally healthy, its class I molecules will display self peptides resulting from normal turnover of self proteins. If a cell has been infected by a virus, then viral peptides, as well as self peptides will be displayed by its class I MHC molecules. Each individual has allelic differences in the peptide-binding clefts of the class I MHC molecules; this is why different individuals within a species have different sets of viral peptides.

The expression of class II MHC molecules is constitutively by antigen presenting cells- macrophages, dendritic cells and B cells with marked differences. That is, class II molecules cannot be defected on pre-B cells but are expressed on the membrane of mature B cells. Each classical class II MHC molecules is composed of two different polypeptide chains encoded by different loci. Fig. 5-8 further illustrates the factor of expression of class II MHC molecules, as expressed by a heterozygous individual. Thus, a heterozygous individual will express the potential molecules as well as molecules containing a and b chains from different chromosomes (from either parent).

Regulation of MHC Expression

How the level of expression of both class I and class II genes, being different among cell types is regulated, is under extensive research, so for the following regulatory mechanisms have been identified:

- **Transcriptional regulation:** CIITA (an MHC II transactivator) and RFX (another transcription factor) bind to the promoter region of class II MHC genes. Several immunodeficiencies may result from the lack of or defects in these transcriptional factors.

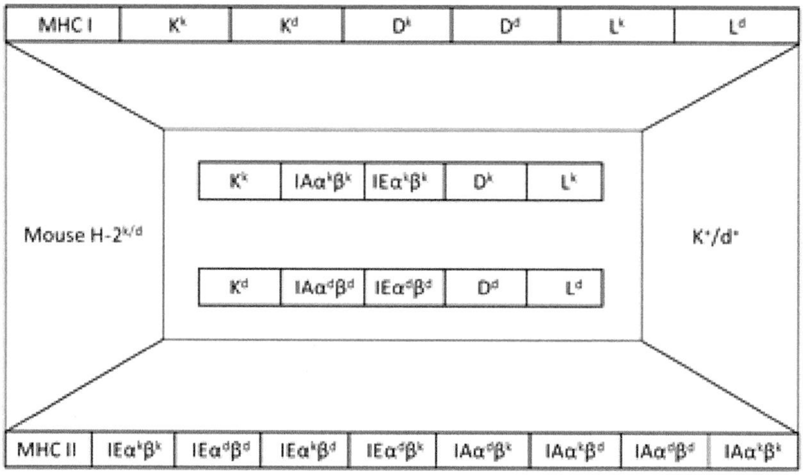

Fig.-5-8. Diagram illustrating various MHC molecules expressed on APCs of heterozygous H-2 $^{k/d}$ Mouse.

- **Cytokinal regulation**: The interferons (α, β and γ) and tumor necrosis factor, each induce the formation of a specific transcription factor that binds to the promoter sequence flanking the class I MHC genes, encoding the class I α chain, β_2-microglobulin. Class II MHC molecules are influenced by the proteasome submits (LMP) and the transporter submits (TAP).
- **Effects of infection**: Infections caused by CMV (cytomegalovirus), HBV (hepatitis B virus), Ad 12 (adenovirus 12) reduce a component needed for peptide transport, prevent assembly of class I MHC molecules and their transport to the plasma membrane, cause a pronounced decrease in transcription of the transporter genes (TAP). The blocking of TAP gene products inhibits peptide transport. As a result, class I MHC molecules cannot assemble with β_2-microglobulin or be transported to the cell membrane. Thus, infection decreases the expression of class I MHC molecules, and the viruses evade the immune response.

Influence of MHC Haplotypes on Immune Response

The loci consisting of the MHC are highly polymorphic (may alternative forms of the genes or alleles exist at each locus). The alleles are inherited sets in the loci, referred to as haplotypes. An individual inherits one each haplotype in the MHC loci, from the mother and father. The class II MHC molecules play a major role in immune responsiveness by presenting antigen to T cells. The haplotypes represent variation and thus account for their immune responsiveness. For this very reason, two explanations have been proposed to account for the variability in immune responsiveness observed among different haplotypes:

- Different class II MHC molecules differ in their ability to bind processed antigen. Animals that respond, for example, to antigen A must express at least one class II MHC molecule whose structure facilitates interaction with antigen A. That is the MHC polymorphism within a species will generate different patterns of responsiveness or nonresponsiveness to different antigens. It is because of the polymorphic character of class II MHC molecules that the haplotypes of the class II molecule show the highest affinity for a particular peptide.
- Immune responsiveness can also be caused by the absence of functional T cells, capable of recognizing a given antigen-MHC molecule complex that is T cells bearing receptors that recognize foreign antigens closely resembling self antigen may be eliminated during thymic processing (leaving a hole in the T cell repertoire).

The two explanations suggest that the T cell response to an antigen involves a trimolecular complex- the T cells receptor, antigenic peptide, and an MHC molecule. This accounts for the observed relationship between MHC haplotype and immune responsiveness to exogenous antigens.

MHC and Disease Susceptibility

Certain Viral diseases, disorders of the complement system, neurologic disorders and several different allergies, have been associated with particular MHC alleles. The proposition is that some HLA alleles occur at a much higher frequency among those suffering from certain/above diseases than in other individuals not suffering from those diseases. The measurement of the frequency of that particular HLA allele gives indication of the relative risk of the individual to that of the disease. Iron absorption is lessened by a disease named hereditary hemochromatosis (HH) commonly prevalent in North Americans of European descent. Usually seen at 40-50 years of age, the disease if not caused, results in hepatic cancer, liver failure severe diabetes and myocardial infarction. Diagnosis and treatment of HH are gaining prominence in research. Table 5-1 lists some significant associations of HLA alleles with potential diseases.

However, the relationship between MHC alleles and the development of diseases is complex; the existence of associations should not be regarded as leading factors for the purpose. Multiple genetic and environmental factors also have roles in the development of disease, especially autoimmune diseases, with the MHC playing a role, but not exclusive. Genomic mapping techniques have been successfully and effectively used to establish the association of HLA alleles and susceptible diseases (since the loci is analysed to assess their contributions in the development of diseases). Complex interactions among the genes at multiple loci may determine the effectiveness of the immune response to a pathogen. The genetic analysis of diseases may therefore highlight the susceptibility of various diseases to the role of MHC alleles. MHC diversity may predispose or prevent a species from being infected by pathogens. A reduction in MHC molecules may result in loss of autoimmune responsiveness or non-responsiveness in otherwise increase in MHC molecules (as antigen presenting). Thus, there is a high level of linkages between MHC genes and infectious diseases.

Table 5-1: Some significant disease and there associated HLA alleles

Disease	*Associated Alleles*
Rheumatoid arthritis	DR4
Reiter's sysndrom	B27
Reactive arthritis (*Yersinia Salmonella, Gonococcus)*	B27
Myasthenia gravis	DR3
Multiple sclerosis	DR2
Insulin-dependent diabetes mellitus	DR4/DR3
Hereditary hemochromatosis	A3, B14, A3/B14

Antigen Processing and Presentation

The sequence of events by which a protein antigen is degraded into peptides is called antigen processing. The peptides derived from antigen are displayed within the cleft of a self MHC molecule on the membrane of T cells. The T cells then recognize these peptides as foreign antigens. The display or association of peptides derived from antigen within the interior cell of MHC molecules is called antigen presentation.

Self-MHC Restriction of T Cells

Antigen-specific proliferation of T_H cells occurs only in response to antigen presented by macrophages of the same MHC haplotype. This is called self-MHC restriction. This phenomenon was experimentally showed by A. Rosenthal and E. Shevach, the results of which are given in Fig. 5-9. These experiments confirmed that the $CD4^+$ T_H cells is activated and proliferated only in the presence of antigen-pulsed macrophages that share class II MHC alleles.

Thus, antigen recognized by the $CD4^+$ T_H cell is class II MHC restricted. In the experiments of R. Zinkernagel and P. Doherty, it was shown that the T_C cells killed only synergic virus infected target cells in mice immunized with lymphocytic choriomeningitis (LCM) virus. That the virus-infected target cells must share class I molecules encoded by the K or D regions of the MHC suggested that $CD8^+$ T_C cells are class I MHC restricted. Substantially speaking $CD4^+$ T_H cells are class II MHC restricted and $CD8^+$ T_C cells are class I MHC restricted.

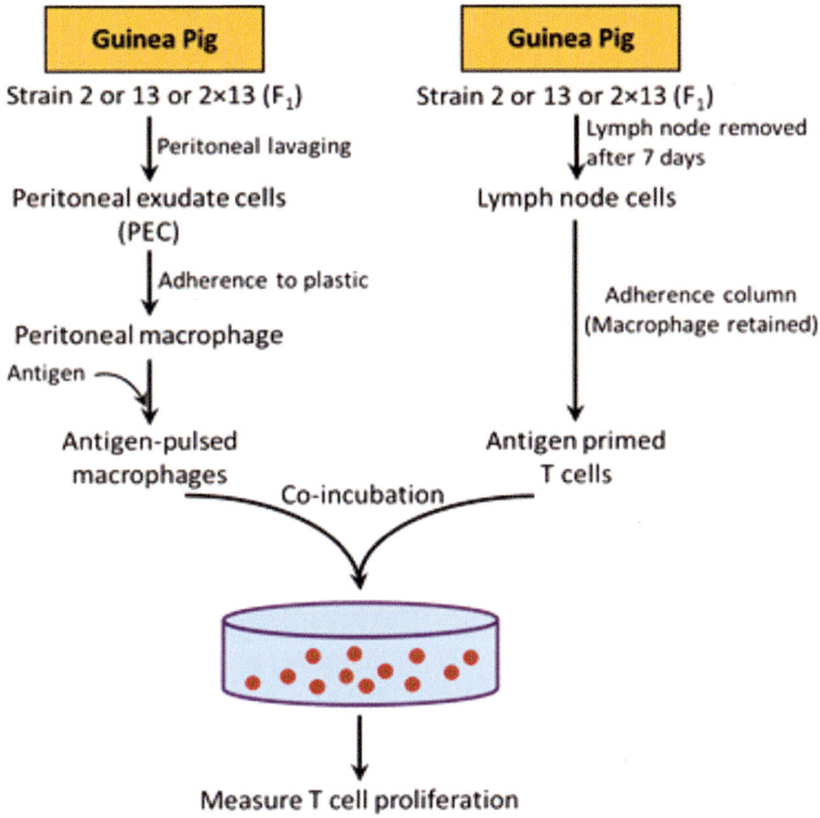

Fig.-5-9. Experimental demonstration of self MHC receptor on T cells.

Role of Antigen-Presenting Cells

Until findings by K. Ziegler and E.R. Unanuae, the dogma prevailed that antigen recognition by B and T cells was basically similar. T_H cell activation by bacterial protein antigens was prevented by a treating the antigen-presenting cells with paraformaldehyde prior to antigen exposure. If the antigen presenting cells were allowed to ingest the antigen and were fixed with paraformaldehyde, one to three hours later T_H cell activation occurred. This suggests that during the interval of one to three hours the antigen presenting, cells had processed the antigen and had displayed it on the membrane suited to activate T cells.

In subsequent experiments antigen presenting cells were treated with glutaraldehyde and then incubated with native ovalbumin or with ovalbumin already subjected to partial enzymatic digestion. It was found that the digested ovalbumin interacted with the glutaraldehyde-fixed antigen-presenting cells, and activated ovalbumin-specific T_H cells but the native ovalbumin could not activate them. So, it was concluded that antigen processing involves the digestion of the protein into peptides which are recognized by the ovalbumin-specific T_H cells.

Experimental evidences corresponding to the conditions that activate T_H cells are further demonstrated in Figure 5-10. In experiments to identify the proteins of influenza virus that were recognized by T_C cells, A. Townsend and his colleagues found that internal proteins of the virus, such as matrix and nucleocapsid proteins, were often recognized by TC cells better than the more exposed envelope proteins. Further, when noninfected target cells were incubated in vitro with synthetic peptides corresponding to sequences of internal influenza proteins, these cells could be recognized by TC cells and subsequently lysed just as well as target cells (target cells are cells that have been infected by a virus or some other intracellular microorganism) that had been infected with the influenza virus. Substantially speaking, antigen processing is a metabolic process that digests proteins into peptides, which can then be displayed on the cell membrane with class I or class II MHC molecules.

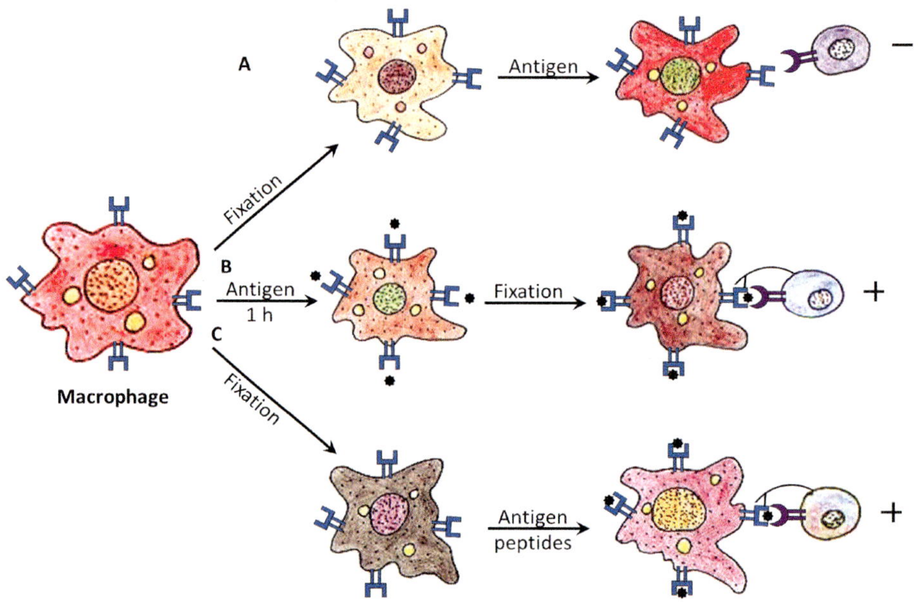

Fig.-5-10. Experimental demonstration of necessity of antigen processing for T-helper cell activation.

Complexes between antigenic peptides and MHC molecules are formed by the degradation of a protein antigen in one of two different antigen-processing pathways. Those cells that process and present peptides associated with class II MHC molecules are called antigen-presenting cells and those that process and present peptides associated with class I MHC molecules are called target cells. Three cell types function in antigen presentation-dendritic cells, macrophages and B-lymphocytes. These cells are called professional antigen-presenting cells and function for a longer period of time than several other cells which function during a sustained inflammatory condition.

Pathways of Antigen Processing and Presentation

The immune system is challenged by intracellular and extracellular antigens. The intracellular antigens are called endogenous and extracellular antigens are called exogenous. The secreted antibodies eliminate extracellular antigens and cytotoxic T-lymphocytes eliminate intracellular antigens. Two different antigens presenting pathways are activated by the immune system; endogenous (intracellular) antigens are processed in the cytosolic pathway and presented on the membrane with class I MHC molecules, and exogenous (extracellular) antigens are processed in the endocytic pathway and presented on the membrane with class II MHC molecules (Fig. 5. 11). L.A. Morrison and T.J. Braciale with two clones of T_C-cells specific for the influenza virus suggest that class I and class II MHC molecules present antigenic peptides derived from different processing pathways. In their experiment, one clone was a typical T_C line that recognized influenza hemagglutinin (HA) associated with a class I MHC molecule and the other an atypical T_C line that recognized influenza HA associated with a class II MHC molecule.

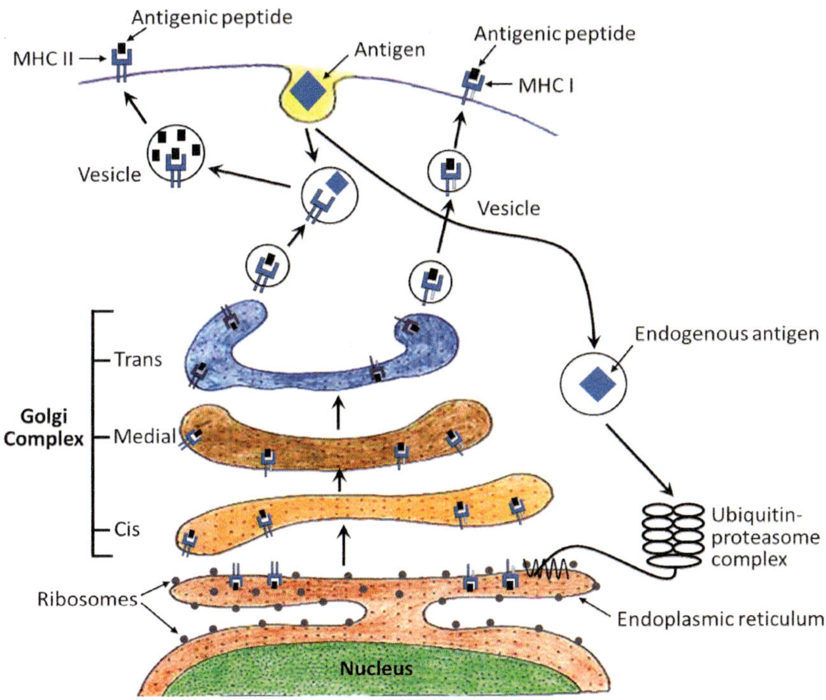

Fig.-5-11. Overview of both type of pathways of antigen processing and presentation.

In another set of experiments, target cells that expressed both class I and class II MHC molecules were incubated with infectious influenza virus or with UV-inactivated influenza virus. The inactivated virus retained its antigenic properties but was no longer capable of replicating within the target cells. Lysis of the target cells was determined by incubating them with the class I restricted or class II restricted T_C cells. These observations and further reports provide evidence that the association between T cell function and MHC restriction is not absolute. Further, the class II restricted T_C cells responded to target cells treated with infectious or noninfectious influenza virions.

Morrison and Braciale in their experiment to determine whether antigen is processed along different pathways for association with class I or class II MHC molecules, as above observed, target cells that had been treated with infectious influenza virions stimulated the class II restricted T_C cells but not the class I-restricted T_C cells. This was the result when emetine was used with infectious influenza virions. In the presence of chloroquine, the target cells treated with influenza virions stimulated the class I restricted T_C cells only.

The results of above experiments show that the mode of antigen entry into cells and its subsequent processing with either the endocytic processing pathway (exogenous antigen) or cytosolic pathway (endogenous antigens) determine whether the resulting antigen peptides will associate with class I or class II MHC molecules. The separate routing of class I and class II MHC molecules to separate intracellular compartments has been further examined in the following sections.

Degradation of Endogenous Antigens in Cytosolic Pathway

Endogenous antigens are degraded within the cytoplasm into peptides that can be presented to T_C cells by class I MHC molecules. The pathway involved in the normal turnover of intracellular proteins is the pathway for degradation of endogenous antigens.

Intracellular proteins are degraded into short peptides by a cytosolic proteolytic system present in the cells. Proteins targeted for proteolysis often have a small protein called ubiquitin. These ubiquitin–protein bonds can be degraded by a multifunctional protease complex called a proteasome (a large–26S cylindrical particle consisting of four rings of protein subunits with a central channel of 10-50 Å. The subunits of proteasome catalyze cleavage of peptide bonds. This prevents lysis of other proteins within the cytoplasm (Fig. 5-12). Peptides bind to class I MHC molecules due to the peptides activities of proteasome through various subunits, encoded with the MHC gene cluster and are induced by increased levels of IFN-γ. Substantially speaking, endogenous antigens are degraded into peptides within the cytosol by a large enzyme complex called the proteasome, which has peptidase activities.

A. Townsend and his colleagues found that there is a mutant cell line called RMA-S cells. These synthesize normal levels of class I a chains and β_2-microglobulin. Peptides might be required to stabilize the interaction between the class I α chain and β_2-microglobulin. They suggested that feeding (transfecting with a gene) of RMA-S cells, which otherwise remain intracellular, enables the predigested cells to be compatible to presentation by class I MHC molecules. This led them to assume that RMA-S cells have defect, in their protein which transports peptides from the cytoplasm to the RER (Rough Endoplasmic Reticulum), where class I molecules are synthesized. RMA-S cells, after transfection with a functional gene encoding the transporter protein, begin to express class I molecules on the membrane, which otherwise remains intracellular. The transporter protein, because it is associated with antigen processing, is a membrane-spanning heterodimer consisting of two proteins TAP1 and TAP2 (transporter antigen processing) belonging to the family of ATP-binding cassette proteins. These proteins are found in the membranes of many cells, including bacteria and mediate ATP-dependent transport of amino acids, sugars, ions and peptides. Thus, TAP transport peptides that will interact with class I MHC molecules. There are LMP2 and LMP7 genes adjacent to the TAP1 and TAP2 respectively. These genes are polymorphic (different allelic forms) which accounts for observed variation in responses to different antigens. Within the RER the peptides assemble with class I MHC molecules (Fig. 5-13). The binding of peptide stabilizes the association between the class I a chain and b-microglobulin. Peptide class I complexes are transported from the RER through the Golgi complex to the plasma membrane. Peptides not bound by class I molecules are rapidly degraded.

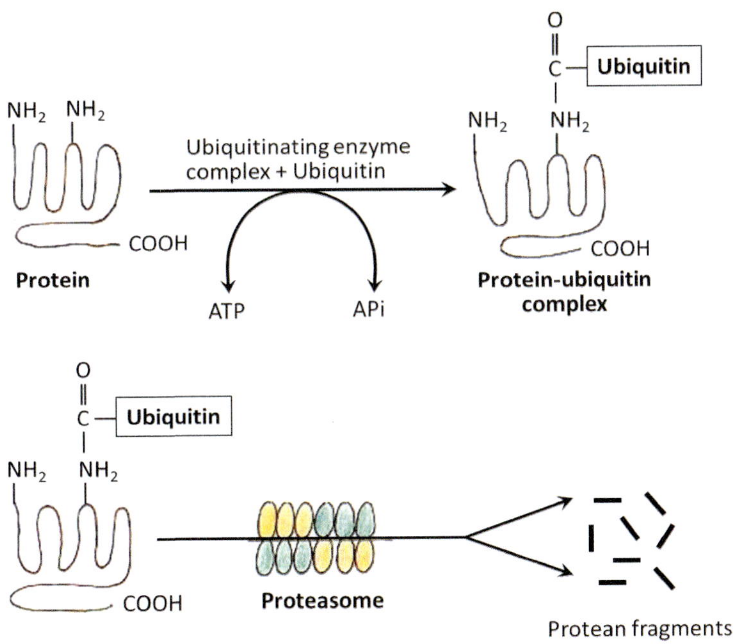

Fig.-5-12. Cytosolic proteolytic system for degradation of intracellular proteins.

Degradation of Exogenous Antigens by Endocytic Pathway

APCs (antigen presenting cells) can internalize antigen by phagocytosis or endocytosis. An antigen after being internalized is degraded into peptides by moving through the endocytic processing pathway consisting of three acidic compartments ranging in pH 6.0 - 6.5 to pH 5.0 - 6.0 (endosomes) and pH 4.5 – 5.0 (lysosomes). The mechanism by which antigen having been internalized moves from one compartment to the other has not been put forth viz. endosomes move to become lysosomes and there are small transport vesicles that carry antigens from one compartment to the other. Antigens are degraded into the compartments encountering hydrolytic enzymes, hydrolases, proteases, glycosidases phospholipases phosphatases etc. and adapting to the pH in various compartments.

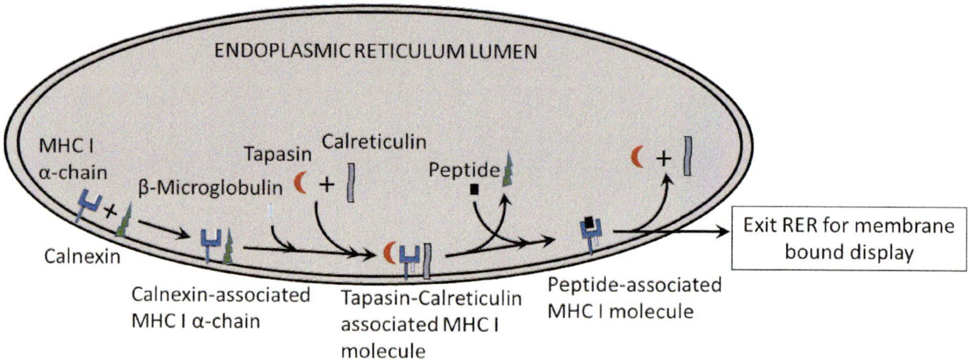

Fig.-5-13. Assembly and stabilization of class I MHC molecules.

Control of Peptide Binding to Endocytic Vesicles

In the RER, there is a trimeric protein called the invariant (Ii) chain. A class II MHC molecule associates with the invariant chain in the process of synthesis within the RER. The invariant chain trimer are preset in the RER, which interacts with the peptide-binding cleft of the class II molecules, resisting any endogenously derived peptides from binding to the cleft. This is how antigen-presenting cells are controlled from binding to the same set of antigen peptides. This mechanism has been demonstrated in transfection experiment with cells that lack the genes encoding class II MHC molecules and the invariant chain. Under immunofluorescent labeling of such cells transfected only with class II MHC genes demonstrated class II molecules localized within the Golgi complex. Substantially speaking, a sort of signaling device in the cytoplasmic tail of invariant chain controls the routing of class II molecules to the endocytic vesicles.

Binding of Peptides with Class II MHC Molecules

As a result of the enhancement of the proteolytic activity in each successive compartment (endosomes to lysosomes), the invariant chain is degraded. A peptide assumes such a compact stable form that once it is bound on the membrane, it is not easy to replace a class II bound peptide with another peptide at physiologic conditions. The transportation of the peptide-class II complexes to the plasma membrane thus poses an interesting study.

After degradation of the invariant chain, as referred to above, a short fragment of it called CLIP (for class II-related invariant chain peptide) remains bound to the class II molecules, thereby resisting any premature binding of antigenic peptide. The separate antigen presenting pathways for both kinds of antigens (endogenous and exogenous) has been shown in the Fig. 5-11. In the MHC complex of humans, there are two nonclassical and nonpolymorphic class II molecule called HLA-DO and HLA-DM which is not expressed at the cell membrane but mainly found within the endosomal compartment. DO is expressed by B cells and DM is expressed by thymus. A reaction catalyzed by HLA-DM allows antigenic peptides to displace CLIP, likely to be inhibited by HLA-DO. The peptide-class II complexes are then transported to the plasma membrane.

Presentation of Non-peptide Antigens

It has long been understood that only peptide antigens are presented by the antigen presenting cells and activate antigen specific immune responses. However, it has been observed that bacterial infection activates a series of events leading to the activation of cell-mediated immunity. The MHC class I molecule present hydrophobic, protease-resistant and periodate-sensitive non-proteinaceous antigen of *Listeria monocytogenes* to murine $CD8^+$ T cells. Sensitivity to periodate strogly suggests that the antigen molecule, which is recognized of T cells must contain carbohydrate moiety. Physical and biochemical analysis of antigen molecule of Listeria monocytogenes reveals that it is glycolipid containing lipoteichoic acids. Similarly, many of the antigens have been characterized in the *Mycobacterium tuberculosis* such as mycolic acid and lipoarabinomannan, which are recognized by the ab T cells. Now, it has established that T cells are able to recognize non-peptide antigens, which are presented in association with CD1 present on antigen presenting cells (APCs), NK cells, etc.

During 1989, Porcelli and coworker found that a molecule called CD1 play important role in antigen presentation. In 1994, Beckman and coworker proposed that CD1 molecules can present lipids to specific T cells. Since then a multitude of progress have been made and now it has been well established that there are a number of non-peptide molecules which are processed and presented to the T cell receptor with the help of CD1 molecule. Thus, CD1 family can be considered as relatively "young" family of antigen-presenting molecules.

CD1 has a unique tissue distribution, with CD1a, -b, and -c present on thymocytes but not peripheral blood T cells. Human CD1 are also present on professional antigen-presenting cells including dendritic cells, Langerhans cells, and B cells. The structural homology of CD1 with class I and class II molecules shows its antigen presenting ability to T cells. However, the amino acids encoded in the predicted antigen binding site, the a1 and a2 domains, are extremely hydrophobic. Like MHC class I proteins, a domain of CD1 molecules are noncovalently associated with b2-microglobulin. Furthermore, the human CD1 gene family consists of five nonpolymorphic genes, CD1 A, -B, -C, -D, and -E, mapped to a cluster on chromosome 1, which display an intra-exon structure similar to MHC I genes, and encode proteins with significant homologies to MHC I and MHC II molecules. Although MHC I and MHC II molecules are confirmed to present peptides, the binding groove of the CD1 molecules is hydrophobic and enfolds alkyl chains of lipid antigens, thereby allowing interactions of polar chains with the TCR.

CD1 antigen presentation pathway

CD1-restricted T cell recognition of non-peptide antigens involves an intracellular presentation pathway (Fig.-5-14). It has novel aspects, although it shares some features common with the classical MHC class II pathway. It has been found that the macrophage mannose receptor is involved in the uptake of LAM through binding with its mannose core. The mannose receptor delivers LAM to late endosomal and lysosomal vesicles, as well as to the MHC class II antigen loading compartment (MIIC). The uptake of LAM into endosomes and its presentation to LAM reactive T cells are blocked by reagents that interfere with the mannose receptor. In the MIIC compartment, the mannose receptor, LAM, and CD1b co-localize, suggesting that this is the place where LAM is loaded onto CD1b molecules. CD1b is endocytosed at the plasma membrane in coated pits and coated vesicle structures, transits to early endosomes, and is then delivered to the MIIC. In striking contrast to other MHC molecules,

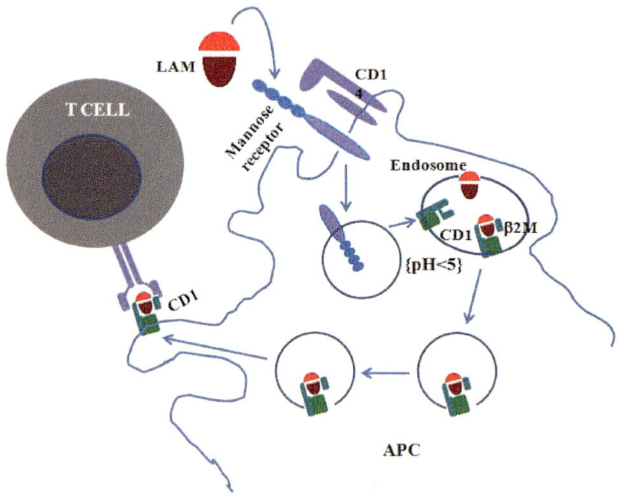

Fig.-5-14. Mechanism of non-peptide antigen presentation in macrophages. Non-peptide antigen presented in association with CD1 molecule is recognized of TCR present on T cells and therefore T cells are activated.

CD1b endocytosis and trafficking to the MIIC compartment is directed by an YXXZ (where Y is a tyrosine, X is any amino acid, and Z is a hydrophobic residue) motif contained in the cytoplasmic tail of the molecule. The endosomal localization motif is essential for antigen presentation by CD1b, confirmed that trafficking of CD1b through lysosomal compartments is required for loading of antigen into CD1b.

This novel antigen presentation pathway bridges two arms of the immune system. Innate immunity pertains to those cells which are preprogrammed to respond in certain ways, including macrophages, natural killer cells, and mast cells. The mannose receptor, a pattern recognition marker, is part of the innate immune response. In this manner, the mannose receptor is able to traffic antigen first recognized by the innate immune system to T cells, which are part of the acquired immune response since their selection and expansion involves the development of memory. In that manner, the CD1 antigen presentation pathway links together the innate immune system and the acquired immune response.

Antigen recognition by γδ T cells

It is also evident that some antigens other than this are capable of activating γδ T cells. γδ T cells comprise 5% of T lymphocyte subset in T lymphocyte pool and are abundant in the inner lining of body, where it plays a pivotal role in protection against infections resulted into pathogen exposure of the inner lining of the body. γδ T cells were shown to recognize isopentenyl pyrophosphate derived from pathogenic microbes including Mycobacterium tuberculosis, isobuyl amine from various bacteria and plants and other related structures in the isoprenoid family (Fig.-5-15) referred to as phosphoantigens or phosphorylated and nonphosphorylated alkylamines. The ability of γδ T cells to recognize these antigens may occur in a manner independent of antigen-presenting molecules and mainly dependent on Vγ 9Vδ 2 TCR. Upon interaction with these phosphoantigens, the gd T cells were found to expand very rapidly in vivo(up to 45% of all T cells in the peripheral blood). Nucleotide conjugates of IPPP and related compounds also stimulate Vγ 9Vδ 2 T cells.

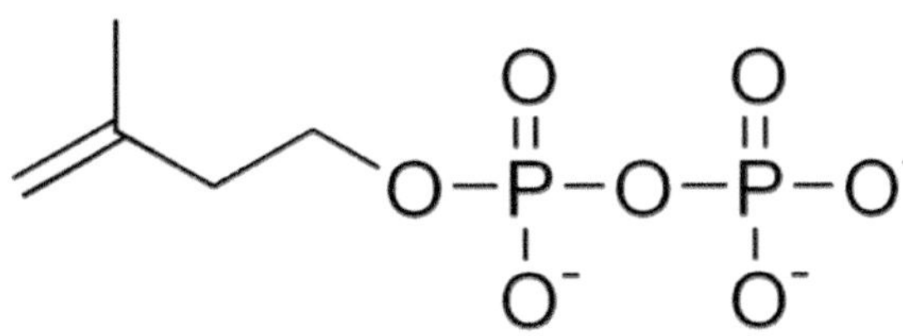

Isopentenyl pyrophosphate (IPP)

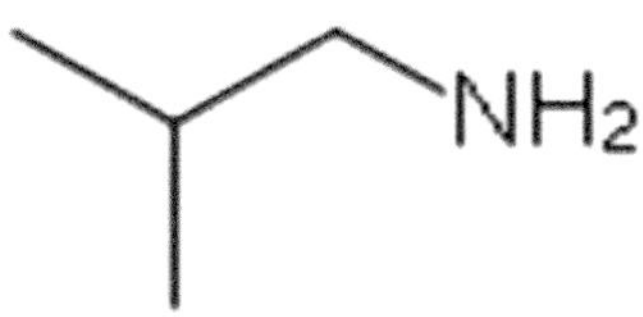

Isobutyl amine

Fig.-5-15. Structure of member of isoprenoid family.

The unusual chemical structure of these nonpeptide antigens suggests that they are presented to γδ T cells in a very distinct manner. Recent studies suggest that recognition of secreted phosphoantigens by human γδ T cells is APC independent and did not require processing. pyrophosphate antigen presentation did not require the expression of known antigen-presenting molecules, including classical MHC class I, MHC class II, CDla, CD1 b, CD1 c, or other β2 microglobulin-dependent molecules. Internalization of antigen is not required. Aldehyde-fixed APCs, which are unable to take up antigens for intracellular processing act efficiently as accessory cells for the response to IPP or other bacterial phosphoantigens. These antigens could not be "pulsed" onto accessory cells and needed to be present continuously during T cell stimulation. Thus, it can be concluded that prenyl pyrophosphate antigens or other phosphoantigens are presented through a novel extracellular pathway of presentation that does not require antigen internalization or processing. As such, they represent a newly recognized class of T cell antigens distinct from peptides and superantigens presented by the MHC systems, and distinct from the lipid antigens presented by the CD1 system. Thus, this pathway allows for the rapid recognition of a variety of naturally occurring small phosphorylated non-peptide antigen of bacteria by γδ T cells which confirm that γδ T cells play a very important role in the early response to bacterial infection.

CHAPTER – 6

T CELL MATURATION TCR-ANTIGEN-MHC INTERACTION T CELL ACTIVATION AND DIFFERENTIATION

Progenitor T cells mature in the thymus and they are activated by the interaction of the T cell receptors with an antigenic peptide in the groove of an MHC molecule. This is known as TCR-antigen-MHC interaction. Following activation, T cells proliferate and differentiate into various types of effector cells and memory T cells. It is this TCR-antigen-MHC interaction (MHC restriction) that distinguishes antigen recognition by T cells only.

Stages of Maturation of T Cells in the Thymus

Thymus is the MHC restricted repertoire of T cells, developed as a result of the selection of self MHC molecules. There are a number of events whereby T cells develop and mature in the thymus, viz. expression of a T cell receptor (Fig. 6-1), TCR rearrangement (RAG-1 and RAG-2) in proper timing etc.

In the early stage when the progenitor T cells enter the thymus, they do not express either surface molecules (CD4 or CD8) or proteins (RAG-1 and RAG-2) which are required for T cell rearrangement. CD4 and CD8 are thus not noticed in the early stage of maturation of T cells. In fact at the early stage, these cells lack CD4 and CD8 surface molecules and are known as double negative cells (DN). Other surface molecules c-kit, CD44 and CD25 mark further development of DN population. At this time, the cells continue to proliferate and CD44 expression is reduced to mark the beginning of the rearrangement of TCR genes. In the thymus developing T cells are known as thymocytes. These thymocytes make productive rearrangement of both γ and β-chain genes and develop into double negative CD3⁺ γδ T cells. This development can be observed in mice by day 14 of gestation. By 17 or 18 days of gestation, there starts a decline until birth. Double negative thymocytes begin to arrange the TCR β-chain genes. The TCR β-chains join a 33 kDa glycoprotein known as the pre-Tα chain and combine with the CD3 group. This complex is known as the pre- T cell receptor (pre-TCR). The pre-TCR has the β-chain and a pre-Tα chain. Some intrathymic ligand is recognized by the pre-TCR and a signal is transmitted through the CD3 complex which activates a protein tyrosine kinase and Lck. There are several effects of signal transduction through the pre-TCR.

- Rearrangement of the TCRα chain is increased.
- Rearrangement of TCR β-chain genes is subsequently suppressed. It results in allelic exclusion of the b-chain gene.
- A double positive state i.e., $CD4^+8^+$ is induced.
- Ensures formation of productive TCR β-chain rearrangement. Thymocytes, which express the β-chain, are selected for further expression and maturation.

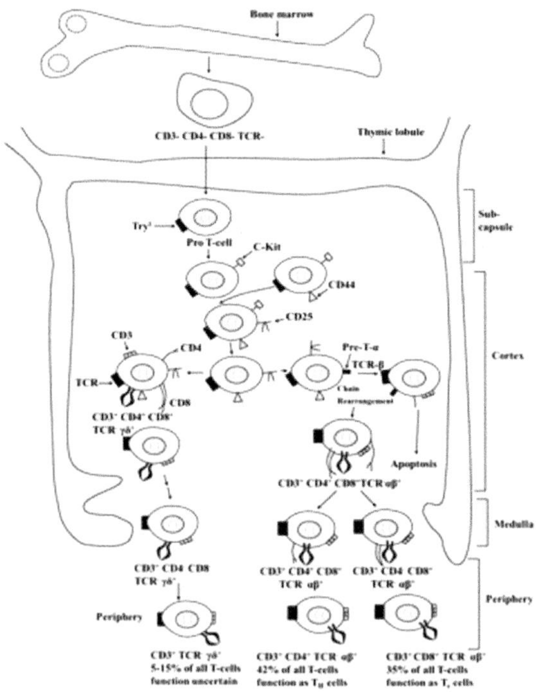

Fig.-6-1. Proposed pathway for murine T cell development in the thymus.

Double positive thymocytes ($CD4^+8^+$) begin to proliferate with the transmission of a signal through the pre-TCR. At the stages of this proliferation, although both the RAG-1 and RAG-2 genes being transcriptionally active, TCR α-chain gene rearrangement is absent (because the RAG-2 protein undergoes rapid degradation in proliferating cells). However, the α-chain gene rearrangement starts after stoppage of proliferation by double positive thymocytes. This proliferative phase contributes to T cell diversity by generating a clone of cells with a single TCR β-chain rearrangement. A greater diversity is resulted in each of these clone cells and different α-chain genes rearrange. At day 17 both CD3 and the ab T cell receptor appear and attain maximal levels by the time of birth. About 98% of all thymocytes fail to mature either for the reason of time of thymic selection or failure to make a productive TCR gene rearrangement. The αβ TCR-CD3 complex survives in thymic selection and develops into single positive thymocytes - $CD4^+$ or CD8+. These single positive thymocytes migrate to the periphery of the thymus for development. The exact pathway is however not yet confirmed.

Positive and Negative Selection of T Cells by Thymus

A TCR repertoire is generated by the random gene rearrangement within the TCR germ-line DNA combined with junctional diversity- 1×10^{15} for the αβ receptor and 10×10^{18} for the gd receptor. The mature T cells recognize only foreign antigen combined with self MHC molecules, and thymocytes, whose receptors exhibit self MHC restriction nature. The interactive T cell population with self-MHC molecules is, by nature left to chance, so those molecules, which fail to recognize a foreign antigen are rendered useless in defense of host. If the αβ TCRs expressed on the mature T cells of a given individual bind to self-MHC, it is known as positive selection. Cells failing to bind to self-MHC are destroyed within the thymus by apoptosis. Since, 98% or more of all thymocytes die within the thymus. Failure to attain positive selection (recognition of foreign antigen combined with self-MHC molecules is known as negative selection.

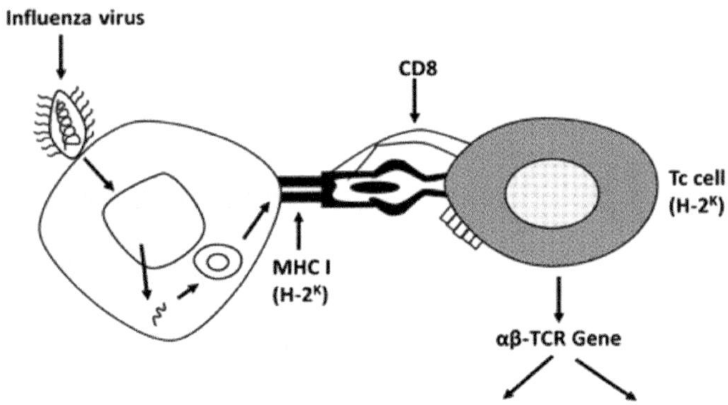

T-lymphocytes	H-2K Transgenic Host	H-2d Transgenic host
CD4+CD8+ TCR αβ+ T-cells	+	+
CD4-CD8+ TCR αβ+ T-cells	+	+

Fig.-6-2. Experimental demonstration of influence of the host haplotype on T cell maturation in mice with H-2b class I- restricted TCR on T cell specific for influenza virus.

T cell receptors are prone to accumulate with MHC molecules at the point of contact between thymocytes and epithelial cells within the thymic cortex. Thus, the positive selection occurs in the cortical region of the thymus. In other words, binding of thymocytes to class I or class II MHC molecules within the thymus leads to positive selection of MHC restricted T cells. Ada Kruisbeck in her experiment found that if the class I or class II MHC molecules are blocked by antibody, developing thymocytes fail to bind to the self MHC molecules on the thymic stromal cells preventing positive selection. The process of positive selection has further been observed with transgenic mice. In the experiment, rearranged αβ-TCR genes (derived from a $CD8^+$ T cell clone specific for influenza antigen) plus H-2K class I MHC molecules were injected into fertilized eggs of two different strains of mouse - one with the H-2K haplotype and one with the H-2d haplotype. From figure 6-2, it comes out that interaction between T cell receptors on immature thymocytes and self MHC molecules is required for positive selection.

In the population of MHC restricted thymocytes, there are some cells with high affinity receptors. These cell undergo negative selection by an interaction with bone marrow derived APCs (dendrite cells and macrophages) in the medulla, and die by apoptosis, because of self reactive T cells allowing maturation of T cells specific for foreign antigen plus self MHC molecules. Deletion of thymocytes reactive with self antigen plus MHC molecules, in experiments with transgenic mice, has further suggests that any difference in the selection of thymocytes in male and female transgenic are related to the presence or absence of H-Y antigen. H-Y reactive thymocytes are self reactive in the male mice and are eliminated. Cells which are not self-reactive escaped elimination.

By positive selection, T cells mature in the thymus, and by negative selection, they are eliminated in the process of apoptosis. In this context a question poses that what determines for a cell to undergo positive or negative selection since the entire repertoire of MHC restricted T cells is never eliminated. Two hypotheses have been forwarded to resolve the issue, which are as follows:

- Positive and negative selection depends on the differences in strength of the signal received by the thymocyte during interactions between its TCR and the MHC-peptide complexes of antigen presenting cells. The strength of the signal is determined by the avidity of MHC-peptide-TCR interactions. High avidity interactions send strong signals and low avidity interactions send weak signals. All thymocytes with TCRs that bind to the diverse collection of peptide-MHC complexes are selected for survival and those bearing receptors that react strongly with peptide-MHC complexes tend to die by apoptosis. That is to say that only thymocytes, whose TCRs bind weakly to self-MHC-peptide complexes on antigen-presenting cells in the thymus undergo positive selection. This is a quantitative hypothesis.
- Another hypothesis takes into account the nature of the signal delivered by TCR rather than its strength. Accordingly, some MHC-positive complexes deliver a weak signal while some deliver an optimum signal (complete). Weak signals to the receptors result in positive selection. This is quantitative hypothesis.

By both the above hypotheses, the role of peptides in thymic selection is established. Peptides not only induce stable MHC expression but also express in the selection process. The avidity hypothesis (as per **i**) above) was investigated by use of different concentrations of peptide on the fetal thymic organ cultures from mice that lacked the peptide transporter TAP-1 and were transgenic for a single TCR specificity. At low concentrations, few MHC molecules bound peptide and the avidity of the TCR-MHC interaction was low and $CD8^+$ cells appear in significant numbers. On raising the concentrations, the number of peptide-MHC complexes and the avidity interaction increased. This confirms the hypothesis of avidity.

The hypothesis of signals was investigated using fetal thymus taken from TCR transgenic mice that locked TAP-1. Three sets of FTOC- no peptide, antagonist peptide and agonist peptide were used. It was observed that if no peptide was added, no $CD8^+$ T cells bearing the transgenically encoded TCR were found. With antagonist peptide positive selection of T cells bearing the transgenic TCR occurred. The Oddition of agonist peptide to the FTOC caused elimination of thymocytes bearing the transgenic TCR. Thus, since MHC complexes of the agonist and antagonist peptides interact with the same TCR yielding different outcomes, the hypothesis is confirmed.

However, it is under further investigation, whether both the mechanisms (avidity and signaling) operate in positive or negative selection or one. Other issue of importance is which subset in the selection of $CD4^+8^+$ or $CD4^-8^+$ T cells. The models have been proposed to explain this phenomenon instructional model and stochastic model. Instructional model suggests that the multiple interactions between the TCR, $CD8^+$ or $CD4^+$ co-receptors, and class I or class II MHC molecules instruct the cells to differentiate into either $CD8^+$ or $CD4^+$ single positive cells, respectively. That is to say that a class I MHC-specific TCR together with the CD8 co-receptors would generate a signal that is different from the signal induced by a class II MHC specific TCR together with the CD4 co-receptor. The Stochastic model suggest that CD4 or CD8 expression is switched off randomly with no relation to the specificity of the TCR and thymocytes whose TCR and remaining co-receptor recognize the same class of MHC molecule, they mature.

T cell Receptor

T cell receptor (TCR) differs from the B cell receptor. The T cell receptor is membrane bound. T cell receptor is specific not for antigen alone but for antigen combined with a molecule encoded by MHC. The T cell receptor is associated on the membrane with a signal-transducing complex named CD3 similar in function to that of the Ig-α/Ig-β complex of the B cell receptor. T cell receptor molecules specifically recognize the MHC-antigen complex.

R.M. Zinkernagel and P.C. Doherty in 1974 demonstrated that T cell recognize antigen as also MHC molecule. Since T cells are membrane bound they recognize antigen only when presented on the membrane of a cell by a self MHC molecule. That is recognition of antigen by T cell is restricted to its presentation by a self

-MHC molecule. This renders T cells a specific nature in recognition of antigen as like B cells (antigen binding receptor). The dual-receptor model suggests that a T cell has two separate receptors one for antigen and one for class I or class II MHC molecules. The altered self model suggests that a single receptor recognizes an alteration in self MHC molecules induced by their association with foreign antigens.

The concept that T cell receptor is specific for both an antigen and an MHC molecule, posed a question for isolation of T cell receptors in terms of their structure. It was assumed that there should be significant structural difference in the receptor specific for antigen and on MHC molecule. For this large number of monoclonal antibodies were produced and then they were screened to find one that was specific to a clone, say clonotypic. In 1982, The T cell receptor was identified by J.P. Allison using this very model. It was found by him that some antisera bound to ab heterodimer from all the clones were not specific; as some were T cells expressing ab heterodimer as also gb heterodimer and differing in various organs and from species to species. However, ab T cells are abundant.

Structure of T Cell Receptors

Amino acid sequencing of αβ and γβ TCR heterodimers resemble TCR as immunoglobulins. There are two domains in TCR - variable (V) and constant (C) like those in immunoglobulins. A TCR molecule resembles a Fab fragment. A detailed concept of TCR structure may be developed by the schematic presentation as in Figure 6-3. As in immunoglobulins, there are light and heavy chains, and the T cell molecules have three hyper variable regions (HV3). A cysteine residue in TCR chain forms a disulfide link with the other chain, anchored by a transmembrane region of 21 or 22 amino acids. The amino acid residues are positively charged due to which the TCR heterodimers interact with chains of the signal-transducing CD3 complex. At the carboxyl terminal end of the TCR chain there is a short cytoplasmic tail of 5-12 amino acids. The antigen binding site of TCR is oriented in the V domain.

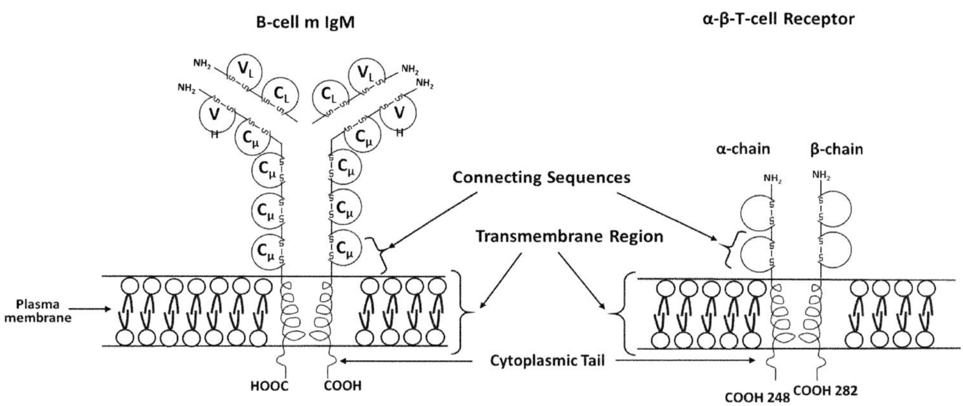

Fig.-6-3. Schematic diagram illustrating the structural similarities between αβ-T cell receptor and membrane-bound IgM on B cells.

TCR Genes- Organization and Rearrangement

Owing to striking similarity between the structure of immunoglobulins and TCR cells the later is called the super family of Ig. Immunoglobulin is a major secreted cell product, but the T cell does not secrete its antigen binding receptor, mRNA of which represents minor fraction in mRNA pool. This is a hindering factor in isolation of mRNA/TCR genes. Hedrick and Davis manipulated this hindrance and proceeded to isolate the membrane bound polyribosomal mRNA from a T_H cell clone and used reverse transcriptase to synthesize 32P labeled cDNA probes. Next they hybridized B cell mRNA with their T_H cell [32P] cDNA, thereby removing or subtracting all the cDNA that was common to all B cells and T cells.

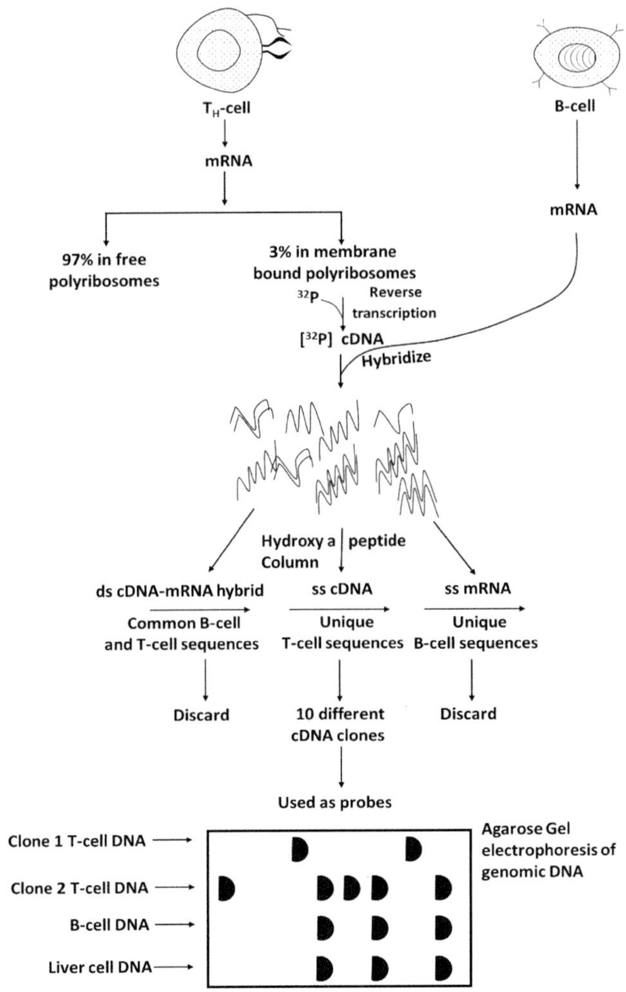

Fig.-6-4. The flow-chart outline of production and identification of cDNA clone encoding T cell receptor. (Adapted from S. Hedrick. 1984. Nature 308:149; with permission).

Thus, the rationale that T cells and B cells should express many genes in common because they are derived from a common progenitor cell came true. Further, since the αβ T cell receptor have constant and variable regions, its genes should undergo DNA rearrangements like the Ig genes of B cells, the above investigators isolated genomic DNA from T cells, B cells, liver cells and macrophages and subjected each DNA sample to Southern-blot analysis, using the 10[^{32}P] DNA probes to identify unique T cell genomic DNA sequences. Fig. 6-4 shows the result of this investigation. With subsequent identification of cDNA clones encoding a-chain, γ chain and δ chain, it was confirmed that TCR gene represent a gene sequence that rearranges and exposes in T cells only as membrane-bound protein.

TCR germ-line DNA is organized into multigene families, corresponding to the α, β, γ and d chains. In the mouse these chain gene segments are located on chromosomes 14, 6 and 13 respectively, and between the Vδ and Jα segments is located the d chain on chromosome 14. Thus arrangement prevents expression of αβ TCR in a given T cell from γδ TCR. The organization of the TCR multigene families in humans is similar to that in mice, but the number of segments differs. Table 6-1 is amply illustrative of the organization of TCR multigene families in humans.

Table 6-1: TCR multigene families in humans

		Number of gene segments			
Gene	**Chromosome location**	**V**	**D**	**J**	**C**
α chain	14	50	-	70	1
β chain	7	57	2	13	2
γ chain	7	14	-	5	2
δ chain*	14	3	3	3	1

Adated from P. A. H. Moss et al. 1992. *Annu. Rev. Immunol.* 10:71.

Variable-Region Gene Rearrangements

Just as B cells rearrange immunoglobulin germ-line DNA, T cells rearrange the variable region. TCR gene segments to form functional genes that encode either the TCR α and β chains or the TCR γ and δ chains. Rearrangement of the TCR α and β chain gene segments results in VJ joining for the a chain and VDJ joining for the β chain. The constant region germ-line DNA encoding the TCR α and β chains is simpler than the Ig heavy chain germ-line DNA (multiple C gene segments encoding distinct isotype with different effector functions)- TCR α-chain DNA has a single C gene segments and β chain DNA has two C gene segments.

Mechanisms of TCR DNA Rearrangements

Functional receptor genes are arranged on the TCR germ-line DNA like that of Ig gene rearrangements. All of the TCR gene rearrangements follow the one turn/ two turn joining only as for the Ig genes. Similarly, the mechanisms of Ig-gene and TCR gene rearrangement has been observed in studies with SCID mice (lacking functional T and B cells). The recombinase enzyme system plays an important role in each cell lineage so that only correct receptor DNA is rearranged. The rearrangement of gene segments in a T- or B cell generates a DNA sequence specific and unparallel to that cell and its progeny. This DNA sequence is specific for each T cell clone.

The clonal population can also be studied from the specific DNA, patterns of which serve as markers. Selection of rearranged TCR DNA may be used as a diagnostic tool when abnormal lymph modes persist (a result of infection causing inflammation or proliferation of a cancerous lymphoid cell). If TCR sequences show no unique/unparallel fragments, the inflammation may be simple and caused by some infection. But with a detectable unique/unparallel DNA fragment in the cell population, there may be proliferation of a cancerous lymphoid cell, as all such cells would contain the same TCR DNA sequence produced by DNA rearrangement in the parent cell.

Allelic Exclusion of TCR Genes

The β-chain gene segments are allelic and organized into to clusters. It is less rigorous for a-chain genes. In the event of productive rearrangement set for one b-chain allele the rearrangement of the other b-chain allele is inhibited. In the event of non-productive rearrangement the thymocyte attempts a second rearrangement. In experiments with transgenic mice, it has been observed that a productively rearranged αβ TCR transgene does not rearrange and express the β-chain genes. But the a-chain genes sometimes are expressed at various levels in place of the already rearranged α-chain transgene. This proves that β-chain genes are less rigorous for α-chain genes. The structure of TCR genes and their rearrangement can be perfectly seen in Figure 6-5.

Generation of TCR Diversity

The V gene segments are fewer in TCR germ-line DNA than Ig germ-line DNA. A large number of random gene combinations are produced by combinational joining of variable-region gene segments. This happens in both the TCR chains as well as in Ig heavy and light chain genes. The difference in V gene segments in TCR germ-line and that in Ig germ-line is due to greater number of J segments in TCR germ-line DNA. Besides, TCR diversity is also generated due to location of one turn (12 bp) and two turn (23 bp) recognition signal sequences in TCR β and δ-chain DNA, being different from Ig heavy chain DNA. Under one turn/two turn recognition signal sequences, it is possible for a Vβ gene segment to join directly with a Jβ or a Dβ gene segment, producing a (VJ)β or (VDJ)β unit. Further alternative joining of d-chain gene segments generate similar units. Likely one Dδ can join with another gene segment producing a (VDDJ)δ unit. This mechanism renders TCR gene segments, a considerable diversity as compared to Ig heavy chain DNA.

The TCR-gene rearrangement exhibits the following features:

- Increase in diversity by encoding several alternative amino acids at each junction, called junctional flexibility,

- ♦ Addition of some nucleotides at the junctions between some gene segments in the process of rearrangement, known as palindromic region nucleotide fusion. the rate is six nucleotides per junction generating about 5461 combinations.
- ♦ Addition of N-region nucleotides, encoding all the TCR chains.

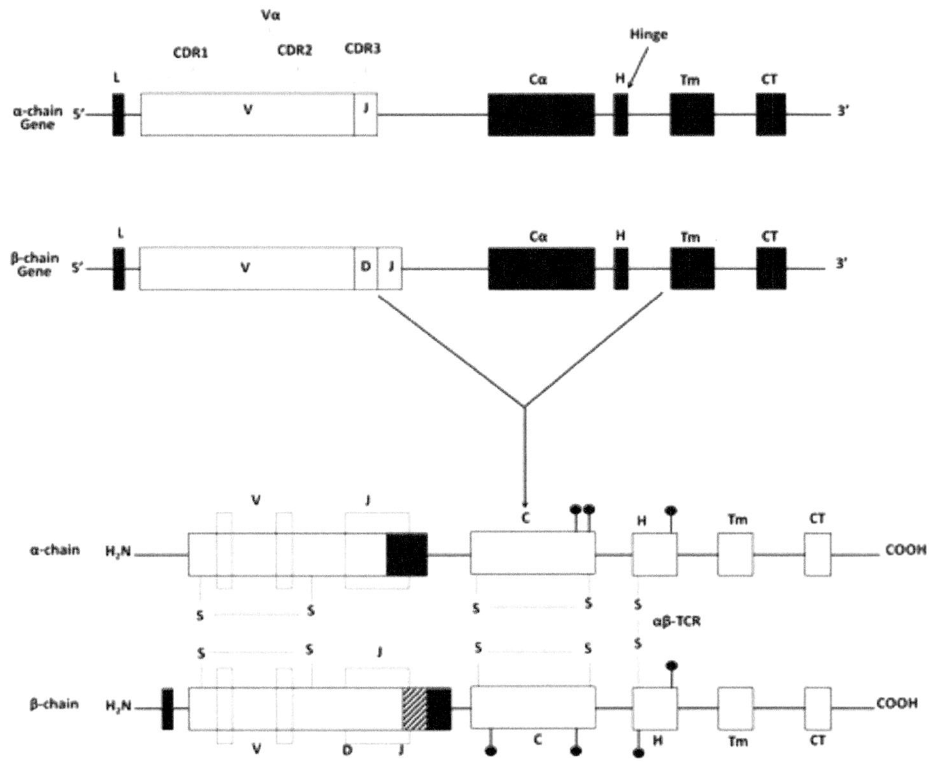

Fig.-6-5. Schematic diagram of rearranged αβ-TCR gene and encoded domains. V= variable, c= constant, J= joining, H= hinge, Tm= transmembrane and CT=cytoplasmic tail.

The nature of TCR diversity is adapted to render it variability needed to recognize diverse peptides while restricting the recognition of MHC molecules. That the TCR for its diversity encode a limited number of CDR regions through V gene segments i.e. CDR1, CDR2 and CDR3. However, there is no somatic mutation in T cells.

TCR-CD3 Membrane Complex

The association of T cell receptor with CD3 (a complex of five invariant polypeptide chains) is called TCR-CD3 membrane complex. This implies that the T cell receptor is associated with other membrane molecules. This characteristic of TCR is similar to that of B cells, corroborated by experiments in which it was observed that fluorescent antibody to the receptor causes addition of another membrane protein i.e. CD3 due to close location of TCR and CD3 in the T cell membrane (J.P. Allison and L. Lanier). Fig. 6-6 shows the TCR-CD3 complex. As stated above, CD3 is a complex of five invariant polypeptide chains associated to form three dimers; a heterodimer of gamma and epsilon chains (γε), a heterodimer of delta and epsilon chains (δε) and a homodimer of two zeta chains (ζζ) or a heterodimer of zeta and eta chains (ζη), the lattermost encoded by the same gene. These polypeptide chains differ in their carboxyl-terminal ends because of difference in RNA splicing of the primary transcript. The CD3 dimers, as above, are needed for expression of the T cell receptor and for signal transduction. The transmembrane region of all the CD3 polypeptide chains contains a negatively charged aspartic acid residue, which enable the CD3 complex to interact with one or two positively charged amino acids in each TCR chain. The cytoplasmic tail of each CD3 chain contains a common sequence known as the immunoreceptor tyrosine based activation motif (ITAM) which is involved in signal transduction after interaction of a T cell receptor with a peptide MHC complex. The TCR-CD3 complex mediates recognition of antigen MHC complexes.

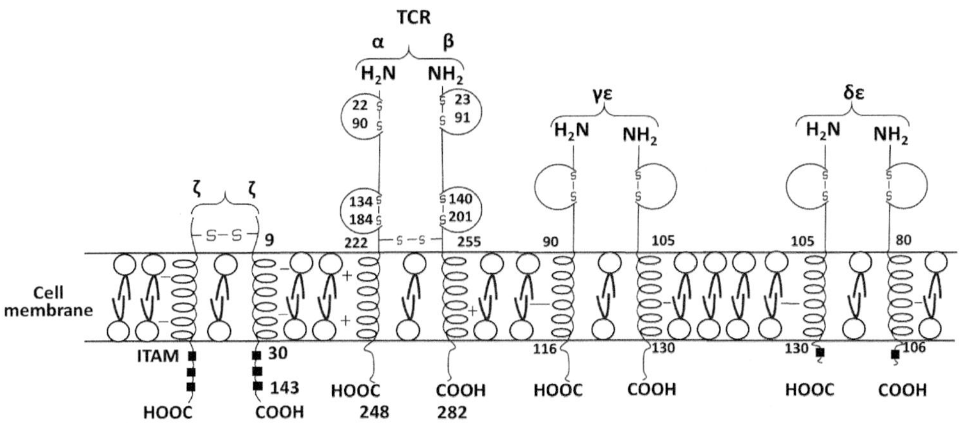

Fig.-6-6. Schematic diagram of TCR-CD3 complex which constitutes the T cell antigen binding receptor and different subunits (ε, γ, δ, εε and ζζ) of CD3 molecule.

Role of T cell Accessory Molecules in TCR Signaling

The interaction between T cell receptor and peptide-MHC complex is strengthened by several other T cell membrane molecules. The T cells accessory membrane molecules (e.g. CD2, CD28, CD45 R, LFA-1 etc.) have a role in signal transduction. Some of the accessory membrane molecules serve both the interaction and signal transduction. Table 6-2 shows some selected T cell accessory molecules which are involved in adhesion, interaction and signal transduction functions.

As accessory membrane molecules of T cells, CD4 and CD8 function largely as cytotoxic cells, CD4 recognize antigen in combination with class II MHC molecules.

CD8 recognize antigen in combination with class I MHC molecules. Because of their abilities to recognize peptide MHC complex and signal transduction, they are also called co-receptors. During antigen recognition CD4 molecules being adjacent to the β2 domain of the class II MHC molecule form a complex TCR-CD4 and MHC, yielding cross linkage of the TCRs on a T cell causing signaling events, through cytoplasmic domains. However, the roles played by these co-receptors, but for experimental evidence against their capacity of recognizing peptide MHC complex and signal transduction, are speculative so far.

Table 6-2: Some important accessory TCR molecules Function

Name	Ligand	Function	
		Adhesion	Signal transduction
CD4	Class II MHC	+	+
CD8	Class II MHC	+	+
CD28	B7	?	+
CTLA-4	B7	?	+
CD45R	CD22	+	+
CD5	CD72	?	+
CD2 (LFA-2)	CD58 (LFA-3)	+	+
LFA-1	ICAM-1 (CD54)	+	?

The affinity of the bond between TCR and MHC-peptide complex is increased by the co-receptors (CD4 and CD8 etc.). Among the co-receptors other than CD4 and CD8, CD2, LFA-1, CD28 and CD45R do not interact with the peptide MHC complex, rather they bind independently to other ligands **(Table 6-2** and **Fig. 6-7 b)** on antigen presenting cells or target cells. Cytokines or cytotoxic substances are transferred during closer contact between the interacting cells. The co-receptors help in antigen recognition by T cells.

TCR-Antigen-MHC called ternary complex is essential for an immune response. This means binding of a peptide to the MHC molecule and recognition of this MHC-peptide complex by the TCR through its complimentary determining regions (CDRs). TCR by itself can recognize and bind the specific antigen. The TCR is situated across the MHC molecule diagonally relative to the long dimension of the peptide. The CDR3 loops of the TCR α and β-chains meet in the core of the peptide and the CDR1 loop of the TCR α-chain is situated at the N terminus of the peptide, and TCR β-chain is located at the C terminus of the peptide.

The CDR2 loops are in contact with the MHC molecule, CDR2α is over the a2 domain alpha helix and CDR2β over the α1 domain alpha helix. Looking down into the MHC cleft shows the peptide as buried under the TCR being invisible from this angle. This is the three dimensional structure of the ternary complex. The structures of the peptide binding regions in class I and class II molecules are similar. This three dimensional structure of TCR molecule is similar to that of Ig (Fig. 6-3).

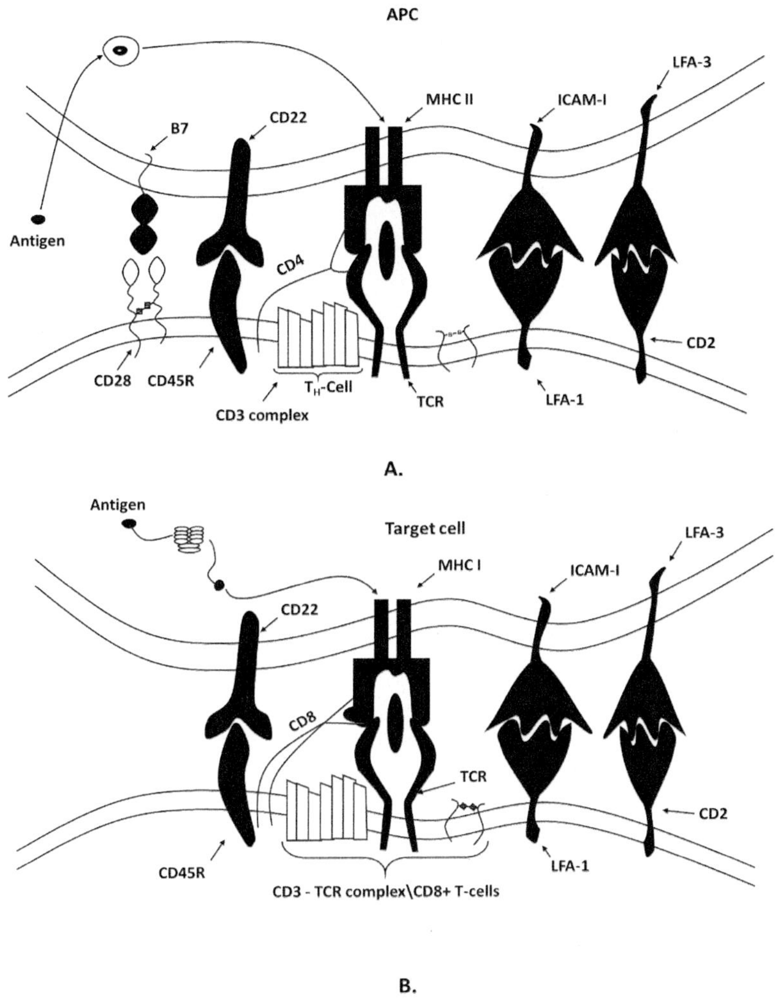

Fig.-6-7. Role of co-receptors in T cell (T_H/T_C) activation. (a) Diagrammatic representation of the interaction between T cell receptor present on T-helper cell and peptide –MHC II complex on APCs along with interaction of various accessory molecules, and (b) Diagrammatic representation of interaction between T cell receptor present on T-cytotoxic cell and peptide-MHC I complex on target cell along with the interaction of various accessory molecules.

Alloreactivity of T Cells

MHC molecules function as histocompatibility antigens, as they are polymorphic in nature. This very polymorphic nature of MHC molecules renders unique sets of MHC molecules in individuals of the same species (individuals of the same species differ genetically or are allogeneic). The reactions of MHC molecules may be termed as alloreactivity, since T cells respond to allografts and reject them.

The alloreactivity of T cells is particular in two ways; first the T cells recognize a foreign MHC molecule and respond to it directly, and secondly the percentage of alloreactive T cells is very high. There are experimental evidences to show that a particular T cell receptor is specific for a foreign antigenic peptide plus a self MHC molecule as also for certain allogeneic MHC molecules. T cell receptor may recognize both an allogeneic MHC molecule plus allogeneic peptide provided they are identical in structure. To determine the cause of high percentage/frequency of antigen specific and alloreactive T cells, three congenic strains ($B_{10,}$ $B_{10A,}$ B_{10S}) were immunized with DNP-OVA. After that from each strain 20-40 antigen specific clones were isolated and tested for alloreactivity against cells expressing different MHC haplotypes in the absence of the specific antigen. It was found that in the range of 19% to 44% allogeneic cells were responded by the antigen specific clones. The findings indicate that the receptors of antigen specific T cells not only recognize antigen plus a self MHC molecule but also cross react with various allogeneic peptides-allogeneic MHC complexes. Figure 6-8 shows the mechanism of T cell reactivity which accounts for higher percentage of alloreactive T cells.

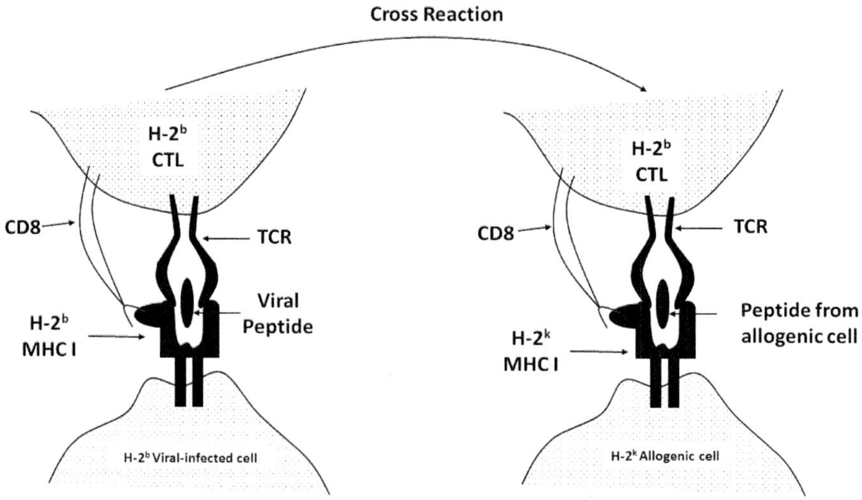

Fig.-6-8. Possible explanation of T cell cross-reactivity with antigen peptide plus self MHC molecules.

T_H Cell Activation

The interaction of the TCR-CD3 complex with a processed antigenic peptide bound to a class II MHC molecule on the surface of an antigen-presenting cell, causes activation of T_H cell, the major factor in immune response (humoral or cell mediated). The resting T_H cell is activated by the interaction with an antigen and a cascade of biochemical events push T_H cells to enter the cell cycle (G_0 to G_1 transition), wherein they proliferate and differentiate into memory cells or effector cells. Numerous genes are also activated following interaction of T_H cells with antigen. These genes have been categorized as per time course taken in their expression viz. immediate genes, early genes and late genes. Table 6-3 gives a detailed account of these gene products, which are the result of signal transduction pathways activated by the encounter between the TCR and MHC-peptide complexes. Some common features in signal transduction are:

- protein elements undergo phosphorylation and dephosphorylation,
- molecules known as adapter proteins play a major role in the signal-induced assembly of receptors in key signaling pathways. Adapters promote the assembly of membrane receptors and other cytoplasmic molecules,
- the role and intensity of signal transduction is influenced by enzyme cascades,
- transmission does not occur in absence of proper signals,
- cyclic nucleotides, calcium ion, diacylglycerol and inositol triphosphate diffuse to other sites in the cell and evoke changes,
- hydrophobic lipoidal signals such as steroids are bound by intracellular receptors,
- guanosine nucleotides (guanosine triphosphate and guanine diphosphate) act as molecular switches through the large proteins known as G proteins composed of α, β and γ- subunits. They participate in the activation of an important proliferation-inducing signal-transduction cascade triggered by binding of ligands to their receptor.

Table 6-3: Gene products following interaction with antigens

Gene Products	*Localization*	*Functions*
	IMMEDIATE	
c-Fos	Nucleus	Nuclear binding
c-Jun	Nucleus	Transcription factor
c-Myc	Nucleus	Oncogene
NF-AT	Nucleus	Transcription factor
NF-κB	Nucleus	Transcription Factor
	EARLY	
IFN-γ	Secreted	Cytokine
IL-2	Secreted	Cytokine
Insulin Receptor	Cell Membrane	Hormone Receptor
TGF-β	Secreted	Cytokine
TNF-β	Secreted	Cytokine
IL-4	Secreted	Cytokine
IL-5	Secreted	Cytokine
IL-6	Secreted	Cytokine
c-Myb	Nucleus	Proto-oncogene
GM-CSF	Secreted	Cytokine

Gene Products Localization		Functions
	LATE	
HLA-DR	Cell Membrane	Class II MHC molecule
VLA-4	Cell Membrane	Adhesion molecule
VLA-1-VLA-5	Cell Membrane	Adhesion molecule

TCR Coupled Signaling Pathways

An increase in the intracellular Ca2+ concentration is the result of TCR mediated/coupled signaling. T cell is activated on the recognition of antigen-presenting cells by the TCR and MHC-peptide complexes. TCR consists of a mostly extracellular heterodimeric ligand binding unit, α and β chains of αβ TCRs and a predominantly intracellular signaling unit. The signaling unit combined with each ligand-binding unit is composed of CD3 complex (CD3γ, CD3δ and CD3ε) and a homodimer of ζ chains. These are essential components of signal transduction, as well for surface expression of the TCR. TCR-MHC-peptide interaction triggers signaling process in which a protein tyrosine kinase comes into association with the cytoplasmic tails of the TCR complex. CD4 in class II restricted T cells or CD8 in class I restricted T cells are translocated to the vicinity of TCR.

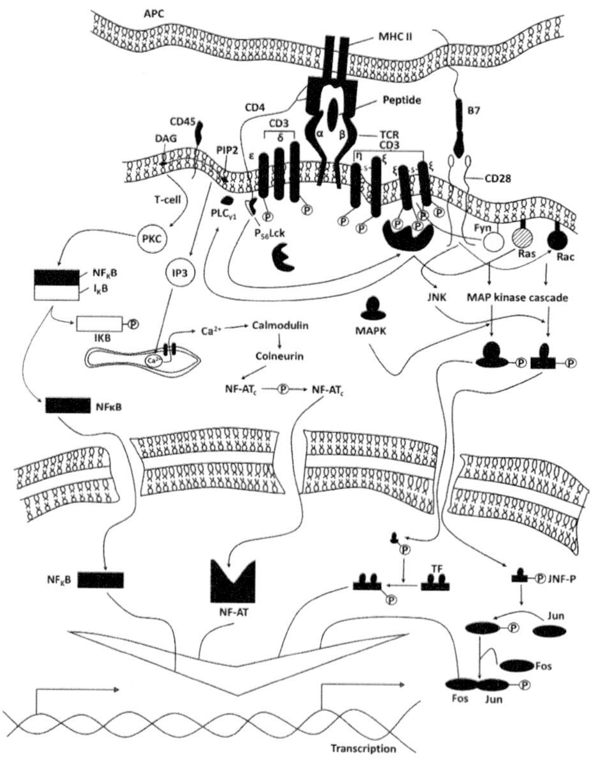

Fig.-6-9. Overview of signal transduction during T cell activation.

The protein tyrosine kinases are required for interaction with other molecules, initiating phosphorylation by Fyn and Lck at two sites. Phosphorylation at an inhibitory site turns off enzyme activity and phosphorylation at activation site sets in activity. However, until phosphate has been removed from the inhibitory site, enzyme activity does not occur. The removal of phosphate from the inhibitory site of Fyn and Lck is performed by CD45, a transmembrane tyrosine phosphatase found in all leukocytes (leukocyte common antigen). The co-receptors of CD3 complex are also phosphorylated by Fyn and Lck. Fig. 6-9 is amply illustrative of the signal transduction during T cell activation. There is another member of the signaling complex - a 70 kDa polypeptide called ZAP (zeta associated protein)-70.

The activation of the ZAP-70 by protein tyrosine kinase leads to the activation of many pathways as shown in the Figure 6-9. Signaling pathways are also initiated by cyclic nucleotides, calcium ion, DAG (diacylglycerol) and IP3 (inositol triphosphate) - IP3 triggers an increase in intracellular cytosolic Ca2+ and the subsequent activation of a calmodulin dependent phosphatase called calcineurin which dephosphorylates the inactive cytosolic form of the T cell specific nuclear factor NF-AT, and DAG activates protein kinase C (PKC) which then phosphorylates various cellular substrates and mediates release of the NF-κB (nuclear factor-kappa-κ B), activating various genes in the nucleus, including IL-2 gene. In T_H-cell activation, only NF-AT is T cell specific.

The Co-stimulatory Signal

As is known, interaction of an antigen peptide with the TCR-CD3 complex activates a naïve T cell (TCR mediated), the interactions between CD28 and T cell members of the B7 family (immunoglobulin super family having a V like domain and a single C like domain). The ligand for B7 are CD28 and CTLA-4 (CD152); signaling through CD28 deliver a positive co-stimulatory signal to the T cell, whereas signaling through CD152 down regulates the activation of T cell. That the signals through CD152 are inhibitory it accounts for lymphocyte homeostasis. The process of lymphocyte homeostasis has been tested on experiments with knockout mice, as T cells proliferate and lead to lymphadenopathy, splenomegaly and death at 3 to 4 weeks after birth. However, the signal transduction pathways employed by CD28 and CD152 (CTLA-4) are still under intensive investigation to ensure the exact molecular targets of Co-stimulation of T cells.

Clonal Expansion versus Clonal Anergy

Non-recognition or non-responsiveness of an antigen peptide-MHC complex by T cell is called clonal anergy and recognition leads to activation of T cell or clonal expansion. In other words anergy is a state of inactivation of cells to proliferate in response to a peptide-MHC complex. Clonal expansion or clonal anergy is dependent upon the co-stimulatory signals, such as those produced by interaction of CD28 with B7. CD28 acts on T_H cells and B7 acts on antigen presenting cells. On receiving TCR mediated signals, the T cell becomes anergic, as also by incubating with normal APCs, because APCs in presence of the Fab portion of anti CD28 block the interaction of CD28 with B7 (Fig. 6-10). The in vivo and in vitro roles of clonal anergy are under active study.

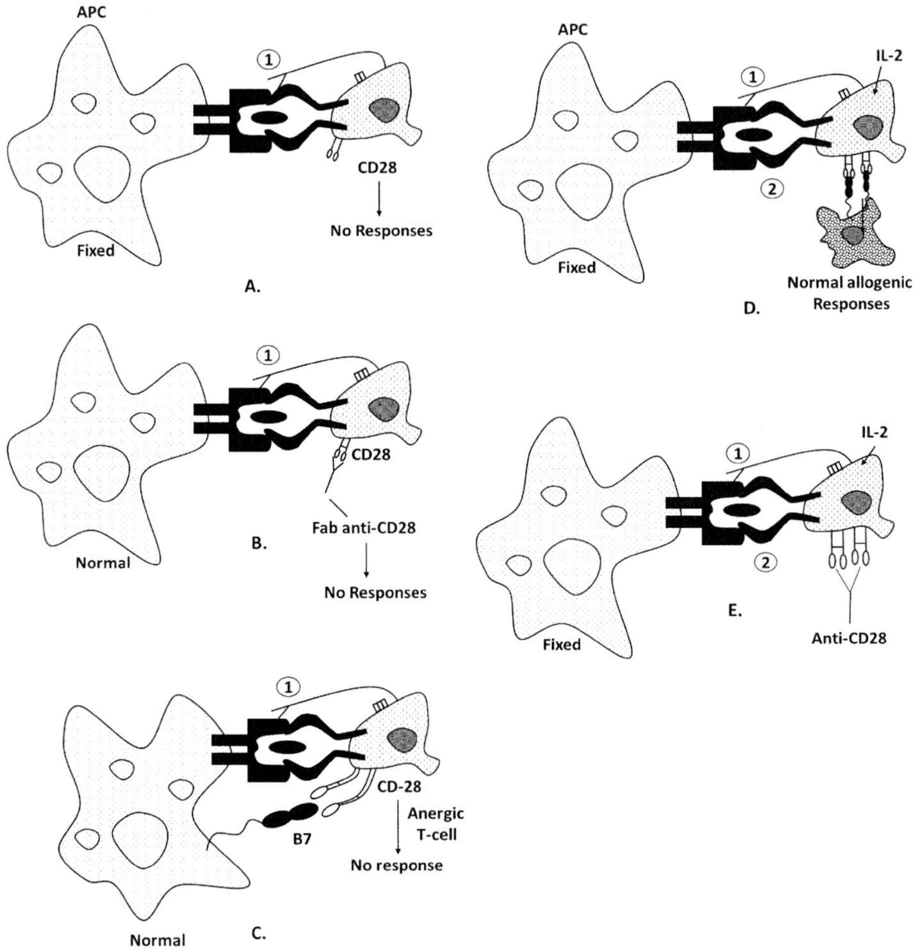

Fig.-6-10. Illustration of clonal anergy and clonal expansion in five different experimental settings.

Superantigen-Induced T Cell Activation

Viral or bacterial proteins binding simultaneously to the Vb domain of a T cell receptor and to the a-chain of a class II MHC molecule are called superantigen, being exogenous and endogenous. Exogenous superantigens are soluble proteins secreted by Gram positive bacteria and endogenous superantigens are cell membrane proteins encoded by certain viruses which infect mammalian cells. Both these superantigens bind particular Vβ sequences in T cell receptors and crosslink the TCR to a class II MHC molecule. Exogenous and endogenous superantigens bind outside of the TCR antigen binding cleft, so any T cell expressing a particular Vb sequence is activated by a corresponding superantigen. As a sequence, a significant percentage of the total T_H population may be affected. Thus the effect is polyclonal and is systemic toxic, resulting in over production of T_H cell cytokines. T cell maturation in the thymus is also effected by the superantigens induced T cell activation, because all the thymocytes bearing a TCR Vb domain corresponding to the superantigens specificity provoke negative selection. In the negative selection, T cells bearing the Vβ6, Vβ7, Vβ8.1, or Vβ9 domain in the inbred AKR mouse strain are eliminated by the mediation of viral proteins called minor lymphocyte stimulating determinants (MIs). But, the thymocytes bearing the Vβ3 or Vβ5 domains are not affected by these MIs superantigens.

T cell Differentiation

T cells that express CD4 recognize antigen combined with a class II MHC molecule and generally function as T_H cells, and those T cells that express CD8 recognize antigen combined with a class I MHC molecule generally function as T_c cells. There are about 90-95% T cells in the periphery of the thymus, which have not yet encountered antigen, enter the circulation as resting cells in the G_0 stage of the cell cycle. Continually circulating between the blood and lymphatic system, these naïve T cells, if not encountered by an antigen in a lymph node, exit through the efferent lymphatic, to join the blood through thoracic duct. This draining of naïve T cells (from the blood to the lymph nodes) takes place every 12-24 hours in to and fro system. This drainage is meant to let the T cell specific for an antigen, recognize it. The life span of these naïve T cell specific for an antigen, recognize it. The life span of these naïve T cell is about or more than 5-7 weeks, if not activated by an antigen.

Generation of Effector and Memory T Cells

If a naïve T cell is encountered with an antigen-MHC complex on an antigen presenting cell or target cell, it is activated and within 48 hours enlarges into a blast cell and begins undergoing repeated rounds of cell division. After activation, T cell enters into the G1 phase of the cell cycle concomitantly inducing transcription of the gene for IL-2 and the a-chain of the high-affinity IL-2 receptor. As a result, the activated naïve T cells begin to proliferate and differentiate (Fig. 6-11). The activated cells divide 2-3 times per day for 4-5 days, generating a large clone of progeny cells, which differentiate into memory or effector T cell populations. Effector cells are short lived and carry functions like cytokine secretion and cytotoxic killing activity - Memory T cells are long lived antigen activated T cells which intensively react to the same antigen with heightened activity, known as secondary response, the primary being the first recognition of an antigen-MHC complex on an appropriate antigen-presenting cell or target cell. Memory T cells express the same membrane molecules as effector cells. Memory cells are different in pattern than naïve or effector T cells. However, certain cell surface characters greatly differs from other T cell sub-types (such as presence of CD45 RO), which are under active study.

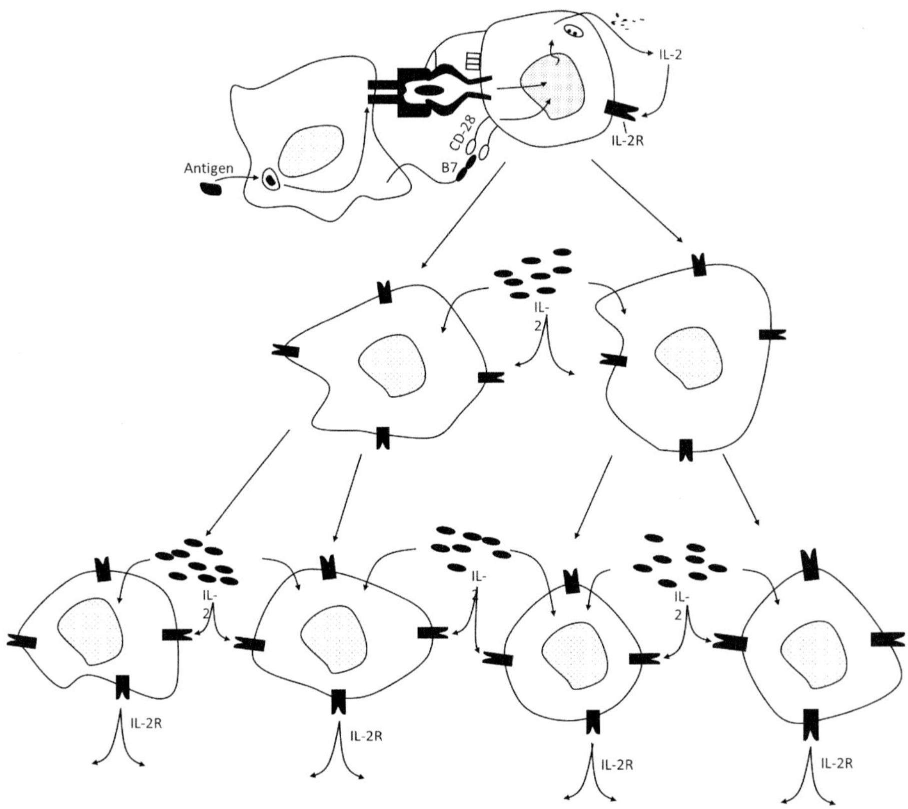

Fig.-6-11. Activation of T-helper cell requires both signal I (primary signal) and co-stimulatory signal (secondary signal). Activation results in the expression of IL-2 and the IL-2 receptor that further induces T cell proliferation.

Cell Death and T Cell Populations

In the formation of T cell repertoire and in the regulation of mature T cell populations, the antigen signaled apoptotic death of T cells is a central characteristic of T cell differentiation. Activation of mature T cells in the periphery causes them to proliferate and differentiate into effector cells populations. The population of effector cells, once the pathogen has been eliminated, is prone to produce high levels of cytokines, so death of T_H cells is necessary to avoid potential harm to the host immune system. This is essential because failure of apoptosis causes defective lymphocyte homeostasis. Apoptosis is attributed two phases- induction and effector. A variety of signals may initiate an induction pathway to cell death- cross linkage of the T cell receptor by antigen or artificial stimuli such as treatment with glucocorticoids, exposure to x-rays and a rise in cytosolic Ca^{2+}.

Fas are a protein of tumor necrosis factor receptor (TNFR) family and are readily expressed on T cells. Fas l is a membrane resident molecule closely related to the cytokine tumor necrosis factor a (TNFα). The engagement of Fas with Fas

L (Fas ligand) triggers cell death of T cells and maintains homeostasis of T cell population. TCR mediated apoptosis of mature T cells outside of the thymus is called activation-induced cell death (AICD), which is also induced by Fas pathway. Fas and Fas L are the important mediators of the activation-induced cell death of mature T cells besides interactions of other member of the TNFR family with ligands of the TNF family.

Peripheral gd T Cells: A Paradigm in Immunology

In the T cell population, there are gd T cells, which are found as intraepidermal lymphocytes (not in human) constituting upto 1% of total epidermal cells and intraepithelial lymphocytes (IELs). These cells profoundly express γδ TCR-CD3 complex. Some cells express only CD4 so called as γδ $CD4^+$ T cells while intraepithelial lymphocytes express both CD4 and CD8 on their cell surface. The gδ T cells remain fixed in the epithelial tissues expressing different Vγ and Vδ gene segments. These V gene-segments reflect a specialization of T cells which respond to certain types of antigens found at the periphery of the intestine.

In immunity, gδ T cells play an important and unique role. They can mediate tumor cell lysis in non-MHC restricted manner i.e. killer cells. The function of γδ T cells is to respond to a mycobacterial antigen called purified protein derivative (PPD), belonging to a group of highly conserved proteins (produced by cells in response to sudden increases in temperature or other environmental stresses). Called heat-shock proteins. They can also be produced by internal stresses such as inflammations caused by viral infections and malignancies. Due to their nature (heat-shock), PPD can eliminate damaged cells as also microbial invaders. Since γδ T cells populate at the epithelial sites, they can manage a surveillance system and monitor the external milieu of the epithelial cells. The spread of cancer or infection of the internal milieu is also prevented by the presence of γδ T cells, providing a cell-mediated immune system.

CHAPTER – 7

B CELL GENERATION ACTIVATION AND DIFFERENTIATION

In vertebrates, B cells are produced in the bone marrow as an orderly sequence of Ig gene rearrangements. B cells are activated by the interaction with antigen, corresponding to the antigen specific to its membrane-bound antibody. The cell proliferates and differentiates to generate a population of antibody-secreting plasma cells and memory B cells. As the B cell activation and differentiation requires an antigen in the periphery, this phase is also called antigen dependent phase of B cell development. Although many B cells are produced in the bone marrow yet not many of these cells mature and in 3 to 8 weeks, they eliminate for want of antigens. This means the diversity of B cells is not accomplished. This chapter deals with developmental pathways of B cell maturation (immunocompetent) activation and differentiation.

B cell Maturation

The naïve B cells, which have not encountered antigen, circulate in the blood and lymph and are carried to the secondary lymphoid organs mostly the spleen and lymph nodes. Lymphoid stem cells differentiate into the earliest distinctive B-lineage cell called pro-B cell expressing a transmembrane tyrosine phosphatase called CD45R (B220). The pro-B cells proliferate within the bone marrow. The stromal cells in the bone marrow provide a microenvironment by which pro-B cells progress into pre-B cell. The presence of stromal cells is thus a condition for the maturation of B cells in the bone marrow as the former while interacting with pro-B and pre-B cells (their interaction is direct) secrete various cytokines, which greatly helps in the developmental process. The pre-B cell and its ligand have several cell adhesion molecules (VLA-4 and VCAM-1) by which they interact with stromal cells. Following contact with stromal cells, a surface molecule called stem-cell factor (SCF) interacts with c-Kit on the pro-B cells. The pro-B cell then begins to divide and differentiates into a pre-B cell, which express IL-7 R (interleukin-7 receptor). Thus, the IL-7 secreted by stromal cells then drives B cell maturation process and at the same time down regulates the expression of adhesion molecules thereby further down regulates maturation processes.

Rearrangement of Ig Genes

B cell maturation is dependent upon the rearrangement of the Ig gene in the pro-B cell. First a heavy chain D_H to J_H gene rearrangement occurs and then a V_H to D_H J_H rearrangement occurs. After completion of heavy-chain rearrangement, the cell is classified as a pre-B cell. Light chain gene rearrangement is also required for continued development of a pre-B cell into an immature B cell. Only one light chain isotype is expressed on the membrane of a B cell to develop it into an antigen specific cell. Heavy chain VDJ sequence and light-chain VJ sequence determine the specificity of the immature B cell to an antigen specific cell by expressing IgM membrane on its surface. Thus, both the light and heavy chain rearrangement is required for the development of pro-B and pre-B cell stages (Fig. 6-1). In the pro-B cell stage the enzyme terminal deoxyribonucleotide transferase (TdT) which catalyses insertion of N-nucleotides at the D_H-J_H and V_H-D_H-J_H coding joints are active, but in the pre-B cell stage, due to occurrence of light chain rearrangement, the TdT expression is not active. In the process of development, immature B cells express low levels of IgD and IgM on the membrane. The membrane is encoded by two mRNAs one at the m chain and the other at the μ chain. Thus the mature B cells express IgD and IgM.

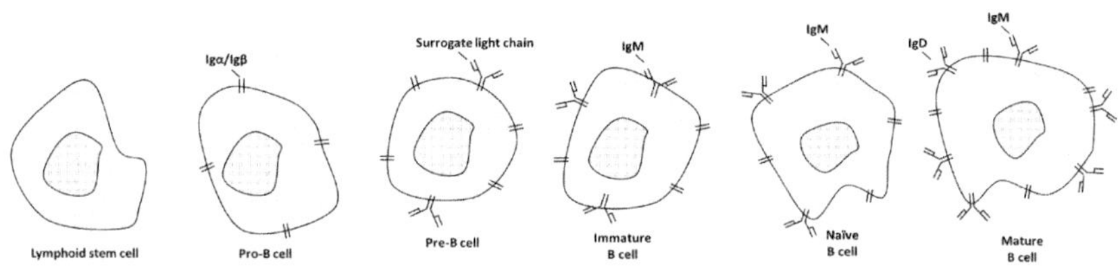

H-chain genes	Germ line	D_H J_H	V_H D_H J_H	V_H D_H J_H	V_H D_H J_H	V_H D_H J_H
Light-chain genes	Germ line	Surrogate Vpre-B & I5 germ line K and I	Surrogate Vpre-B & I5 germ line K and I	V_L J_L	V_L J_L	V_L J_L
RAG-J/2	-	+	+	-	-	-
M-Ig-H-chain	-	-	μ	μ	μ	μ + δ
M-Ig-light	-	Surrogate Light chain	Surrogate Light chain	κ or λ	κ or λ	κ or λ
Transcription factors	Pu.1, Ikaros	BSA (Pax-5), Sox-4, EBF, E2A, Oct-2	BSA (Pax-5), Sox-4, EBF, E2A, Oct-2	BSA (Pax-5), Sox-4, EBF, E2A, Oct-2	BSA (Pax-5), Sox-4, EBF, E2A, Oct-2	BSA (Pax-5), Sox-4, EBF, E2A, Oct-2
Surface markers	c-kit	c-kit, CD45R, CD-19, CD-24, Igα/Igβ, IL-7R (CD-43)	c-kit, CD45R, CD-19, CD-24, Igα/Igβ, IL-7R	c-kit, CD45R, CD-19, CD-24, Igα/Igβ, mIgM	c-kit, CD45R, CD-19, CD-24, Igα/Igβ, mIgM	c-kit, CD45R, CD-19, CD-24, Igα/Igβ, mIgM

Fig.-7-1. Diagrammatic illustration of B cell development and maturation in bone marrow and expression of different molecules and surface markers.

Pre-B Cell Receptor

During B cell development in the pre-B cell the membrane μ chain is associated with an unusual light-chain complex called the surrogate light chain. The surrogate light chain has two proteins – a V like sequence (Vpre-B) and a C like sequence (-l_5) located at distance from l light-chain genes. Pre-B cells capable of expressing membrane bound m heavy chains proceed in association with the surrogate light chains associated with Ig-α/Ig-β called as pre-B cell receptor, to mature B cells. This pre-B cell receptor recognizes some signal that inhibits V_H to DHJ_H gene rearrangement of other heavy chain gene thus induces an allelic exclusion. After that pre-B cell undergoes cell proliferative phase and each of which produces 32 to 64 progeny by the process of cell divisions. The overall population of these progeny pre-B cells increases the diversity of thus developed antibody repertoire.

B cell development is dependent on the transcription factors E2A, EBF (early B cell factor) and BSAP (B cell-specific activator protein) and sox-4 (Fig. 7-1). In the early stages of development of B cell lineage E2A and EBF transcription factors play an important role, whereas BSAP contributes beyond the early stage development of B cell lineage. Thus, some of these factors contribute at the early stages, and some at the later stages of development of B cell.

Cell Surface Markers

At the pro-B cell stage, B cells do not display the heavy or light chains of antibody, rather they express CD45R (a form of protein tyrosine phosphatase found on leukocytes), the signal-transducing molecules Ig-a and Ig-b, and c-kit (found on the surface of pro-B cells). The CD45R, Ig-a and Ig-b, c-kit are surface markers of pro-B cells whereas CD45R, and Ig-a and Ig-bare the surface markers of pre-B cells. When the IgD class of Ig joins IgM on the cell surface, the naïve mature B cells reach to a stage of proceeding to the periphery of the mature B cells and this marks the completion of the developmental process, that completion of developmental process may be ascertained by the techniques of immunohistology and flow cytometry.

IgM and IgD bearing suB cells B-1 and B-2

Some IgM and IgD expressing suB cells appear during fetal life by the display of CD5 in lymph nodes and spleen. These sub cells are called B-1 B cells and B-2 B cells. B-1 cells are also found in the peritoneum. B-1 cells arise from stem cells in bone marrow, and respond to carbohydrate antigens. While more is yet to be known about their function and contribution, if any, in the immune system, it has been observed that they help formation of antibodies.

In mice about 90% of the B cells die within the bone marrow, because they express onto antibodies against self antigens. That cross linkage of mIgM on immature B cells cause the cells to die within the bone marrow. Experiments with transgenic mice have revealed that when immature B cells encounter self-antigens within the bone marrow they are eliminated by apoptosis, a process known as negative selection. Further studies have shown that negative selection of immature

B cells does not always result in their immediate death instead maturation of cells producing self antibodies is arrested. This monitoring is a result of change in the expression of transgene encoded chains - light chain can convert self reactive immature B cells into non-reactive cells to rescue the cells from elimination. Actually this monitoring is better understood by the concept of "editing" of light chain genes. When some immature B cells bind a self antigen, maturation is arrested, the cells up-regulate RAG-1 and RAG-2 expression and further rearrangement of their endogenous light-chain gene starts. Some of these cells succeed in replacing the k light chain of the self-antigen reactive antibody with a l chain encoded by endogenous l-chain gene segments. Then cells begin to express edited mIgM with a different light chain and a specificity that is not self-reactive. As a result, these cells are no longer self reactive and thus evade negative selection and leave the bone marrow as non-self reactive mIgM B cells.

Activation and Proliferation of B cell

After B cells leave the bone marrow, encounter with antigen causes their activation, proliferation and differentiation in the periphery. In case antigens are not encountered, naïve B cells die by apoptosis. Activation induced by antigens is called antigen-driven/induced activation, as a result of which humoral response is generated. B cell activation may occur as dependent upon T_H cells or independent of T_H cells. The nature of the antigen determines whether T_H cells will participate in the activation or not. Antigens that can activate B cells without participation of T_H cells are called thymus dependent antigens, and those antigens which activate B cells without participation of T_H cells are called thymus independent antigens. Some bacterial cell wall components (lipopolysaccharides) or polymeric proteins are B cell activators regardless of their antigenic specificity. At low concentrations lipopolysaccharides stimulate the production of antibodies specific to them, but at high concentrations lipopolysaccharides are polyclonal B cell activator (mitogens). Polymeric proteins (e.g. bacterial flagellin) activate B cells by extensively cross linking the mIg receptor. The polymeric proteins activate mature B cells and inactivate immature B cells without participation of T_H cells.

Generation and Transduction of Activating Signals

Naïve cells, unless activated do G_1 not divide and remain out of the cell cycle. Through activation they pass into G1 and S stage. In the S stage moving through G_2 and into mitosis, a cell completes the cell cycle and DNA is replicated. The route of a naïve cell for activation and division, in which it goes from G_0 stage to S stage (of course G_0 stage is the stage of a cell in which it rests) two types of events drive it into successive stages; the event, which makes a cell competent to progress and the event, which drives the cell to develop (division). When multivalent antigen binds and cross links mIg, the two events take place by mediation of thymus-dependent and thymus independent antigens. The occurrence of these events, since by these naïve cells are pushed into cell cycle through activation, (yet to occur) are better known as activating signals (Figure 7-2).

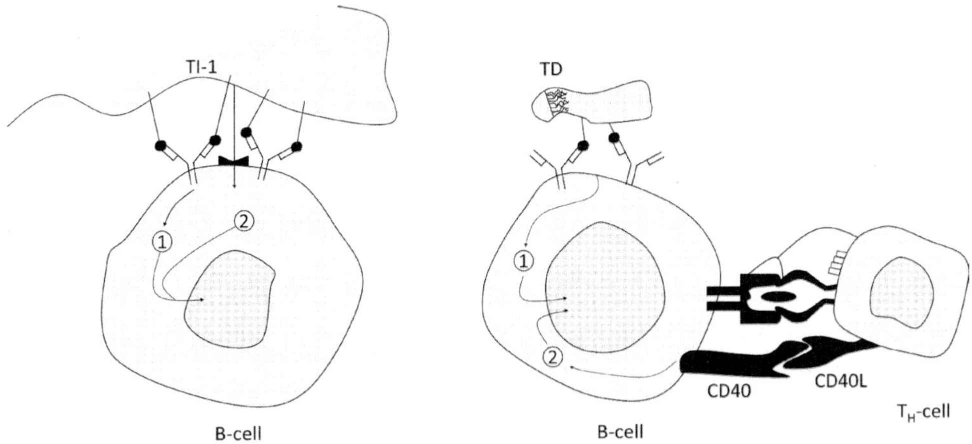

Fig.-7-2. B cell activation involves two distinct signals.

The isotype of mIg have very short cytoplasmic tails to generate a signal by associating with intracellular signaling molecules (tyrosine kinases and G proteins). A question arises as to how engagement of the Ig receptor by antigen could activate intracellular signaling pathways. This is because the membrane Ig is associated with the disulfide-linked heterodimer Ig-a / Ig-b. This association forms the B cell receptor (BCR).

BCR is functionally divided into the ligand-binding Ig molecule and a signal transducing Ig-a / Ig-b heterodimer. The cytoplasmic tails of both the heterodimer contains 61 and 48 amino acids respectively, and 18 residue motif. This 18 residue motif is called immunoreceptor tyrosine based activation motif (ITAM) (Figures 7-3). The cytoplasmic tails of Ig-a / Ig-b transduce the stimulus produced by cross linking of mIg molecules into an effective intracellular signal. However, the following points further elucidate the activation process of B cell / T cell.

- Polypeptides having long cytoplasmic domains are the signal transducers.
- A functional receptor signaling complex is essential for phosphorylation.
- Dephosphorylation is inhibited by CD45.
- Generation of active transcription factors, which stimulate or inhibit the transcription of specific genes in the nucleus.
- A large signaling complex is formed by PTK (protein tyrosine kinase) activity - ZAP70 in T cells and Syk in B cells.

A defective signal transduction in the development of B cells produces an immune deficiency disease characterized by failure to synthesize all classes of antibody, called antibody deficiency disease. This is an inherited disease linked with X chromosome (affects males only) in which the patient's immune system is affected due to profound blockade in the development of B cells and the production of antibodies.

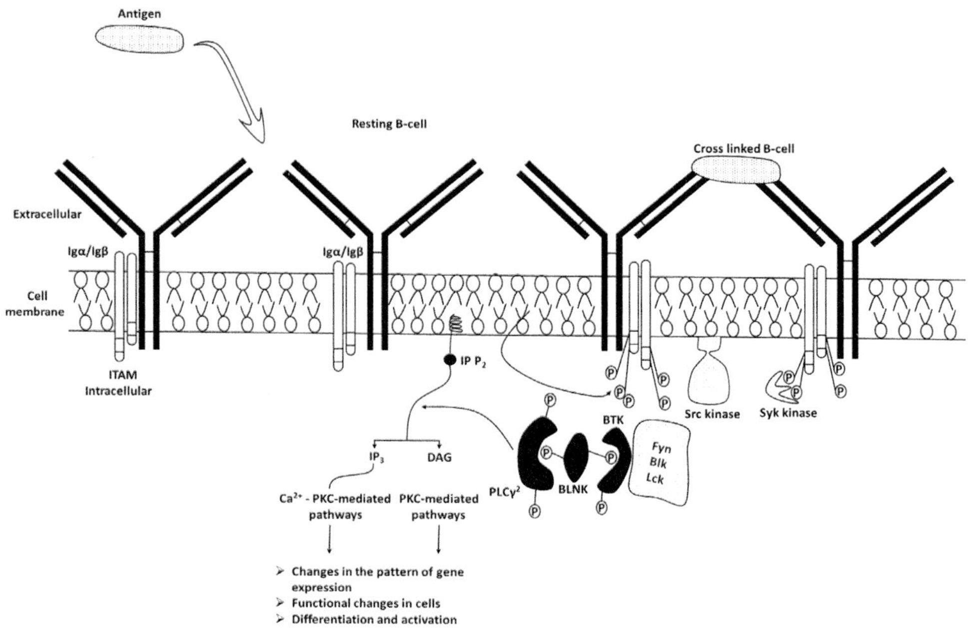

Fig.-7-3. Schematic representation of the activation of B cell receptor (BCR) and signal transduction mechanism through the BCR.

B cell-Coreceptor Complex

Signals through co-receptors can modify the consequences of stimulation through antigen receptors. The component of the B cell membrane, which provides stimulatory signals and capable of modifying the consequences of stimulation, is called B cell co-receptor. B cell co-receptor has three proteins (Fig. 7-4)-

CR2 (CD21)	a receptor of C3d, a breakdown product of the complement system, functioning as a receptor for a membrane molecule, binds to complement coated antigen that has been captured by the mIg on the B cell and cross links the co-receptor to the BCR allowing the CD19 to interact with the Ig-a/Ig-b component of the BCR.
CD19	a member of the immunoglobulin super family with a long cytoplasmic tail and three extracellular Ig fold domains, permits to bind a number of signaling molecules including the PTK (protein tyrosine kinase), Lyn.
TAPA-1 (CD81)	a transmembrane protein that spans the membrane four time.

The co-receptors thus deliver signals to the BCR complex and serve to amplify the activating signal transmitted through the BCR. This is how the low affinity B cells responds to low concentrations of antigen and develops into antibody of high affinity. The B cell co-receptors play an important role in the interaction of different components of immune system.

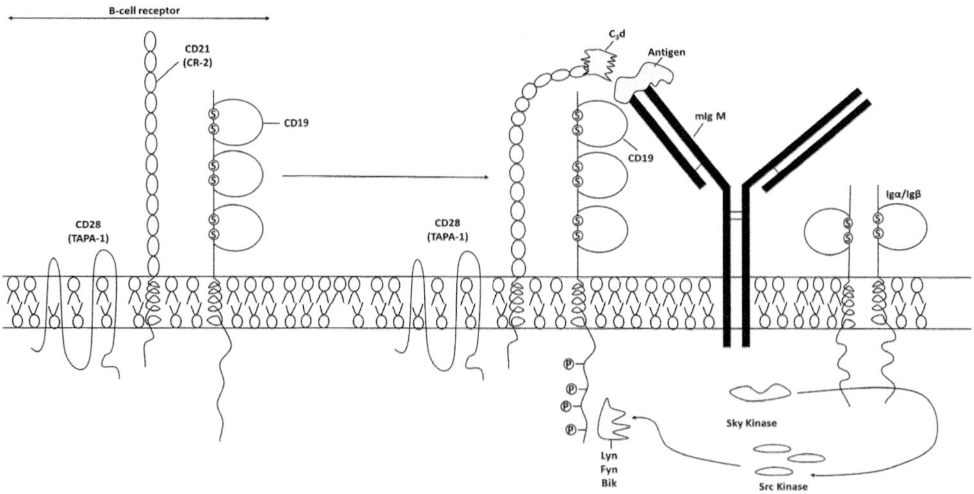

Fig.-7-4. Assemblage of three membrane bound molecules in B cell receptor complex.

Role of T_H Cells in Humoral Immune Response

Activation of B cell is a complex process conducted by thymus dependent and thymus independent antigen. Only binding of antigen to the mIg does not cause signaling for activation of B cell, but interaction with membrane molecules on the T_H cell is required. Two events occur following binding of antigen by mIg on B cells - the antigen is internalized by receptor-mediated endocytosis and processed within the endocytic pathway into peptides, and secondly the antigen binding also initiates signaling through the BCR, which induces the B cell to up regulate a number of cell-membrane molecules (class II MHC, co-stimulatory ligand and B7). The antigenic peptides produced within the endocytic pathway associate with class II MHC molecules and are presented on B cell membrane to T_H cell.

The T_H cell is thus activated. Once a T_H cell recognizes an antigenic peptide displayed by a class II MHC molecule on the membrane of a B cell, the two cells interact. This phenomenon is called contact between a T cell and B cell, resulting in coverage of a broad area of membrane within the two cells (Fig. 7-5). Now, this contact not only leads to the directional release of T_H cell cytokines, but also to the up regulation of CD40 L (CD154), which then interacts with CD40 on B cells. This provides signal from the T_H cell to the activation of B cell. Interaction of CD40 L and CD40 on the B cell delivers a signal to the B cell resulting in changes in gene expression. The transcription factor NF-kB is also activated. The role of CD40 L in B cell activation has been shown in experiment with naïve B cells (incubated with antigen and plasma membranes prepared from activated T_H-cell clones or from resting TH-cell clones), wherein only the membranes prepared from activated T_H cells induced B cell proliferation. Thus, the role of contact between the membranes of B and T cells has been proved. Although activated T_H cells cause proliferation of

B cells, yet they fail to differentiate unless cytokines are also present. This suggests that both a membrane contact and cytokines are needed to B cell proliferation. The contact results in directional release of cytokines toward the interacting B cell. After activation, the B cell begins to express membrane receptors for various cytokines (e.g. IL-2, IL-4, IL-5, etc.). The signals produced by these cytokine-receptor interactions help B cell proliferation and can induce differentiation. During an experiment C. Goodnow demonstrated the process of negative selection of mature B cells in the periphery by the transgenic system. He induced two groups of transgenic mice - single transgenics carrying a transgenic encoded antibody against her egg white lysome (HEL) and double transgenics carrying the anti-HEL transgene and a HEL transgene linked to the zinc-activated metallothionine promoter (to elevate the serum HEL concentration). In both the single and double transgenics, 60% - 90% of the B cells expressed anti HEL membrane bound antibody.

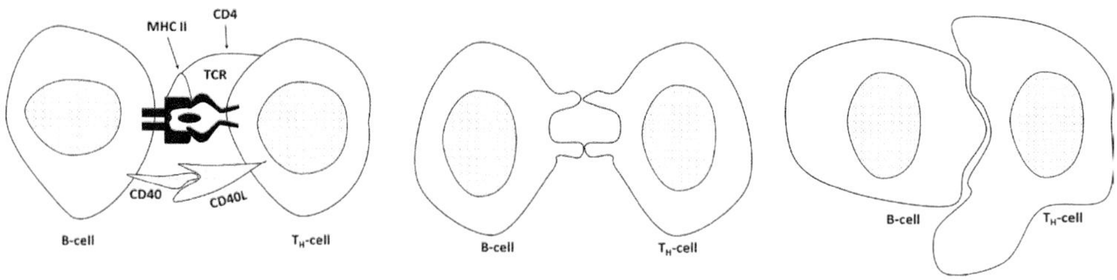

Fig.-7-5. Diagrammatic representation of initial contact between B cell and T cell, and ligand binding (upper panel).

The Humoral Immune Response

When an exogenous antigen comes in contact with an individual, a primary response characterized by a lag phase, during which naïve B cells undergo clonal selection, subsequent clonal expansion and differentiation into antibody secreting plasma cells and memory B cells. The lag phase is dependent upon the nature of antigen, and the nature determines the rise and decline of lag phase. The activation of naïve lymphocytes is called primary response, and activation of memory lymphocytes is called secondary response of the host individual. This response to an exogenous antigen is called the humoral response. As a result of humoral response, antibodies of various isotypes are produced (secreted), which differ in their ability to mediate specific effector functions, by remaining in the life of the individual. Memory lymphocytes activated during the primary response differ from naïve lymphocytes in respect of lag period i.e. memory lymphocytes have a higher affinity for the antigen, lasts longer, larger in population (the memory B cells). This higher affinity provides in host a higher magnitude of defense against infection. That the antibody response with which a memory lymphocyte is encoded at an initial stage (say by vaccination of a particular antigen) can elicit a response of higher magnitude and affinity to a subsequent exposure or vaccine strain encountered later.

Immunization with small organic compounds or haptens conjugated with large proteins (carriers) is called hapten-carrier conjugates, which induces a humoral immune response consisting of antibodies responsive to hapten epitopes and to unaltered epitopes on the carrier protein (see Table 7.2). Hapten-carrier conjugates provide an interesting study in immunology. Under such studies it is again confirmed that B cell activation is dependent upon T cell or T cells help in activation of B cells. Experiments have revealed that both T_H and B cells must recognize antigen determinants on the same molecule for activation of B cell. Thus, it is interaction of T and B cell, which causes recognition of antigens. In various studies of hapten-carrier conjugates, the following have been observed:

- Specific antibody generation is possible when a hapten is chemically coupled with a large carrier molecule.
- Anti-hapten response is possible when immunization is done with the same hapten-carrier conjugate (carrier effect).
- If the populations of hapten and carrier are separate (immunized with Dinitrophenol and Bovine serum albumin - DNP-BSA, but DNP-BCG (Bovine gamma globulin) is a mutually unrelated conjugate), an anti-hapten response ensues.
- B cell response is not limited to the hapten determinant rather some B cells react to epitopes on the carrier.

Table 7-1: Common hapten-carrier conjugates used in immunologic research

Hapten-carrier acronym	*Hapten*	*Carrier protein*
DNP-BGG	Dinitrophenol	Bovine gamma globulin
TNP-BSA	Trinitrophenyl	Bovine serum albumin
NIP-KLH	5-nitrophenyl acetic acid	Keyhole limpet hemocyanin
ARS-OVA	Azophenylarsonate	Ovalbumin
LAC-HGG	Phenyllactoside	Human gamma globulin

Generation of Humoral Response in Lymph Nodes

Any antigen can be filtered up to the extent of 90% by a lymph node carried into it by the afferent lymphatics. While percolating through the cellular mechanism of a node, an antigen encounters any of the three types of antigen presenting cells; in the paracortex interdigitating dendritic cells, macrophages scattered throughout the node, or specialized follicular dendritic cells in the follicles or core. The humoral immune response occurs in the peculiar micro-environments within a lymph node (Fig. 7-6). The lymph nodes are penetrated by the antibodies through the afferent lymphatics or through high endothelial cells. The activation of B and T cells occurs in the paracortex (heavily populated with T cells, macrophages, and interdigitating dendritic cells).

The interdigitating dendritic cells express high levels of class II MHC molecules in the paracortex. Within 1-2 days of trapping of antigen in the paracortex, T_H cells are activated and proliferated. At the same time, naïve B cells from the B cell-rich cortex migrate to the T cell-rich region of the lymph node. On entering the T cell region, a B cell that is presenting antigen can interact with a T_H cell specific to the antigen trapped from a T-B contact. This results in activation of B cell. Within 3-4 days of antigen entry in the paracortex, small foci of proliferating B cell are formed at the edges of the T cell-rich region. The small foci differentiate into plasma cells secreting IgM and IgG isotopes. Most of the antibodies are produced during this period of response. Thus, a few activated B cells and T_H cells migrate from the foci to the primary follicles. These all events are called formation of microenvironment favorable for interactions between B cells, activated T_H cells and follicular dendritic cells. Follicular dendritic cells do not arise in bone marrow, neither express class II MHC molecules nor present antigen to CD4+ T cells.

As a result of the formation of microenvironment, the antigen-antibody complex developed as described above, is retained for long periods of time on the cell membrane. The B cells are amply activated during this period of time and the period is known as that of secondary response of immune system. During the period of secondary response, follicular dendritic cells release small membrane-derived particles, 0.3-0.4 m in diameter (heavily coated with immune complexes called iccosomes). The iccosomes are bound by mlg on B cells bearing antibody with specificity for antigen and endocytosed.

The endocytosed antigen is then processed and presented on class II MHC molecules allowing the B cells to function as effective antigen presenting cells for activated T_H cells (Fig. 7-7). The formation of T-B contact follows as a consequence thereof B cell and T_H cell activate and proliferate. Some such activated B cells and T_H cells migrate towards the core of the secondary follicle, forming the germinal center. In the initial stage of the formation of germinal centers, activated B cells continue to proliferate within the secondary follicle, known as centroblasts. Centroblasts have large size expanded cytoplasm, diffuse chromatin and lack membrane Ig. Centroblasts mostly remain in the one edge of the follicle called dark zone.

The centroblasts give rise to progeny called centrocytes expressing membrane Ig. The increasing number of centrocytes move from the dark zone to a region containing numerous follicular dendritic cells called light zone. In the light zone helper T cells also interact with centrocytes. Centrocytes differentiate in the light zone giving rise to two types of progeny - small memory B cells and large plasmablasts. The plasmablasts move from the germinal center to the medulla of the lymph node, where they develop into plasma cell and begin to secrete antibodies. At this very time, some memory B cell leave to lymph node through the efferent lymphatic vessel and recirculate to the other parts of the body and some memory B cells remain in the vicinity of the germinal center.

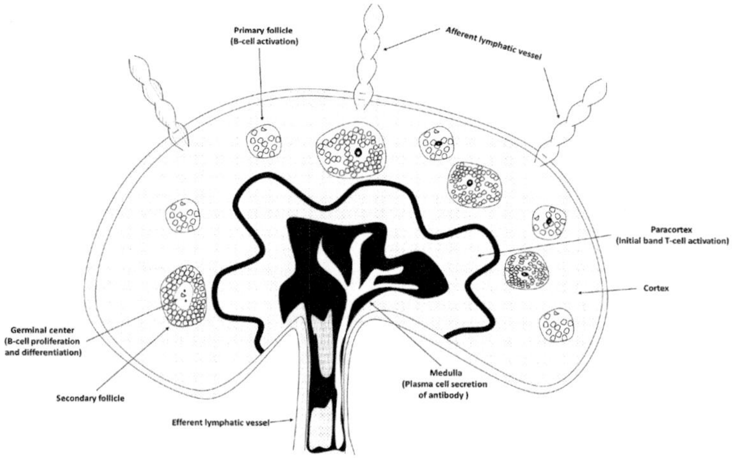

Fig.-7-6. Diagrammatic view of peripheral lymph node showing various anatomic sites

Centrocytes die by apoptosis within the basal light zone of germinal centers and are phagocytosed by an unusual type of macrophage called the tingible-body macrophages. Tingible-body macrophages specialize in phagocytosis of lymphoid cells. Centrocytes die of apoptosis because they cannot bind the antigen presented by the follicular dendritic cells. By apoptosis those cells, which do not recognize the antigen at all, and bear mIg with low affinity for the antigen are eliminated.

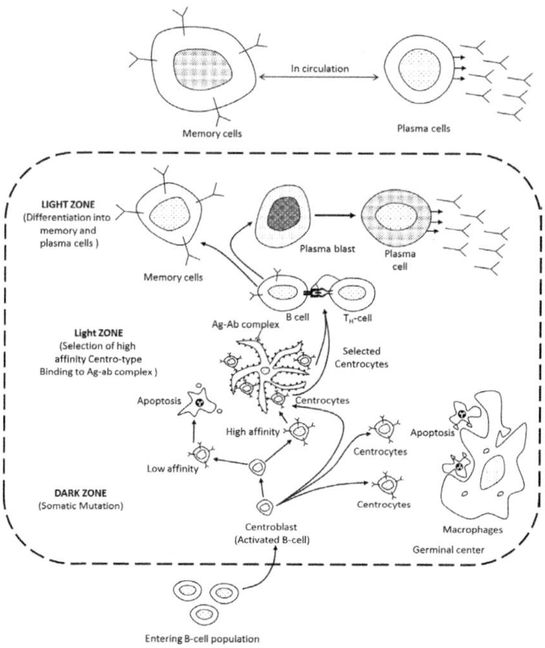

Fig.-7-7. Overview of cellular events within the germinal center.

Role of Germinal Centers in B cell Differentiation

Affinity maturation, class switching, and formation of plasma and memory B cells, all three events that happen during B cell proliferation and differentiation within germinal centers.

Humoral Response versus Affinity of Antibodies: During the course of humoral response, the average affinity of the antibodies is increased to 100 or more times. The increase in the average affinity of the antibodies was first observed in an experiment by H.N. Eisen and G.W. Siskind on rabbits. The rabbits were grouped in two; the first group was immunized with 5 mg dose of DNP-BCG (hapten-carrier complex) and the second group was immunized with 250 mg of DNP-BCG. At intervals of 2, 5 and 8 weeks, the affinity of the serum anti-DNP antibodies produced in response of the antigen was measured. It was found that the first group showed an increase in the average affinity of the antibodies to 140 times during 2-8 weeks, but the affinity of the second group did not show increase for more than 5 times in 2 weeks. This suggests that with low immunizing dose plasma cells (arising from B cells with high-affinity membrane antibody) are the major source of antibody whereas with high immunizing dose plasma cells express both high-affinity and low-affinity membrane antibody.

Humoral Response versus Somatic Hypermutation: Further works suggest that somatic hypermutation during proliferation of centroblasts and antigen selection of high affinity centrocytes lead to affinity maturation, so that the average affinity of the induced serum antibodies increases during the course of thymus dependent humoral immune response. The germinal centers are the sites of somatic hypermutation. A comparison of the mutation frequencies in B cells isolated from germinal centers with those from areas of intense B cell activation outside the germinal centers, has shown that somatic hypermutation of heavy and light-chain variable-region genes probably occurs when centroblasts proliferate in the dark zone of the germinal center. The mutation rate during somatic mutation is about 10-3 base pair/division which is a million time greater than the normal mutation rate for cells. The heavy and light-chain V(D)J segments total about 700 base pairs means that after two cell divisions, nearly every centroblast acquires a mutation in either heavy or light-chain variable regions. Some of the B cells while leaving the germinal center as memory B cells or plasma cells return to the dark zone of the germinal center for further rounds of division and mutation. In this process of returning of some of the B cells to the dark zone of the germinal center, further selected in the light zone, as a result of which antibodies of higher affinity are produced. Thus, the somatic hypermutation has high rates and precise targeting in the immune system. This is an extraordinary process, wherein determination of molecular basis of this process is still a matter for further investigation in immunology. B cells having high-affinity receptors for the antigen are subjected to positive selection and those with low affinity are subjected to negative selection and eliminated in the germinal center.

Humoral Response versus Selection of Centrocytes: By negative selection a centrocyte dies in the germinal center and by positive selection a centrocyte emerges from the germinal center. The selection of centrocytes in the light zone of the germinal center is dependent upon the ability of the membrane Ig molecules or the centrocyte to recognize the antigen and bind it. Antigens are displayed by the follicular dendritic cells. A centrocyte whose membrane Ig binds an antigen gets a signal. Centrocytes who get such signal survive and those failing to get signal die. However, centrocytes with receptors of high affinity are likely to get such signals and bind an antigen. Signals are generated by interaction with a $CD4^+$ T_H cell together with engagement of CD40 on the B cell (centrocyte) by CD40 L on the helper T cell. Centrocytes capable of getting signal either from T_H cell or the antigen-membrane Ig survive, while those failing to get signal undergo apoptosis. This selection feature is an important factor of the germinal center.

Humoral Response versus Class Switching: As is known, class switching induced by several cytokines viz. IL-4, IL-13 etc. allows any given V_H domain to associate with the constant region of any isotype (Fig. 7-8). Consequently antibody specificity remains constant while the biological effector activities of the molecule change. Membrane interaction between CD40 and CD40 L on the T_H cell is a must for the induction of class switching. Failure of this interaction causes an immunodeficiency syndrome in which T_H cells fail to express CD40 L resulting in failure of formation of germinal centers and antibodies.

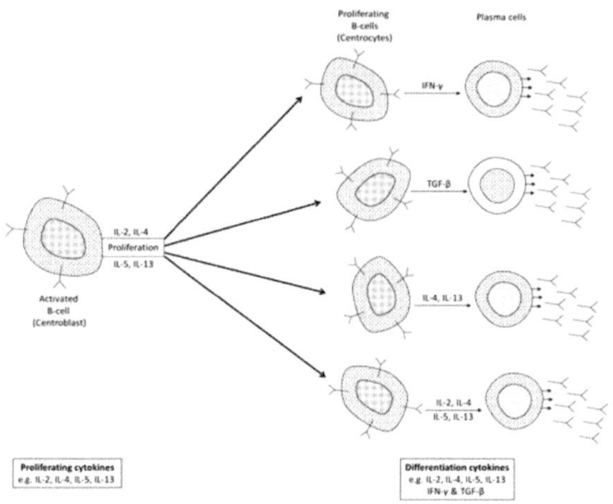

Fig.-7-8. Effects of various cytokines on B cell proliferation, activation and differentiation.

Formation of Plasma and Memory B cells: Small B cells (centrocytes) having high affinity membrane Ig interact with antigen expressed on the follicular dendritic cells. Centrocytes with low affinity mIg are eliminated by apoptosis. Thus the centrocytes, which have high affinity mIg undergo positive selection and differentiate into plasma cells and memory B cells in the light zone of the germinal center. A change in RNA is involved in the formation of plasmablasts for synthesis of the membrane form (higher levels of transcription factors). Memory cells are also formed of high affinity centrocytes in the germinal center, expressing IgG, IgA and IgE molecules, more than IgD. Table 7.3 overviews the comparison of naïve and memory B cells.

Table 7-2: Comparison of naïve and memory B cells

Properties	*Memory B cell*	*Naïve B cell*
Membrane marker *Immunoglobulin*	IgD, IgM	IgD, IgM (?), IgG, IgA, IgE
Complement receptor	low	high
Anatomic location	spleen	bone marrow, spleen, lymph node
Life span	short-lived	may be long-lived
Receptor affinity	lower average affinity	higher average affinitydue to affinity maturation*
Adhesion molecule	low ICAM-1	high ICAM-1

Factors in B Cell Development

The following transcription factors influence the development of B cell:

- B cell specific activator protein (BSAP) expressed only by B-lineage cells influences all the stages of development of B cells to the formation of memory B cells and plasma cells. BSAP binds to the l5 and Vpre B-genes of the surrogate light chain, the J-chain gene of polymeric IgM and the 3a heavy-chain (in germ line DNA).
- Transcription factors- NF-kB, EtsL, c-Jun, Ikaros, Oct-2, Pu.1, EBF, BCF and E24 (DNA binding proteins) promoting or preventing pro-B cells development.

Regulation of the Immune Effector Response

Effector response is essential for reaction of an antigen in its elimination or potential threat to life. If upon encounter with an antigen (self antigen or pathogens), immunity is not developed (dependent upon specific recognition of antigen by antigen reactive T or B cells), this state of unresponsiveness is called tolerance. This process of recognition or response to antigens regulates immune effector system.

There are two routes through which immune response or recognition of antigens occurs; the humoral route and the cell-mediated route. The humoral route eliminates extracellular bacteria and bacteria progeny, whereas the cell-mediated route eliminates infected cells, tumor cells and foreign grafts. Secreted antibodies bring about humoral response, and therefore both antigen and antibodies are well involved in regulatory circuit of humoral immune response whereas specific and nonspecific cells contribute to cell mediated immunity and are regulated by complex repertoire of cytokines and other effector mediators.

CHAPTER – 8

CYTOKINES

In response to a number of stimuli, the white blood cells and other cells of the body secrete a low molecular weight proteins or glycoproteins which regulate the complex interactions among lymphoid cells, inflammatory cells and hematopoitic cells. An effective immune response depends upon the interactions of lymphoid cells, inflammatory cells and hematopoitic cells. The group of low molecular weight proteins or glycoproteins secreted by the white blood cells in response to some stimuli is called cytokines. In this manner, cytokines are effector mediator of glycoprotein nature secreted of lymphocytes and monocytic cells. However, earlier, those cytokines, which are secreted by lymphocytes, are termed as lymphokine, and similarly, those cytokines, which are produced and secreted by monocytes and macrophages are termed as monokines. Some of which are also termed as interleukin to represent its source and target. Chemokines are also a group of cytokines that has a specialized function to drive cell chemotaxis. Cytokines play an important role in cell to cell communication and in the onset of effective immune response. Cytokines regulate the development of immune effector cells and exert direct effector functions. Cytokines are however, locally acting components of immune system. The structure and biological activity of cytokine and their receptor, signal transduction by cytokine receptors, cytokine related diseases and therapeutic uses of cytokines or their receptors, have been covered in this chapter.

The Structure of Cytokines and their Receptors

Cytokines are composed of a molecular mass of less than 30 kDa and exert its effect on various immune and non-immune cells by autocrine, paracrine and endocrine fashion. Cytokines are large group of glycoproteins, members of which may belong to one of the following families:

- The hematopoitin family,

- the interferon family,
- the chemokine family, or
- the tumor necrosis factor (TNF) family.

To achieve effector function, cytokines need to bind to its receptor to communicate signals for biological effects (however, a cytokine is able to induce a plethora of biological processes in a cell). The cytokine receptors are expressed by variety of cells and therefore a cytokine is able to induce biological processes in many cell types at the same time. The cytokine receptors are transmembrane proteins, which are categorized under 5 families. The amino acid sequences of these family members are considerably different, but all of them have a high degree of a-helical structure. The b-sheet structure is either absent or very little. The molecules have similarities in polypeptide fold and helical region (four helices connected by a loop run parallel to one another).

The receptors of cytokines belong to one of the following families:

- Ig superfamily,
- TNF family,
- Hematopoietin (class I),
- Interferon (class II),
- Chemokine

The class I and class II cytokine receptor families are composed of more than one type of polypeptide chain and help in ligand binding. Class I cytokine receptors share common signal-transducing subunits with other members of the same family. The redundancy and antagonism exhibited by some cytokines are due to sharing of signal-transducing subunits among receptors. There are three sub-families of class I cytokine receptor (Fig. 8-1)-GM-CSF, IL-2 and IL-6. IL-2 and its receptor play an important role in the clonal proliferation of T cells, occurring in three forms - low-affinity monomeric, intermediate-affinity dimeric, and the high-affinity trimeric. Signal transduction by the IL-2 receptor requires both b and g-chains, but only the trimeric receptor containing the a chain and binds IL-2 with high affinity. Most of the cytokines involved in immune responses are activated by regulation of the expression of their receptors, e.g. only activated T_H cells express the high affinity IL-2 receptor. Since clear mechanisms operate in the immune response, specificity is maintained by regulation of the expression of cytokine receptor on cells.

When cytokines bind to receptors on the membrane, three actions are initiated. On binding to receptor on the membrane of the same cell, cytokines exert autocrine action. On binding to receptors on a target cell in the vicinity of the producer cell cytokines exert paracrine action, and in distant parts of the cell exert endocrine action. Cytokines are generally secreted after activation of a particular cell and live for a few hours to a few days only. The autocrine and paracrine actions refer to short lived activation of a particular cell.

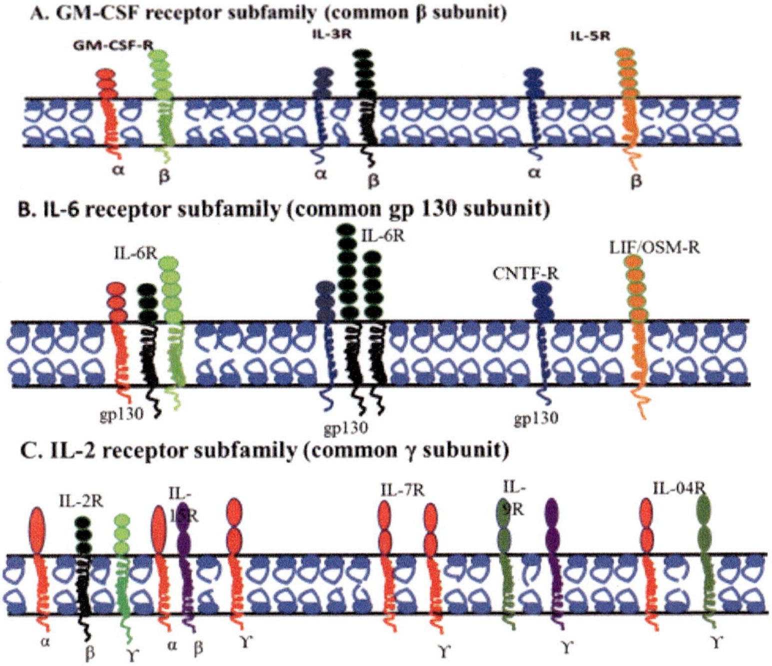

Fig.-8-1. General diagram of GM-CSF, IL-6, IL-2 sub form of class I cytokine receptors. All members of the subfamily have common signal transduction subunit such as α-subunit in GM-CSF-R, IL-3R and IL-5R. (a) gp130 subunit in IL-6R, IL-11R, CNTF-R and LIF/OSM-R, and common γ-subunit in IL-2R, IL-15R, IL7R, IL-9R and IL-4R.

The Biological Activity of Cytokines

Cytokines are mainly secreted by T_H cells and macrophages. The interaction of antigen with macrophages and the subsequent activation of resting T_H cells lead to release of numerous cytokines. By this activation of T_H cells, a complex network of interacting cells is generated in the immune system. Cytokines function in an antigen-nonspecific manner. In other words cytokines affect whatever cells they encounter bearing appropriate receptors and being in a physiological state that allows them to respond. Table 8.1 deals with the principal biological activities.

Table 8-1: Functions of TH1 and TH2 cells and their cytokine secreted

Cytokines	T_H1	T_H2	*Functions*
IL-2	+	-	Induces cell proliferation
IL-3	++	++	Induces histamine secretion
IL-4	-	++	Induces class switching to IgG1 and IgE and MHC II expression
IL-5	-	++	Induces class switching to IgA

Cytokines	T_H1	T_H2	*Functions*
IL-10	-	++	Decreases class II MHC expression
IL-13	-	++	Inhibits inflammatory response
IFN-γ	++	-	kill viral cells, increases expression of MHC class I and II molecule
TNF-β	++	-	Enhances phagocytic activity
GM-CSF	++	+	Differentiation and proliferation of cells

Role of Cytokine Receptor in Signal Transduction

As a result of the interaction of a cytokine with their receptors (either class I or class II family receptors), a series of protein tyrosine phosphorylation ensue. Interferons are capable of inducing cells to block the replication of a number of viruses. That is IFNs have antiviral activities. When IFN-g interacts with its receptor IFN R, signal transduction begins. This may be due to cytokine-induced dimerisation of receptors. As a result of this dimerisation successive reactions are generated and transcription of specific genes is initiated. The resulting gene products mediate the biological effects associated with various cytokines. The following salient features of the IFN-g system are the basis of the unifying signaling device.

- There are two polypeptide chains in the IFN-γ receptor-
    - an a-chain for signal transduction and cytokine binding, and
    - a β-chain for signal transduction only.
- The naïve protein tyrosine kinases (PTKs) are combined with the above chains of IFN-γ receptor.
- An other cytokine receptor known as IL-6 combines with a new family of PTK, known as Janus Kinases (JAK). The JAKs combine with the above chains of IFN-γ receptor in cells without binding of cytokine. The janus kinases or JAKs are present as JAK1, JAK2, JAK3 and Tyk (tyrosine kinases). The different JAKs are active for signaling of different cytokines such as signaling by IL-4 through the IL-4 receptor, a heterodimer IL-4 Rα and IL-13 Rα1 activates JAK1 while signaling by IL-13 activates JAK2 Table 8-2 deals with the type of signaling intermediates and STAT.
- Cytokine binding induces the association of the two separate cytokine receptor subunits i.e. JAK1 associated with the a chain phosphorylates JAK2 and the JAK 2 associated with the β chain phosphorylates JAK 1. In this way, JAKs are activated.
- The activated JAKs phosphorylate specific tyrosine in the receptor chains of the complex. As in 3, IL-4 signaling activates JAK1 and in turns, JAK1 phosphorylates tyrosine residues of IL-4 Rα subunit of IL-4 receptor while IL-13 signaling through IL-4 Rα-IL-13 Rα1 complex (receptor for IL-13) activates JAK2, which in turn phosphorylates tyrosine residues of IL-13 Rα1 cytoplasmic tail. The phosphorylation of tyrosine residues of the receptor subunits forms docking sites for signal transducers and activators of transcription (STAT).

STAT is a member of a family of transcription factors, which after docking to the phosphorylated tyrosine residues of receptor subunit becomes phosphorylated and thereof dimerises to form STAT homodimer. STAT dimer is then translocated to the nucleus and initiates transcription. Specific STATs play essential roles in the signaling pathways of a wide variety of cytokines. However, every cytokine is not associated with the activation and translocation of STAT (Fig. 8-2).

Table 8-2: Some important cytokines and their signaling intermediates

Cytokines	*Signaling intermediates and STAT*
IL-2	JAK1 & JAK3, STAT5
IL-4	JAK1 & JAK3, STAT6
IL-6	JAK1, STAT3
IL-10	JAK1, JNK, STAT3
IL-12	JAK2, JNK, STAT4
IFN-γ	JAK1 & JAK2, STAT1
Adapted from Bach, Aguet, and Schreiber, 1997, Annu. Rev. Immun. 15:563.	

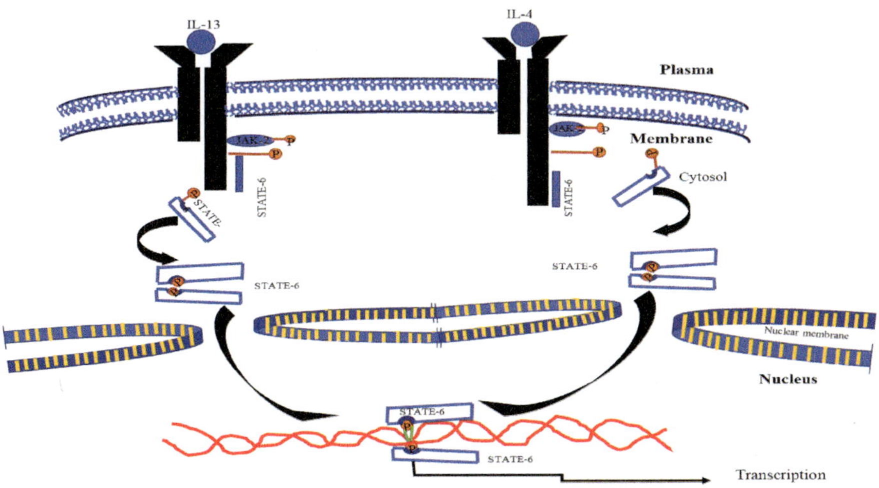

Fig.-8-2. Overview of a proposed pathway of cytokine signaling induced by IL-4 and IL-13 to activate signal transducers and activators of transcription (STAT), which finally leads to initiation of transcription.

However, not every cytokine signaling lead to the activation and nuclear translocation of signal transducers and activators of transcription (STAT), cytokines produced by usually TH1 cells leads to the activation and nuclear translocation of nuclear factor kappa B (NF-κB) that leads to the transcription of immunerelevant genes important for effective immune response. The NF-κB activation pathway is better represented by the signaling of lipopolysaccharides (LPS) and heat shock proteins (HSPs) through its receptor known to be TLR2 and TLR4 (toll like receptor) (Fig. 8-3). IFN-γ signaling leads to activation of NF-κB, and therefore it is necessarily interrelated with TLR2/TLR4 signaling pathway, however not much are known about it. NF-κB is a heterodimer of p65 and p50 proteins and found to be associated with inhibitory kappa B alpha (IκBα) unit in inactive state in the cytosol. The activation of NF-κB needs the dissociation of from NF-κB. A kinase known as IKK α and IKK β phosphorylates the IκBα unit and makes dissociated from NF-κB. NF-κB is then translocated to the nucleus and initiates transcription. IκBα, after dissociation, is degraded by proteasome system in the cytosol.

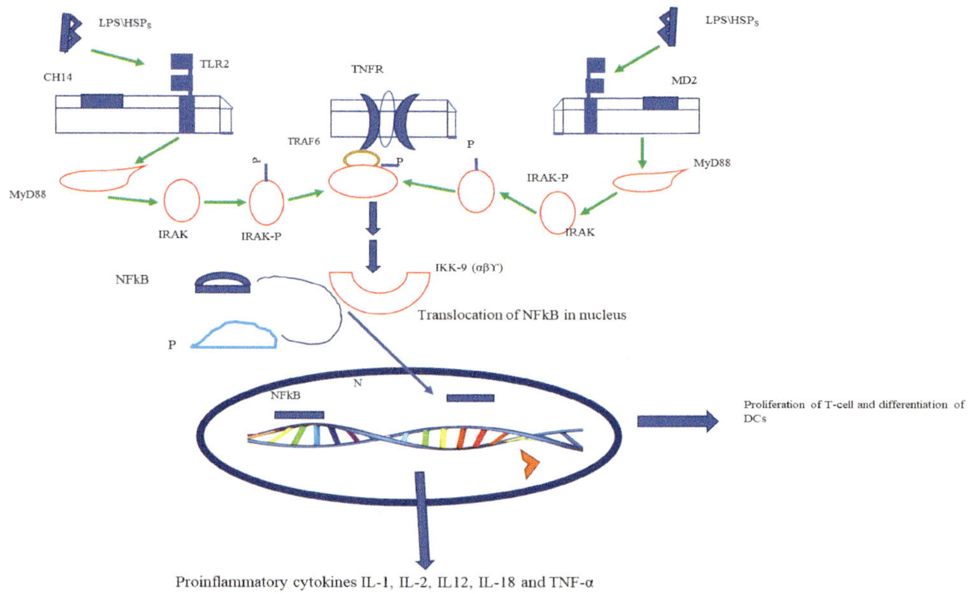

Fig.-8-3. Diagrammatic representation of the proposed mechanism of NF-κB activation and translocation to the nucleus that leads to the transcription of inflammatory cytokines. However, the NF-κB activation mediated by TH1 cytokine is not yet elaborated, but it seems speculative that in the case of cytokine signaling, IKK α and IKK β are also activated to phosphorylates IκBα. The figure is adapted from P. Deepak, S. Kumar and A. Acharya. 2006. European Journal of Inflammation. 4(1):1.

The STAT pathway and NF-κB pathway is interrelated. It has been found that T_H2 cytokines usually activate STAT pathway and down regulate NF-κB pathway whereas T_H1 cytokines usually activates NF-κB pathway and down regulate STAT pathway. However, these pathways are interlinked, are not well elucidated to date. It has been assumed that probably these are regulated at gene level.

Cytokine Signaling Inhibitors

Cytokine inhibitors are present in the bloodstream and extracellular fluid. These inhibitors are soluble and arise from enzymatic cleavages of the extracellular domain of cytokine receptors. The soluble cytokine binding proteins may inhibit or lesser the inflammatory effects of the cytokines. Consequently, the virus is provided with a selective advantage, since some viruses also produce cytokine-binding proteins or cytokine mimics. The molecules produced by these cytokine mimics allow the virus to manipulate the immune response in ways that aid the survival of the pathogen. This is an interacting characteristic of some viruses to combat host immunity.

Cytokine Secretion by T_H1 and T_H2 subsets

The type of immune response made to a particular antigen depends upon the cytokine secretion patterns. T-helper cell subpopulations have two subsets; T_H1 and T_H2 (Recently another subsets; TH0 and TH3 have been identified. T_H0 subset produces both T_H1 or type I and T_H2 or type II cytokines whereas T_H3 subset produces either cytokines at a time). T_H1 cytokines activates B cells to produce opsonization-promoting IgG and exercise inflammation and tissue injury. T_H2 stimulates B cells to produces profuse IgM and IgE by inducing immunoglobulin class switching resulting in allergic reactions.

T_H1 subset is made to respond to viral infections and intracellular pathogens by specially shifted patterns of cytokine secretion. IFN-g (a cytokine) activates macrophages, by which microbial activity is increased and cytokines such as IL-12, induces antibody class switching. IL-12, IL-2, IL-4, IL-6, IL-7, IFN-γ, TNF-α, TNF-β and LIF have some of the soluble cytokine receptors, expression of which has been increased after T_H1 stimulation as a consequence of/for increased biological effects. TNF-β and IFN-γ (cytokines) mediate inflammation, and because of their secretion. T_H1 cells are usually associated with inflammatory phase and responsible for the outcome of cell-mediated immune response, which are essential for immunity against viral infection and tumor cell killing.

T_H2 subset secretes IL-4, IL-5 and IL-13 which induces production of IgE (induces IgE class switching) and induce T_H2 immunity very effective in nematode expulsion in helminthic infections (roundworm). The roundworm (helminth) infections induce T_H2 responses and evoke anti roundworm IgE antibody. The antibody bound to the worm binds to the Fc receptors of eosinophils, forming an antigen-specific bridge between the worm and the eosinophils. The attack of the eosinophils on the worm is triggered by the cross linking of the Fcε - bound IgE (however, responsible for allergy).

Thus, the two subsets of cytokines (T_H1 and T_H2) are grossly aggressive to microbial and helminth infections. The in vivo outcome of these two subsets is illustrative of the nature of response - i.e. T_H1 responds to intracellular pathogens and T_H2 responds to allergic and helminthic infections. The Imbalance of these cytokines in the internal cytokine milieu (T_H1/T_H2 cytokine dichotomy) of the host body leads to many pathophysiology condition viz. asthma, multiple sclerosis, and malignancies.

T_H1 and T_H2 subsets Cytokine Cross Regulation

Responses by T_H1 and T_H2 cells are dependent upon cytokine receptors such as IL-12R and IL-4R. That TH1 cells require IL-12 and IFN-γ for their optimum development. IFN-γ stimulates IL-12 production by activating macrophages and dendritic cells. IFN-γ is generated by stimulation of T cells and activated NK cells, which in turn activated by IL-12.

T_H2 cells require cytokine receptor IL-4. This is proved by observation that knocking out the gene that encodes IL-4 the development of T_H2 cells is prevented.

T_H1 and T_H2 subsets are cross regulated i.e. the growth of one subset inhibits the development of the other subset (Fig. 8-4). And further these cytokines have opposing effects on target cells other than T_H subsets. IFN-γ secreted by the T_H1 subset promotes IgG2 a production by B cells but inhibits IgG1 and IgE production. On the other hand IL-4 secreted by T_H2 subset promotes production of IgG1 and IgE and suppresses production of IgG2a. Thus, the phenomenon of cross regulations explains that there is an inverse relationship between antibody production and delayed hypersensitivity i.e. when antibody production is high, delayed type hypersensitivity is low and vice-versa. The interacting patterns of these two subsets will be further detailed in subsequent chapters. The findings of T_H1 and T_H2 subsets have enormously contributed to the understanding of the control of the immune response.

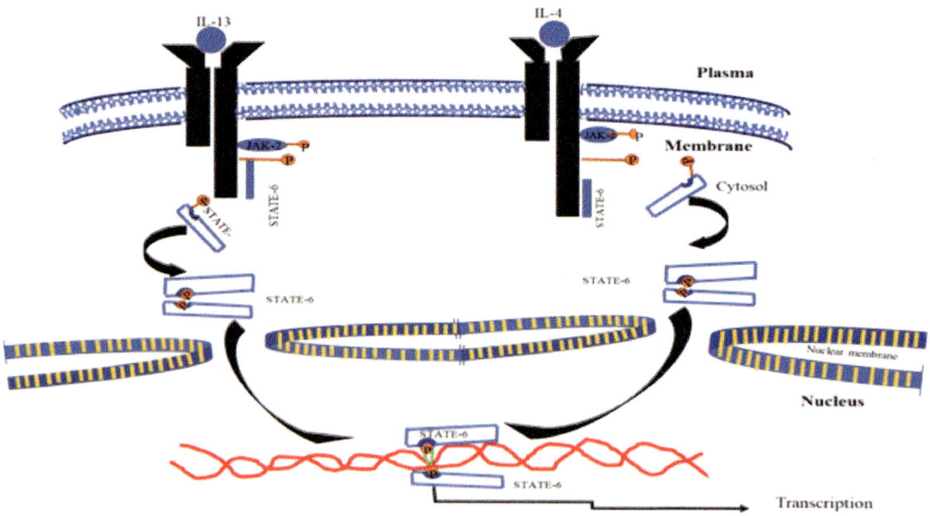

Fig.-8-4. Overview of the regulation of TH1 subunits of T-lymphocytes by cytokines.

In an interesting study of leprosy, in humans, it has been found that it is not a single clinical entity, rather the disease presents as a spectrum of two major forms of disease, tuberculoid and lepromatous, at each end of the spectrum. In tuberculoid leprosy, a cell-mediated immune response forms granulomas, which destroy most of the mycobacteria, leaving only a few organisms in the tissues. Further, in tuberculoid leprosy, skin and peripheral nerves are damaged, and even with the progress of the disease, the patients usually survive whereas in lepromatous leprosy, mycobacteria are widely disseminated in macrophages, causing extensive damage of the skin bone cartilage and nerves. This damage depends upon the balance of T_H1 and T_H2 subsets.

In tuberculoid leprosy, the immune response is characterized by a T_H1 type response with delayed-type hypersensitivity and a cytokine profile composed of high levels of IL-2, IFN-γ and TNF-β. In lepromatous leprosy, there is a TH2 type response with high levels of IL-4, IL-5 and IL-10 i.e., there is diminished cell-mediated immunity and increased production of serum antibody. A similar balancing of T_H1 and T_H2 subsets has been found in AIDS. The T_H1 activity is initially high in AIDS but with its progression T_H2 response occurs. In addition, some pathogens may influence the activity of the T_H subsets. T_H1 activity is suppressed by cross-regulation with the presence of Epstein-Barr virus producing vIL-10 acting like IL-10 and cellular IL-10. The vIL-10 reduces the cell-mediated response to the Epstein-Barr virus and renders it a chance of survival. Thus, the balancing between T_H1 and T_H2 determines the progression of some diseases, as stated above.

Abnormalities Caused by Cytokines or their Receptors

Abnormalities in the expression of either cytokines or their receptors may result in various diseases such as bacterial toxic shock, certain lymphoid and myeloid cancers, and Chagas' disease.

Bacterial toxic shock: In bacterial toxic shock, blood pressure drops, besides fever, diarrhoea and widespread blood clotting. The bacterial toxic shock is a fatal disease and its treatment is exorbitantly expensive. This disease develops because bacterial cell-wall endotoxins stimulate macrophages to overproduce cytokines IL-1 and TNF-α by the same mechanism as shown in figure 8-3 by LPS. Patients dying of meningitis, the levels of TNF-α were found very high than those who escaped death and recovered. In experimental studies, it has been found that monoclonal antibody to TNF-α or IL-1 activity are fairly reversible i.e., neutralization of TNF-α or IL-1 with monoclonal antibodies may prevent the development of toxic shock. Based on these findings there has emerged hope to the treatment of bacterial shock. Superantigens bind directly to class II MHC molecules apparently outside of the antigen binding cleft. Similarly, superantigens bind to an exposed region of the b pleated sheet on the side of the T cell receptors, well away from the sites that bind normal antigenic peptides and can activate large number of T cells irrespective of their antigenic specificity. Superantigens can cause several diseases like bacterial toxic shock. Diseases caused by such superantigens result in excessive production of cytokines, inducing systemic reactions such as fever, widespread clotting and shock (superantigens are namely *Streptococcus pyrogenes, Staphylococcus aurous, Mycoplasma arthritis*).

Lymphoid and Myeloid Cancers Plasmacytosis: Cardiac myxoma cells, myeloma, plasmacytoma cells and cervical and bladder cancer cells secrete IL-6. The production of IL-6 stimulates cell proliferation, exhibiting a massive fatal plasma cell proliferation. This plasma cell proliferation is called plasmacytosis (a high rate plasma cell proliferation). Plasmacytosis may cause development of cancer.

Chagas' disease: Chagas' disease is caused by the protozoan Tripanosoma cruzi, resulting in severe depression of immunity. The immune reactivity of T. cruzi has the been observed by culturing peripheral-blood T cells. T cells are not activated in presence of *T. cruzi* as the contact between lymphocytes and protozoa are enormously inhibited. Further studies in this respect are aimed at finding solution to the treatment of leukemia and other immunity related diseases.

Use of Cytokines and their Receptors in Therapy

The interferons and colony-stimulating factors such as GM-CSF have been found of clinical importance in the treatment of immunity-related diseases. Interferons induce increase in the expression of class I and class II MHC molecules and increase NK cell activity. IFN-γ increases the expression of class II MHC molecules on macrophages and dendritic cells. IFN-α is responsive to leukemia in its chronic phase (wherein granulocytes are massively increased), malignancies, multiple sclerosis (an autoimmune neurologic disease), and chronic granulomatosus disease (CGD). In CGD, the ability of phagocytic cells to kill ingested microbes is severely impaired and patients suffer from recurring infections caused by bacteria and fungi. Treatment with IFN-g has proved superior to administration of antibiotics and surgical drainage of abscesses, since with IFN-γ microbicidal oxidants are profusely generated to reverse the impairment.

However, a number of barriers have prevented the use of many other cytokines for the treatment of diseases. The seemingly under-hand barriers are the maintenance of appropriate dose levels, formulation of effective combinations of cytokines with other drugs or cytokines and regulating cytokine toxicity. The side effects from administration of recombinant cytokines or their receptors, such as fever, chills, diarrhoea, thrombocytopenia, shock, respiratory distress, and coma, are a few warranting considerable counteraction.

CHAPTER – 9

THE COMPLEMENT SYSTEM

Antibody-mediated lysis of bacteria and red blood cells is a result of the interaction of a large and complex group of proteins. The activity of blood serum that completes the action of antibody is called the complement system. It is the major effector of the humoral branch of the immune system. This chapter will deal with the complement components, generation of complement system (activation), biological activities of complement generation and regulation of the complement system.

Components of the Complement System and Some Related Diseases

The complement system is composed of proteins and glycoproteins synthesized mainly by liver hepatocytes. The proteins and glycoproteins are also produced by blood monocytes, tissue macrophages and epithelial cells of the gastrointestinal and genitourinary tracts. These components constitute 5% (by weight) of the serum globulin fraction and most circulate in the serum in functionally inactive forms. By proteolytic change an inhibitory fragment is removed from these components and the complement reaction sequence is started with an enzyme cascade. In this process, if the activated components are unable to react with the next component (sequential charging) they become inactivated and circulate in the serum in the inactive forms. Peptide fragments are formed by activation of a component. The larger fragments bind to the target near the site of activation and the smaller fragments diffuse from the site and can initiate localized inflammatory response by binding to specific receptors. The complement fragments interact with one another to form functional complexes. Complement components are designated by numerals (C1-C9) by letter symbols (e.g. factor D) or by trivial names (e.g., homologous restriction factor). The peptide fragments formed by activation of a component are denoted by small letters. In most cases, the smaller fragments resulting from cleavage of a component is designated "a" and the larger fragment designated as "b" (e.g. C3a, C3b. In the designation of components C2 is an exception: C2a is the larger cleavage fragment).

Inherited deficiencies of the complement components are associated with immune-complex disease and susceptibility to recurred bacterial infections. The deficiencies of these complement components correspond to the specific fragment and affect the pathways of complement activation. Some of the diseases caused by complement deficiencies are given hereunder:

Diseases / Infections	***Deficient Complements (factors)***
Severe bacterial infections immune-complex diseases (e.g. Systemic lupus erethematosus, glomerulonephritis and vasculitis)	C3, C3b
Note: The complement components prevent recurrent infection by mediating a localized inflammatory response and opsonizing the bacteria.	
Neisseria infections (causing recurrent meningococcal and gonococcal infection)	D and properdin
Note: *Neisseria* species are gram-negative bacteria, generally susceptible to complement-mediated lysis or are cleared by the opsonizing activity of C3b.	
Congenital angioedema (localized edema of the tissue often following trauma and if in bowel or upper respiratory tract following abdominal pain or obstruction in the respiratory pathway) and chronic hemolytic anemia (in hemoglobinuria)	C1 Inh (C1 inhibitor) prevents activation of C4 and C2, membrane-bound regulatory components (decay accelerating factor-DAF) and homologous restriction factor (HRF)
Note: In paroxysmal nocturnal hemoglobinuria the glucosyl phosphatidylinositol membrane anchorage is defective resulting in absence of DAF and HRF on the cell membrane. DAF and HRF are inhibitors of lysis of red blood cells.	

In humans, however, with genetically engineered lack of expression of complement components, is not practical as that in mice, to study the controlled situations. As such with partial depletion of a given component, impairment of immune function is not perfectly comprehensible. This fact is a hazard in manipulating congenital C.

Complement Activation

The ultimate result of germination of complement system passing through various pathways is formation of C5b, aimed at membrane attack to pathogens. There are three pathways through any of which complement system may be germinated.

- Classical pathway
- Alternative pathway
- Lectin pathway

Classical pathway: This pathway is activated by the formation of soluble antigen-antibody complexes (immune complexes) or by the binding of antibody to antigen on a suitable target i.e. bacterial cell. This pathogen is also generated by certain subclasses of IgM and IgG (human IgG1, IgG2, and IgG3). Figure 9-1 shows the classical pathway activation in the generation of complement component. C1 in serum is a macromolecular complex consisting of C1q and two molecules each of C1r and C1s held together in a complex C1qr2s2 stabilized by Ca2⁺ ions. The C1qr2s2 may exist in two configurations - S or 8. C1qr2s2 is present in S-shape only when it is free, but when it binds to C1q, it assumes 8-shape (Fig. 9-2). In the IgM, C1q binding sites are not exposed, so the complement cascade cannot be activated by itself. C1q binding is achieved by only two molecules of IgG within 30-40 mm of each other on a target surface or in a complex providing two attachment sites for C1q. In case, the two molecules of IgG end up close enough to initiate C1q binding, more molecules are required to activate classical pathway, whereas a single molecule of IgM bound to a red blood cell can activate the classical pathway. The red blood cell is also lysed.

The binding of IgM and certain subclasses of IgG to soluble antigens or cell-surface antigens activates classical pathway, which results in the generation of C1, C2, C3 and C4 components at different stages. These are intermediates in the classical pathway. C2 and C3 are proenzyme, and C4 is a glycoprotein containing three polypeptide chains: α, β and γ. The reaction sequence generates the enzymatically active C4a2b complex (C3 convertase) which can split C3 into C3a and C3b, and the C4a2b3b complex.

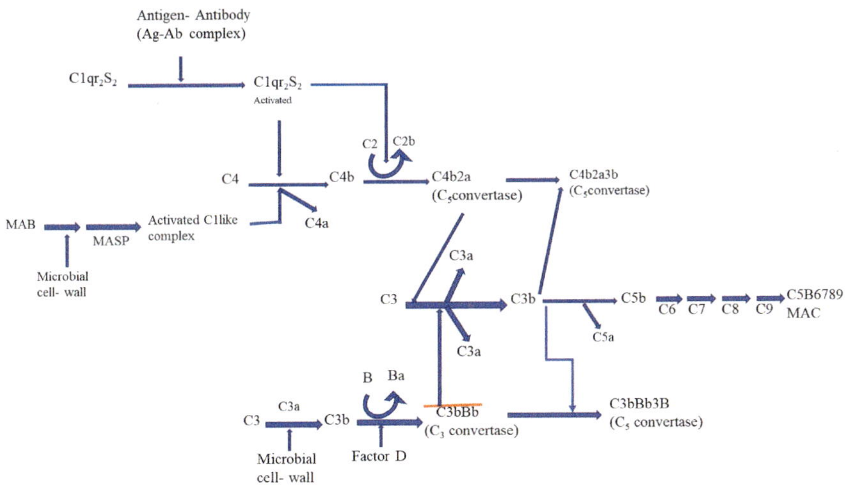

Fig.-9-1. The Figure shows the four possible type of non-covalent bond formation between antibody and antigens. These are hydrogen bonds, ionic bonds, hydrophobic bonds and van der Waals interaction.

Alternative pathway: C4b2a3b complex is called C5 convertase. How this complex changed to a membrane-attack complex will be described in a subsequent section. The alternative pathway is an antibody-independent pathway and a component of the innate immune system. That is to say that alternative pathway is initiated without requiring antibody. This pathway is activated by gram negative and gram positive bacteria. The intermediates in the pathway are pathogens and particles of microbial origin. Non-pathogens too contribute to the initiation of the alternative pathway. The C3 and C5 convertase react into non-enzymatic complexes and bind to the antigenic surface.

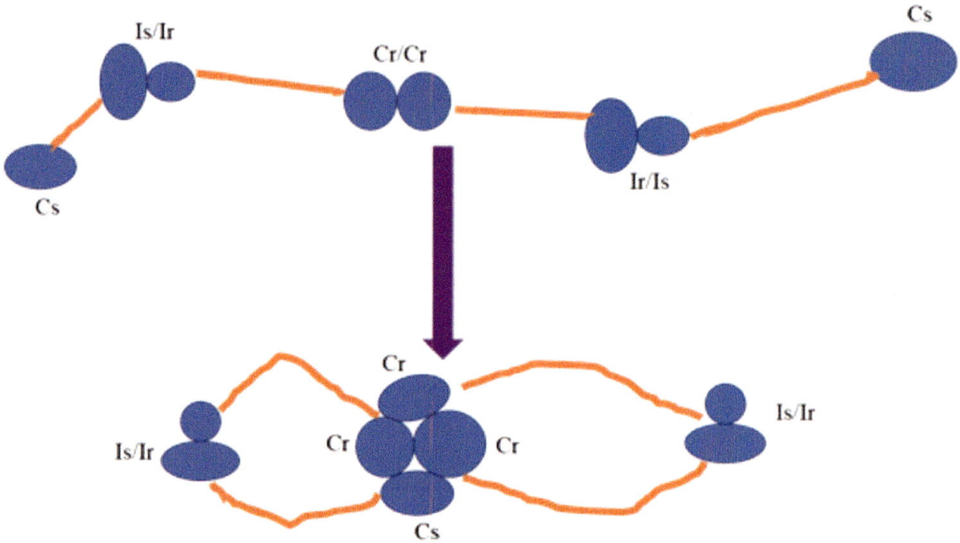

Fig.-9-2. Determination of antibody affinity by equilibrium dialysis.

Lectin pathway: Lectins are proteins that bind to a carbohydrate. This pathway is activated by the binding of mannose-binding lectin (MBL) to mannose residues on glycoproteins or carbohydrates on the surface of microorganisms. The binding of MBL to the surface of a cell or pathogen results in association with serine protease (MASP). MASP resembles C1r and C1s in structure. Since this pathway is also independent of antibody, it is a part of innateimmune system.

Formation of Membrane-Attack Complex

By sequential interaction of C5b, C6, C7, C8, and C9, a macromolecular structure called the membrane attack complex (MAC) develops, which displaces the membrane phospholipids. Consequently, a large transmembrane channel is formed. This channel disrupts the membrane of the target cell and enables ions and small molecules to diffuse through it without any barrier. The components C5, C6, C7, C8, and C9 have different immunologic function. A detail of the complete formation of MAC in relation to active protein/split product (complement components) has been given in Table 9-1. MAC mediates cell lysis by forming a large pore-70-100 Å in the target cell membrane.

Table 9-1: Split product in the formation of membrane attack complex (MAC)

Component	*Split Product*	*Function*
C5	C5a	Acts as anaphylatoxin and mediates
	C5b	chemotaxis and inflammation.
C6	C6	Activation products C5b6 binds to C7
C7	C7	Activation products C5b67 inserts into the lipid bilayer of the target cell
C8	C8	Activation products binds a number of C9 molecules and initiates polymerization
C9	C9	After binding to C5b678, polymerizes to form MAC pore in the target.

Biological Activities of Component Generation

The effective defense mechanism or humoral response is aimed at destruction of invading microorganisms and viruses. The activated complement system is, therefore, accountable for many of the biological activities viz. cell lysis, inflammatory response, opsonization of antigen, viral neutralization, and solubilisation of immune complexes.

Complement mediated lysis: All viruses are susceptible to complement mediated lysis. Herpes viruses, orthomyxo viruses, paramyxo viruses, and retroviruses, all are pathogenic viruses, to be lysed by complement mediated lysis. Escherichia coli, Salmonella, Neisseria gonorrheae, gram-negative bacteria develop resistance to complement mediated lysis, because cell wall lipopolysaccharides (LPS) has long polysaccharide side chains, which prevents insertion of the MAC into the bacterial membrane. But, most gram-negative bacteria are susceptible to complement mediated killing (lysis). There is a thick peptidoglycan layer in the cell walls of gram-positive bacteria. This layer prevents insertion of the MAC into the membrane; therefore gram-positive, bacteria are generally resistant to cell-mediated lysis. The peptidoglycan layer encapsulates the bacteria, hindering interaction between C3b deposited on the membrane and the CR1 on phagocytic cells. In Streptococcus pneumonia (gram-positive bacteria), the peptidoglycan layer encapsulates it, to the insertion of MAC into its membrane. In addition to the encapsulation of bacteria by peptidoglycan layer, various bacteria, viruses, fungi and protozoans contains proteins that can hinder the normal complement cascade on their surfaces, and thereby making the normal complement regulatory proteins C4bBP, CR1 and DAF ineffective. MAC can lyse a red blood cell but not a nucleated cell. This is why; complement-mediated lysis by antibody specific for tumor cell antigen is not effective.

Inflammatory Response: Various peptides generated during formation of the MAC play a decisive role in the development of an effective inflammatory response (Table 9-2). The smaller fragments resulting from complement cleavage, C3a, C4a and C5a called anaphylatoxin bind to receptors on mast cells and blood basophils and induce degranulation with the release of histamine and other pharmacologically active mediators. The anaphylotoxins induce inflammatory an response.

Table 9-2: Biological effects of complement products

Complement product	*Biological effects*
C5b-9, membrane-attack complex	Cell lysis
C3a, C4a, and C5a (anaphylatoxin)	Degranulation of mast cells and basophils*
C3a, C5a	Degranulation of eosinophils, aggregation of platelets
C3a, C5a, C5b67	Extravasation and chemotaxis of leukocytes at the inflammatory site
Bb	Inhibition of monocyte/macrophage migration and induction of their spreading
C3c	Release of neutrophils from bone marrow
C5a	Release of hydrolytic enzymes from neutrophils, increased expression of complement receptors type1 and 3 (CR1 and CR3) on neutrophils
C3b, C4b, iC3b	Opsonization of particulate antigens, increasing their phagocytosis
C3b, C5b-9 (MAC)	Viral neutralization
C3b	Solubilisation and clearance of immune complexes
*Degranulation leads to release of histamine and other mediators that induce contraction of smooth muscle and increased permeability of vessels.	

Opsonization of antigens: The complement system also induces a localized inflammatory response with a buildup of fluid and inflammatory cells, and it facilitates phagocytosis of antigen through its effect as an opsonin. C3b and C4b are the crucial opsonin, by activation of which phagocytic cells and some other cells express component receptors – CR1, CR3 and CR4 that bind to C3b and C4b facilitating phagocytosis of antigens.

Viral Neutralization: Viral infectivity is immensely neutralized in the host complement system through classical pathway. A number of mechanisms come into play in neutralization of viruses e.g., viral aggregates. Viral aggregates reduce the net number of infectious viral particles through C3b component. Two molecules of antibody for virions are enough to neutralize a virus by its binding to the surface of a viral particle. The coating of molecules neutralizes viral infectivity by blocking attachment to susceptible host cells.

Solubilisation of Immune Complexes: Complement also plays an essential role in the clearance of circulating immune complexes. Individuals, who suffer from autoimmune diseases, produce large quantity of immune complexes, which results in tissue damage (as is found in the disease systemic lupus erythematosus SLE). Deficiencies in C1, C2, C4, and CR1 predispose an individual to SLE, wherein deficiency of C4 is the most effective cause. The levels of C3b are reduced with deficiencies of C1, C2, and C4, on immune complexes and hence, their clearance is inhibited, besides proper binding.

Mechanism of Regulation of the Complement System

The complement system requires elaborate regulatory mechanisms to prevent damage to normal tissues. The regulations are exerted by active and passive components of complement system.

Active Components: There are several proteins that inactivate various complement components viz (see Table 9-3),

- the glycoprotein C1 inhibitor forms a complex with C1r2s2 causing it to dissociate from C1q. This prevents further activation of C4 or C2.
- C3b (the C3 convertase enzyme) undergoes spontaneous hydrolysis, and thereby preventing damage to the normal host cells (Fig. 9-3).

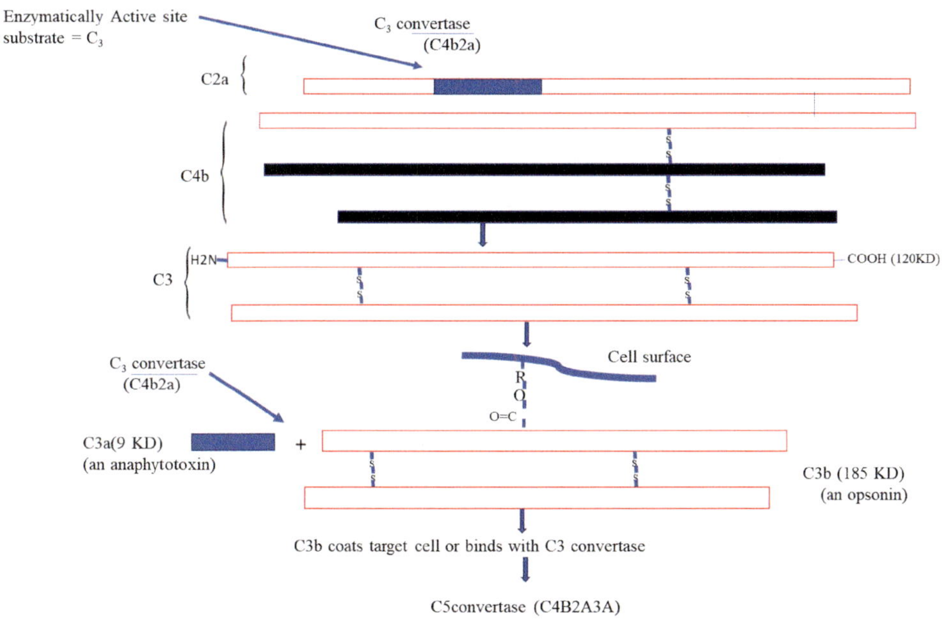

Fig. -9-3. Different steps of C3 activation in the application phage.

- The three RCA proteins;
  - soluble C4b binding proteins (C4bBP),
  - two membrane bound proteins (CR1), and
  - membrane co-factor protein (MCP)

prevent assembly of C3 convertase, and each of the above proteins binds to C4b. This association also prevents binding to C2a **(Fig. 9-4).**

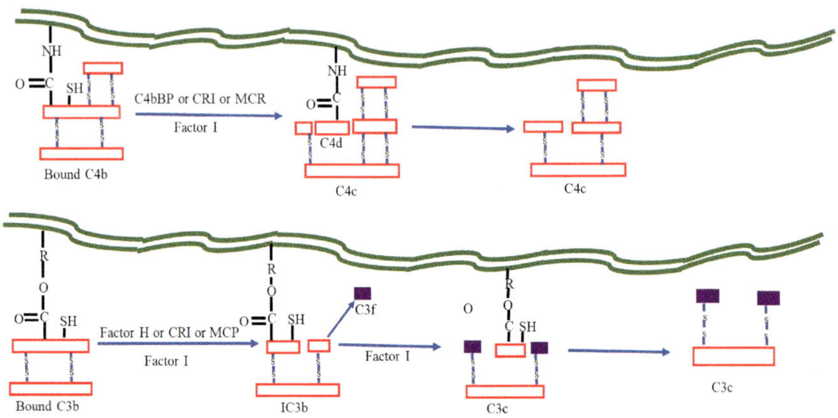

Fig.-9-4. Diagrammatic representation of interaction of the a bound C4b and C3b (a) representing classical pathway (by regulatory proteins of the complement system).

- serum proteins counter binding to nearby cells. A serum protein called as S-protein can bind to C5b67 including a hydrophilic transition and thereby preventing insertion of C5b67 into the membrane of nearby cells. Fig. 9-5 deals with the mechanism of C5 activation and formation of C5 convertase, which are neutralized by S-proteins.

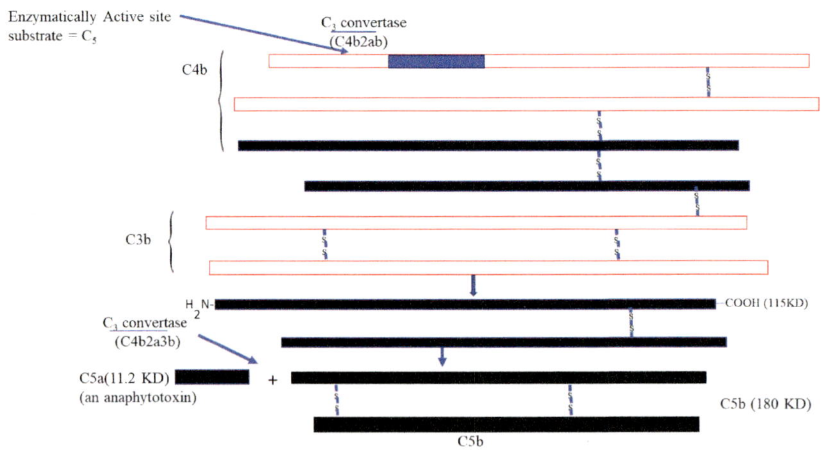

Fig.-9-5. Different steps of C5 activation (cleavage of C5).

Passive Control: Passive control is exerted by the extreme lability of many of the active components, which lose activity unless they react with a subsequent component. The extremely labile components undergo spontaneous inactivation as they diffuse away from the target cells.

Table 9-3: Some important proteins involved in the regulation of the complement system

Protein	*Nature*	*Pathway*	*Immunologic function*
C1 inhibitor (C11nh)	Soluble	Classical	Serine protease inhibitor: causes C1r2s2 to dissociate from C1q
C4b-binding Protein (C4bBP)*	Soluble	Classical and lectin	Blocks formation of C3 convertase by binding C4b; cofactor for cleavage of C4b by factor1
Factor H*	Soluble	Alternativ	Blocks formation of C3 convertase by binding C3b; cofactor for cleavage of C3b by factor1
Complement-receptor type1(CR1)*	Membrane bound	Classical, alternative and lectin	Blocks formation of C3 convertase bybinding C3b or C4b; cofactor for factor1-catalyzed cleavage of C4b or C3b
Membrane- Cofactor protein(MCP)*	Membrane bound	Classical, alternative and lectin	Blocks formation of C3 convertase bybinding C3b or C4b; cofactor for factor1-catalyzed cleavage of C4b or C3b
Decay-accelerating Factor (DAF)*	Membrane bound	Classical, alternative and lectin	Accelerate dissociation of C4b2a2a and $C3\overline{bBb}$ (classical and alternative C3 convertases)
Factor 1	Soluble	Classical, alternative and lectin	Serine protease: cleaves C4b or C3b using C4bBp, CR1, factor H, DAF, or MCP as cofactor
S protein	Soluble	Terminal	Binds soluble C5b67 and prevents its insertion into cell membrane
Homologous restriction factor (HRF)	Membrane bound	Terminal	Bind to C5b678 on autologous cells, blocking binding of C9
Membrane inhibitor of reactive lysis (MIRL)	Membrane bound	Terminal	Bind to C5b678 on autologous cells, blocking binding of C9
Anaphylatoxin inactivator	Soluble	Effector	Inactivates anaphylatoxin activity of C3a, C4a, and C5a by carboxypeptidase N removal of C-terminal Arg

CHAPTER-10

LEUKOCYTE MIGRATION AND INFLAMMATION

The specific feature of leukocytes is that they while circulating throughout the blood migrate into the tissue threatened by infection or tissue injury. This migration of leukocytes into the affected tissues is a well controlled process caused by various molecules. As a result of this migration of leukocytes and inflammation occurs at the site of antigen population, visibly by redness, heat, swelling and pain. Inflammation, therefore, marks the encounter of leukocytes with the antigens, involving mediator molecules and specific physiologic changes. This whole process called inflammatory response is protective to this host organism. This chapter elaborates the inflammatory response.

Lymphocyte Recirculation

Lymphocytes are the major cellular constituents of the lymphoid tissue. They are quite mobile and travel from blood to lymph and then return to blood to ensure coordinated functions of immune system. Lymphocytes circulate continually in the blood lymph and various lymphoid organs. They stay for about 30 minutes, circulating in the blood stream. From the bloodstream about 45% of lymphocytes move to the spleen and about 42% of lymphocytes move to various peripheral lymph nodes. After staying for about 5 hours in the spleen and 12 hours in the lymph nodes, the cycle is completed. During this continual recirculation, 1-2 times per day, maximal numbers of antigenically committed lymphocytes encounter and interact with antigens, whereby inflammatory response or immune response is generated. The population of lymphocytes is aggravated depending upon the population of antigens in a tissue

The lymphocytes migrate from one lymphoid organ to an other and cover the largest distances in their considerable life span. This circulation of lymphocytes increases the efficiency of local immune responses in many ways. Firstly, it allows antigen-induced selection of precursor cells; secondly it enables systemic immunity to be induced by local entry of antigen due to the dissemination of effector and memory cells; thirdly, it facilitates collaboration of T and B lymphocytes in immune responses. This type of migration of lymphocytes from primary lymphoid organ to secondary lymphoid organ where they encounter with an antigen through the blood vascular system and lymphatic system is called lymphocyte trafficking (Fig.-10-1), where as trafficking of lymphocyte to particular tissue or organs site is termed as lymphocyte homing. In other words, lymphocyte homing is the directed migration of subsets of circulating lymphocytes into particular tissue sites such as migration of lymphoid stem cells to central lymphoid organs, migration of mature lymphocytes to peripheral lymphoid organs and migration of lymphocytes to the site of infection or inflammation.

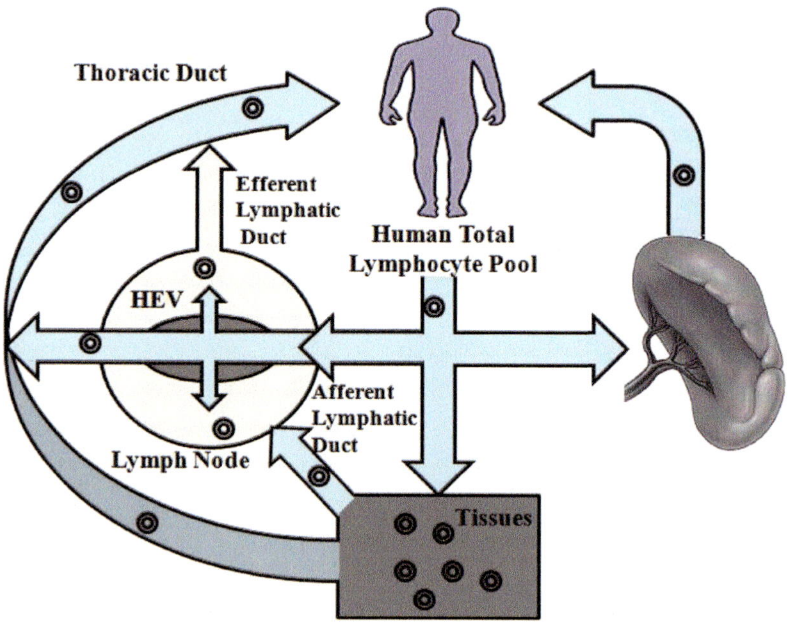

Fig.-10-1. Lymphocyte trafficking through central and peripheral lymphoid organs and tissues.

Although, they travel to and from the lymphoid tissue, their movement is not random. Lymphocytes of different origin have different migratory properties. Lymphocytes obtained from intestinal lymph migrate preferentially through the gut, whereas those obtained from efferent lymph of a peripheral lymph node migrate through peripheral lymph nodes. Now, it has been shown that under strong regulatory mechanisms, lymphocyte trafficking is non-random. Eugene Butcher and his colleagues in 1999 were the first to propose the combinatorial model of multi-step leukocyte trafficking. They have shown that lymphocyte trafficking is determined by the association and sequential operation of adhesion and chemoattractant receptors that control the multi-step processes of lymphocyte homing.

Circulating lymphocytes enter lymph nodes via homing receptors by recognizing corresponding ligands (addressins) on high endothelial venules (HEV) and conventional blood vessels. It is because of the fact that blood cells traverse the post-capillary venule at a velocity of 500μm/s, resulting in a shear force that must be overcome before a lymphocyte can adhere to the endothelium lining and to extravasate into tissue. The migration of lymphocytes (both B and T lymphocytes) is highly dependent on the responsiveness to the chemokines, chemokine gradient across the blood vessels or lymphatic system and the expression of homing receptor like selectin superfamily (P-selectin, E-selectin, and L-selectin) and integrin. However, other molecules such as vascular adhesion protein-1 (VAP-1) on endothelial cells and possibly hyaluronate, also participate in the rolling of leukocytes. The contribution of inflammation-inducible VAP-1 to initial interaction between lymphocytes and vascular endothelium has been recently confirmed in animal models.

High endothelial venules (HEV)

HEVs are postcapillary venules, different from conventional or ordinary venules in several aspects. HEV is composed of a prominent perivascular sheath, a thick basal lamina and a layer of endothelial cells (Fig.-10-2). They are located mainly in the T cell zones, such as the paracortical areas of lymph nodes and the interfollicular areas of Peyer's patches. They are also found in the B cell zones, particularly in the periphery of B cell follicles. Some adhesion molecules are common to both normal venules and HEVs (such as CD31 and ICAM2), while certain molecules are expressed specifically in HEVs.

Fig.-10-2. Ordinary or conventional venule and high endothelial venule (HEV)

Peripheral lymph-node HEVs mainly expresses highly glyc-osylated and sulphated forms of sialomucins, including glycosylation-dependent cell-adhesion molecule 1 (GLYCAM1), CD34, podocalyxin, endoglycan and endomucin, carbohydrate moiety of which known as peripheral-node addressins (PNADs). HEVs express high levels of lymphoid chemokines, whereas normal venules generally do not (venules in the intestine and skin do express very high level of chemokines).

Extravasation process: The process by which leukocytes enter inflamed tissue of peripheral lymphoid organs, adhering to and passing between the endothelial cells (lining the wall of blood vessels) is called extravasation. Endothelial cells express leukocyte specific cell adhesion molecules (CAMS). CAMS are composed of four families of proteins-

- Integrin Family,
- Immunoglobulin Super-family,
- Selection of family, and
- Mucin like Family

The Integrin family is characterized by heterodimeric proteins. Different integrins help binding of leukocytes to different CAM_S and play an important role in extravasation process.

Immunoglobulin super-family is characterized by immunoglobulin like domains and direct lymphocyte entry into mucosa. The domains are named as ICAM-1, ICAM-2, ICAM-3 and VCAM expressed on various endothelial cells.

The Selection of family is composed of molecules named L, E, and P are expressed on vascular endothelial cells. These molecules cause adhering of leukocytes to endothelial cells.

Mucin like family is characterized by a group of heavily glycosylated proteins (serine and threonine rich proteins) on some endothelial cells of lymph nodes.

Selectins and mucin-like CAMS interact with each other, and member of each family are expressed on both leukocytes and endothelial cells. Integrins, expressed on leukocytes, interact with Ig-super family CAMS expressed on endothelial cells.

Extravasation of Neutrophils and Lymphocytes

Neutrophils recognize the inflamed endothelium and adhere strongly to them to protect themselves against sweeping by the flowing blood. After binding to endothelium, neutrophils enter the endothelial layer and migrate into the underlying tissue through four sequential steps- rolling, activation by chemo-attractant stimulus, arrest and adhesion, and lastly transendothelial migration (Fig. 10-3) and (Fig. 10-4).

Rolling is a type of weak binding, in which cytokines and other mediators act upon the local endothelium to which neutrophils remain bound. At this stage selectin and mucin like molecules interact with each other, causing tethering of neutrophils, which are soon detached by the flowing blood. This very process is repeated, causing rolling of neutrophils.

In the course of rolling, neutrophils are activated by chemo-attractive cytokines called chemokines (secreted by local cells involved in the inflammatory response).

The chemokines are IL-8 and macrophage inflammatory protein (M/P-1b). Besides, some breakdown bacterial proteins are also added to the neutrophils enroute tissues, and finally the interaction between integrins and Ig superfamily paves the way to the tissues. The addition of breakdown bacterial proteins and interaction between integrins and Ig super-family constitutes the arrest, adhesion and transendothelial migration of neutrophils for migration into the tissues.

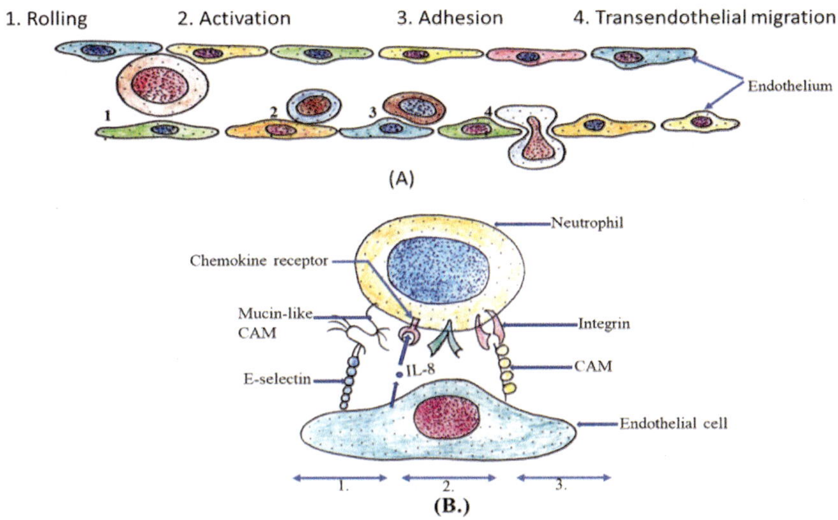

Fig.-10-3. (a) Four different steps in neutrophil extravasation. (b) Initial steps required for secretion of cell adhesion molecules and chemokines (IL-8). This conformational change enables neutrophils to extravasate.

As with neutrophils, extravasation of lymphocytes involves interactions among a number of CAMS. Lymphocytes extravasation is similar to neutrophils extravasation, in their entry into lymph nodes. Cytokines are produced locally in response to antigen capture. The general process of lymphocyte extravasation is however different from that of neutrophil extravasation in terms of "homing receptors". The receptors which direct the circulation of various populations of lymphocytes to particular lymphoid and inflammatory tissues are called homing receptors. Homing of lymphocytes is also called "trafficking".

Until activation, a naive lymphocyte is unable to become an effector cell. Naïve lymphocytes become active by presentation of antigen through dendritic cells. Once naïve lymphocytes encounter antigen trapped in a secondary lymphoid tissue, they become active as lymphoblasts. Activation takes about 48 hours, and during this time the blast cells are retained in the para-cortical region of the secondary lymphoid tissue. At this phase-lymphoblast is known as shut down phase, since blast cells are retained (shut down) in the para-cortical region of the secondary lymphoid tissue. And during this phase, antigen specific lymphocytes cannot be detected in the circulation of blood, leaving the node in the efferent lymph. The shut down phase generates the effector and memory cells. Leaving the secondary lymphoid tissue, lymphocytes converted into effector and memory cells begin to recirculate. This phase of conversion of lymphocytes into effector and memory cells is called shift phase.

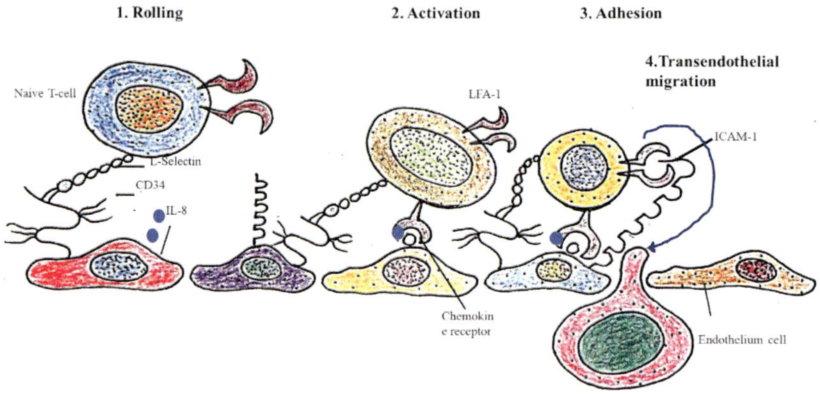

Fig.-10-4. Diagrammatic representation of three distinct steps; rolling, activation and adhesion in extravasation of naïve T cells through the endothelium.

Specific Homing Patterns of Lymphocytes, Effector, and Memory Cells

The homing patterns of naïve lymphocytes, effector and memory cells differ from one another. Some regions of vascular endothelium in post capillary venules of various lymphoid organs are composed of specialized cells with a plump having cuboidal shape. These regions are called high endothelial venules or HEV_S. The development and maintenance of HEV_S in lymphoid organs is affected by cytokines produced in response to antigen capture. HEV_S express a variety of cell-adhesion molecules. Some of these adhesion molecules are distributed in a tissue-specific manner.

The homing pattern of naïve lymphocytes is designed to keep adhesion molecules expressed by HEV_S constantly recirculating through secondary lymphoid tissue whose primary function is to trap blood-borne or tissue-born antigen. This type of homing helps to maximize the possibility of antigen trapping by a lymphocyte recognizing it. It is at these stages that naïve lymphocytes become active and enlarge into lymphoblasts, as described earlier.

Effector and memory cells exhibit tissue-selective homing pattern, where different combination of adhesion molecules- LPAM-1 (a4b7), LFA-1 (aLb2) and ICAMS are joined to them. Table 10-1 shows the cell surface molecules or receptors involved in the leukocyte extravasation. These adhesion molecules interact with ligands present on tertiary extra lymphoid tissue such as skin and mucosal epithelia and at the sites of inflammation. This interaction helps homing of effector and memory cells to the sites of inflammation. Effector and memory cells home through HEVS into peripheral lymph nodes through the afferent lymphatic vessels.

Table 10-1: Cell adhesion molecule and their role in leukocyte extravasation

Adhesion Molecules	*Expression*	*Ligands on endothelium*	*Step involving interaction*	*Main function*
CLA or ESL-1	Effector T cells	E-selectin	Tethering/ rolling	Homing to skin and migration into inflamed tissue
L- selectin	All leukocytes	GlyCAM, CD34 MAdCAM-1	Tethering/ rolling	Lymphocyte recirculation via HEVs to peripheral lymph nodes and migration into inflamed tertiary sites
LFA-1(αLβ2)	Leukocyte subsets	ICAM-1,2,3	Adhesion/ arrest	General role in leukocyte extravasation via HEVs and leukocyte migration into inflamed tissue
LPAM-1 (α4β7)	Effector T cell monocytes	MAdCAM-1, VCAM-1	Rolling and adhesion	Homing of T cells to gut via mucosal HEV; migration into inflamed tissue
Mac-1(αMβ2)	Monocytes	VCAM-1	----------	Monocyte migration into inflamed tissue
PSGL-1	Neutrophils	E- and P-Selectin	Tethering/ rolling	Neutrophil migration into inflamed tissue
VLA-4 (α4β2)	Neutrophils, T cells, Monocyte	VCAM-1, MAdCAM-1, fibronectin	Rolling/ adhesion	General role in leukocyte migration into inflamed tissue
VLA-6 (α6β1)	T cells	Laminin	---------	Homing of progenitor T cells to thymus; possible role in T cell homing to non-mucosal sites

Interaction pattern of Adhesion Molecules (extravasation of lymphocytes)

Naïve lymphocytes initially bind to HEVS by L-selectin (a homing receptor) directing the lymphocytes to specific tissues expressing a corresponding mucin-like vascular addressin such as CD34 or GlyCAM-1. Because of the slow rate of blood flow in the regions of HEVS, the lymphocytes are not dislodged. Chemokine binding to G-protein leads to the activation of integrin molecules on the membrane, and thereby interaction occurs with Ig-super family adhesion molecules, resulting in adherence of lymphocytes to the endothelium. However, more has yet to be understood about transendothelial migration.

Mechanism of Lymphocyte Trafficking

Lymphocyte trafficking or migration involves complex regulation of adhesion in time and space. Trafficking begins with the extension of leading edge called lamellipodia in the direction of motion with the subsequent anteroposterior contraction of microfilament at the trailing edge called the uropod. It involve the regulated polymerization and depolymerization of actin cytoskeleton and regulated anchorage of lamellipodia with endothelial cells via adhesion molecules, typically integrins and drawing the cellular component of cytoskeletal framework by myosin-based interaction designated as contraction wave. This myosin-based contraction wave pulls the cells forward. Before triggering adhesion, lymphocytes are loosely attached to the endothelial cells, known as tethering. It allows lymphocytes to roll on the endothelial wall. The particular process is mediated by the members of the selectin superfamily of the molecules interacting with specific carbohydrate moiety present on the endothelial wall, which together constitutes an cascade. Rolling is, however, very complex processes, which can be meaningfully divided into (1) tethering/loose rolling mediated by a first interaction (requiring presentation of "tethering" receptors on microvilli to allow initial contact, and rapid on and off rates and tensile properties to allow rolling and (2) separate "braking" mechanisms for slowing rolling further to the point that chemoattractant activation and integrin engagement can occur (Fig.-10-5). Here, α4β7 interaction with MAdCAM-1 acts as a brake specifically during naïve lymphocyte interactions in Peyer's patch HEV. However, it has been found that a number of molecules such as CD44, LOX1 and certainly VAP1 may be specialized for braking. Selectin engagement leads to weak adherence to the endothelium of venule, which results into simply rolling of lymphocytes in the venule's lumen (Fig.-10-6). The mechanism of extravasation of neutrophil in the case of inflammatory response generated by bacterial infection has also been illustrated in the First Chapter (see Fig.-10-5 & Fig.-10-6).

Table 10-2: Difference between ordinary or conventional venules and high endothelial venules (HEVs)

	Ordinary Venules	*HEVs*
Endothelium	Flat	Tall & plump
Basal lamina	Thin	Thick
Perivascular sheath	Scanty	Prominent
CD31	+	+
ICAM-1	-/+	++
ICAM-2	+	+

	Ordinary Venules	***HEVs***
E-cadherin	+	+
Sialomucins	-/+	++
MAdCAM-1	-	+
Chemokines	-	+

Both endothelial cells and lymphocytes are known to express a number of molecules responsible for adhesion, which together constitute unique receptor-ligand pairs (Table 10-3). These receptor–ligand pairs such as α4β7/MAdCAM-1, α4β1/VCAM-1, CD44/hyaluronate, LOX1, and VAP1 are differentially regulated on leukocytes and/or endothelial cells, and thus make a point of differential specificity of trafficking. Lymphocyte chemoattractants secreted from within peripheral lymph nodes and their G protein-coupled receptors expressed on lymphocyte subpopulations provide strong signal to stimulate integrin-dependent recruitment of lymphocyte subsets across the venule and also to the lymph nodes.

Moreover, the arrival of Ag and the induction of an immune response in the node greatly increase blood flow and traffic of lymphocytes across HEV, which is coupled with a transient, sharp decrease in recirculating lymphocyte output from the efferent lymphatics. In lymph nodes, lymphocyte traffic across the HEV may increase substantially within 3 hours after antigenic stimulation and by as much as 10-fold over the first 48 hours of the induction of an immune response.

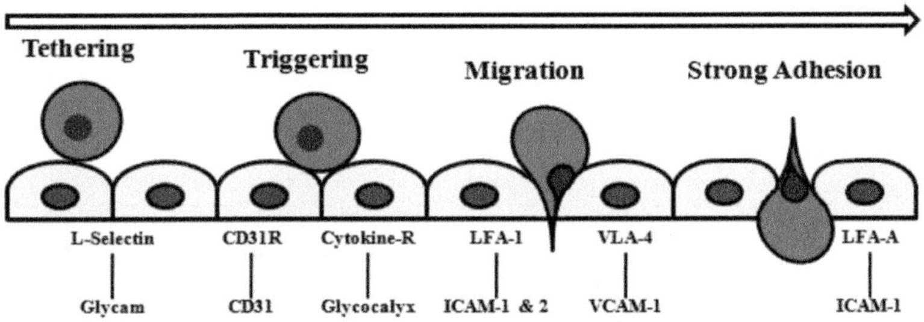

Fig.-10-5. Mechanism of lymphocyte trafficking in the venule (ordinary or HEVs). Firstly, lymphocytes are loosely attached to the endothelial wall (tethering) and rolls on the surface. After triggering the adhesion cascade, lymphocytes strongly attached to the endothelial surface (strong adhesion) and exudes from the inter-cell space (extravasation) to the site of inflammation. All these events need particular receptor-ligand interaction as shown at the bottom of each panel.

Regulation of lymphocyte trafficking and homing

Chemokines and chemokine receptor (CKR) play a central role in regulating lymphocyte trafficking and homing. Chemokines constitute a large family of small (8-12 kDa) proteins. About 40 chemokines form an extensive network of chemotactic signals. It is now clear that non-chemokine molecules such as lipid mediators, pathogen-derived products, antimicrobial peptides, complement products, and other normal constituents of our body are also chemotactic and regulate lymphocyte trafficking. Sphingosine 1-phosphate and homeostatic chemokines regulate lymphocyte localization in lymph nodes and egress to circulation. Upon antigen priming, regulatory T cells rapidly upregulate CXCR5 to migrate and suppress germinal center T and B cells. Plasmacytoid dendritic cells can migrate to lymph nodes through high endothelial venule cells, and chemokines and non-chemokine G-protein-coupled seven-transmembrane domain receptor ligands such as chemerin can regulate pDC migration. Gut dendritic cells and retinoic acid induce the expression of α4β7 and CCR9 on T cells for their homing to the gut. Leukotriene B4 and interleukin-8, known chemoattractants for myeloid cells, are also selective chemoattractants for cytotoxic effector T cells.

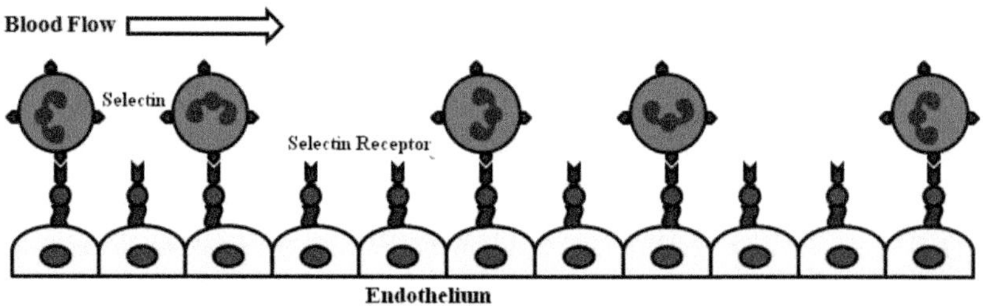

Fig.-10-6. Selectin expression results in a weak binding of lymphocytes to the venules endothelium, which subsequently leads to rolling of lymphocytes across the venule's lumen, rather extravasation to extracellular matrix (ECM).

Table 10-3: Expression of molecules responsible for adhesion and lymphocyte migration through the high endothelial venules.

Lymphocytes →

Endothelial cells ↓	L-selectin	PSGL-1	CLA	A4/β7	A4/β1	LFA-1	CD44	CD-31
Glycam-1	++							
CD34	++							
P-selectin	+	++						
E-selectin		+	++					
MAdCAM-1	+			++	+			
VCAM-1				+	++			
ICAM-1						++		
ICAM-2						++		
Hyaluronate							++	
CD31								++

Among the chemokine family, members of the β-chemokine family are known to direct T lymphocyte migration along a chemokine gradients or chemokine field (chemotaxis) and to induce adhesion to extracellular matrix proteins and/or lymphocyte homing site. The β-chemokines MIP-1α, MIP-1β, monocyte chemotactic protein (MCP)-1, MCP-2, MCP-3, and RANTES reportedly show chemotactic activity for T cells *in vitro*. MIP-1α, MIP-1β chemokines are known to preferentially attract $CD8^+$ and $CD4^+$ T cells, respectively and play an important role during contact hypersensitivity and delayed-type hypersensitivity.

Chemokines have been shown to exert their functions by binding specific Gα1-protein coupled chemokine receptor (CKR) on the lymphocyte surface. In addition, chemokines rapidly increase binding avidity of integrin adhesion receptors by inside-out signaling pathways. This activation process is required to arrest blood-borne lymphocytes inside postcapillary venules of secondary lymphoid organs or non-lymphoid tissue. Under physiological conditions, integrin activation appears restricted to selected chemokines presented on the luminal side of blood vessels. The migration and integrin activation may be interdependent events in some cases such as interstitial B cell migration within the spleen.

CCR9 ligands is responsible for entry of BM-derived pre-thymocytes into thymus through blood vessels located at the cortical–medullary junction (CMJ) in adults (Fig.-10-7). CCR7 is highly expressed in early thymocytes displaying an intermediate phenotype between CD4/8-double negative (DN) 1 and DN2, termed DN1-2.27 In plt/plt or CCR7-deficient mice, DN1-2 cells accumulate at the CMJ while DN3/4 cellularity is decreased. Positively selected DP thymocytes reinduce CCR7 expression, which is required for efficient migration into thymic medulla.

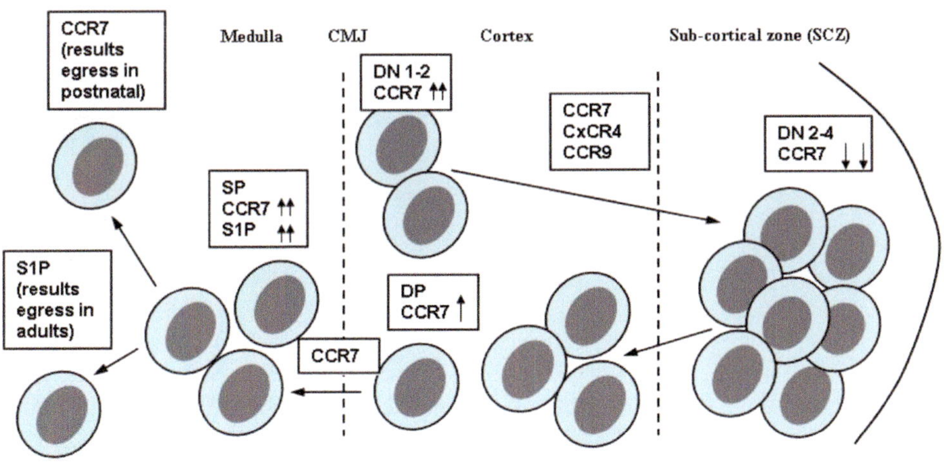

Fig.-10-7. Regulatory mechanisms operated by chemokines in thymocytes egress during postnatal and adult stages.

Overexpression of CCR7 induces premature migration of DP cells into the medulla and CCR7-deficient single positive (SP) thymocytes appear inside cortex instead of medulla. Unique thing is that same CCR7 and CRK guides the migration of lymphocytes from SCZ to CMJ and from cortex to medulla, at different stages of thymocyte development. CCR7 ligands and CXCR5 play a central role in the trafficking of lymphocytes coming from HEVs of secondary lymphoid organs to the red pulp of spleen, into splenic white pulp. Migration of B and T lymphocyte differ; B cell migration into white pulp is integrin dependent, while T cells migration is integrin independent. Another chemokine CCL21 have been suggested to perform chemorepulsive function on double negative (DN) 1-2 cells.

The two chemokines, CCR7 and sphingosine-1-phosphate (S1P) receptor 1 are involved in thymocyte emigration into periphery in postnatal and adult animals respectively. S1P1 is also central to lymphocyte egress from secondary lymphoid organs into efferent lymphatic vessels, highlighting a double role for this receptor for efficient T cell exit from the thymus and secondary lymphoid organs. CCL19, CCL21, CXCL12 and CXCL13 are important motility-inducing factors that provoke migration and movement of interstitial lymphocytes (Fig.-10-8). However, other factors such as thromboxanes A2 (TXA2) may also contribute to interstitial lymphocyte motility. Expression of CCR4 and/or CCR10 characterizes skin-homing effector T lymphocytes activated in peripheral lymph nodes (PLN), whereas the expression of CCR9 and α4β7 in T lymphocytes activated in Payer's patches. CCL17 and CCL25 are preferentially expressed in skin and small intestine venules. Taken together, chemokines clearly regulate differential trafficking and migration of lymphocytes.

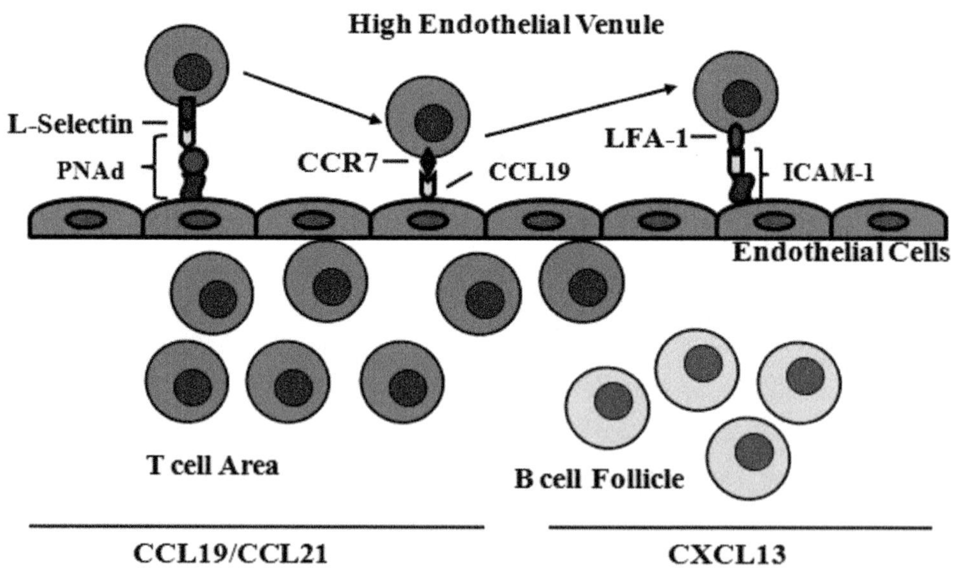

Fig.-10-8. Role of chemokines in the binding of lymphocytes to the endothelium cells of lymph node HEVs and lymphocyte migration from T cell area of secondary lymphoid organs (SLO).

Mediators of Inflammatory Response

The specific aspects of inflammatory response highlight the organizing phase of lymphocyte recirculation. In the development of inflammatory response, a number of mediators have their role to play in the entire process. In this process of inflammatory response, the following mediators play their roles:

- Cytokines
- Chemokines
- Plasma enzymes
- Platelet activators (lipids)

In the immune system, inflammation is used to assemble the components of an active response and these resources at the site of infection. It is in this sense, the inflammatory response characterized by lymphocyte recirculation, organizes the components, which have their impact on the response. The above mediators of an inflammatory response are resources focused by the immune system to the site of infection, through inflammation i.e., body's defense.

Table 10-4: Pleiotropic effects exhibited by IL-1, IL-6 and TNF-α

Effect	*IL-1*	*IL -6*	*TNF- α*
Endogenous pyrogen fever	+	+	+
Synthesis of acute phase protein by liver	+	+	+
Increased vascular permeability	+	+	+
Increased adhesion molecules on vascular endothelium	+	-	+
Fibroblast proliferation	+	-	+
Platelet production	+	-	+
Chemokine induction (e.g.., IL-8)	+	-	+
Induction of IL-6	-	+	+
T cell activation	+	+	+
B cell activation	+	+	+
Increased Ig synthesis	-	+	-

Cytokines: Cytokine inflammatory mediators are IL-1, IL-6, IL-12, TNF-α, IFN-γ, and IFN-β. Some of the effects on inflammatory response mediated by IL-1 IL-6 and TNF-α are listed in Table 10-4. Chronic inflammation is caused by IFN-g and TNF-a. TH1 cells, NK cells and TC cells release IFN-γ and macrophages secrete TNF-α. All three members of the interferon family of glycoproteins (IFN-α, IFN-β and IFN-γ) are released from virus infected cells and confer antiviral protection on neighboring cells, depending on the type of cell infected. Figure 10-9 shows the redundant and pleiotropic function of IFN-γ. IFN-g activates macrophages exhibiting increased expression of class II MHC molecules, increased cytokine production and increased microbicidal activity. As a result, these activated macrophages kill intracellular microbial pathogens. It is worth noting that the accumulation of large number of activated macrophages, exhibiting increased extension of class II MHC molecules, relieve various hydrolytic enzymes, reactive oxygen, and nitrogen intermediates, which cause damage of the surrounding tissue. The cytokine TNF-α secreted by macrophages has a direct cytotoxic effect on tumor cells but not on normal cells. Thus, TNF-α has got this double aspect in the immune system.

Chemokines: Chemokines are a super-family of small polypeptides, mostly containing 90-130 amino acid residues. They are produced by lymphoid and non-lymphoid tissues in connection with the initiation or progress of inflammation. Chemokines cause leukocytes to move into various tissue sites by inducing the adherence of these cells to the vascular endothelium, resulting in high localized concentration. Chemokine action is mediated by receptors that activate large GTP-binding proteins, which are heterotrimeric G-proteins. This initiates signal transduction processes that generate such potent second messengers as cAMP, IP3, $Ca2^+$ and activated small G-proteins, resulting in a variety of modifications in the physiology of the target cell (Fig. 10-10). The signal transduction pathways promote other changes such as the release of granular contents, proteases in neutrophils and macrophages, histamine from basophils, and cytotoxic proteins from eosinophils. Different cells (T_H cells) display different profiles of chemokine receptors bound to them, as recognizing devices.

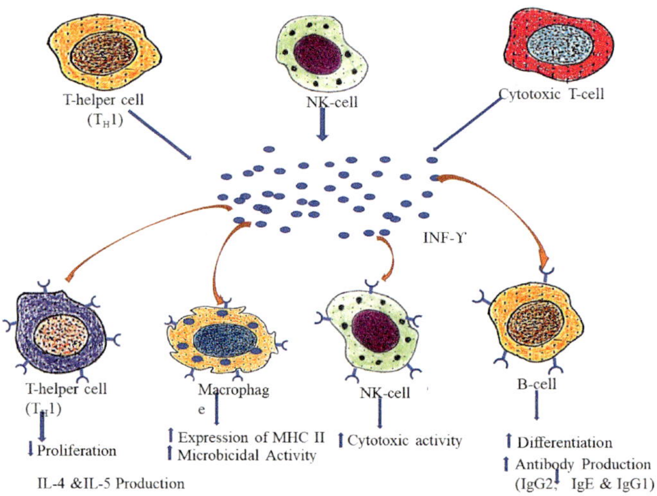

Fig.-10-9. Major secreting cells of IFN-γ and wide effect of IFN-γ on different cell types (Adapted from Science 259:1693; with permission).

Plasma enzymes: Some enzymatic cascades come into play with the damage caused by defensive secretions and releases of body's defense (immune system). The cascades are activated to form a web of interacting systems that generate a number of mediators of inflammation. These plasma enzyme mediators include bradykinin, fibrinopeptides, and plasmin.

After a tissue injury, a plasma clotting factor called Hageman factor is activated, which in turn activates prekallikrein to form Kallikrein, which cleaves kininogen to produce bradykinin. Bradykinin is an inflammatory mediator being a potent vasoactive basic peptide that increases vascular permeability causing vasodilation, inducing pain and contraction of smooth muscle (particular to an injured tissue).

After damage to blood vessels, large quantities of thrombin is released in tissue fluid or plasma, which in turn produce insoluble strands of fibrin or fibrinopeptides. The fibrin strands crisscross one another to form a clot. This clot serves as a barrier to the spread of infection. As a result of clot formation, bleeding is prevented as also spread of involving pathogens into the bloodstream.

The clotting from the injured tissue is removed by an enzyme known as plasmin. Plasmin is a potent proteolytic enzyme, which breaks down fibrin clots and activates other pharmacologically active mediators such as histamine. Vascular permeability and smooth muscle contraction are increased by these mediators and fibrin clots are also degraded into chemotactic products. The complement system is generated, as a sequel to this, resulting in the influx of fluid that carries antibody and phagocytic cells to the site of injury. Antibodies and phagocytic cells are components of the complement system described earlier.

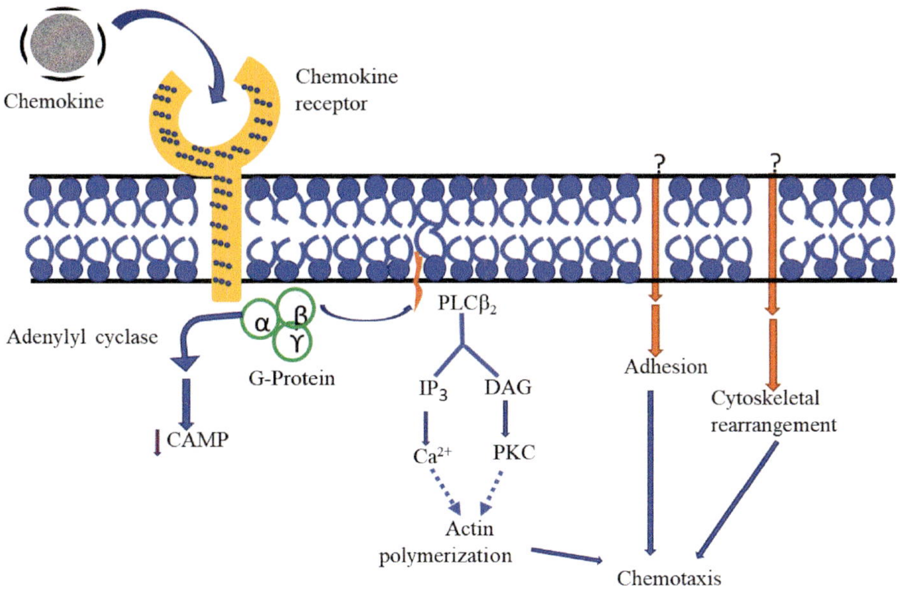

Fig.-10-10. Mechanism of activation by chemokines through chemokine receptors.

Platelet activators: The degradation of phospholipids in the membrane of macrophages, neutrophils and mast cells into arachidonic acid and also-platelet-activating factors, composed of platelet activators. Platelet activators or platelet-activating factor (PAF) exert inflammatory effects.

Arachidonic acid produces prostaglandins and thromboxanes. Prostaglandins increase vascular permeability, vascular dilation and induction of neutrophil chemotaxis. Thromboxanes cause platelet aggregation and contraction of blood vessels. Prostaglandins and thromboxanes are products of the cyclooxygenase pathway, caused by metabolism of arachidonic acid. Metabolism of arachidonic acid by the Lipoxygenases pathway causes formation of leukotrienes. Leukotrienes are produced by monocytes, macrophages and mast cells as a result of which contraction of smooth muscles sets in.

Acute and Chronic Inflammatory Response

When the body is affected anywhere by an internal stimulus (infection) or a tissue is injured, a rapid chain of physiologic response to such stimuli sets in. This physiologic response is called inflammatory response. The inflammatory response may be of two types; acute and chronic. A rapid inflammatory response, which lasts for a short while involving both localized and systemic effects, is called an acute response. It is characterized by a rapid alteration in the levels of several plasma proteins. When tissue damage and wasting of cells is caused by some microorganisms resistant to pathologic conditions and last for a long duration of time, chronic inflammation sets in followed by pathologic consequences. A chronic inflammatory response may accompany allergies, autoimmune diseases, microbial infections, transplants and burns.

Mediators of Acute Inflammatory Response

Among the mediators of acute inflammatory response are neutrophils, macrophage inflammatory proteins (MIP-1a and MIP-1b) chemokines, cytokines promoting increased phagocytic activity and increased release of lytic enzymes into the tissue spaces. Increased release of lytic enzymes and concomitant phagocytic activities are the hallmarks of acute inflammatory response. The release of lytic enzymes and onset of phagocytic activities are localized response against infection or tissue injury. Increased levels of cytokines are known as systemic response in the inflammatory process.

Neutrophils - are the first mediator in an acute inflammatory response. During a period of acute inflammation, neutrophils infiltrate the affected tissue with ten times of normal neutrophils production in the body (per day). Neutrophils leave the bone marrow and circulate within the blood in response to infection or injury. The circulating neutrophils express mucin (PSGL-1) or tetrasaccharides (sialyl Lewisa/x), which bind to E and P-selection (expression of endothelial cells). This binding mediates the attachment or tethering of neutrophils to the vascular endothelium, allowing the cells to roll in the direction of blood flow. Neutrophils are acted upon by chemo-attraction triggering a G-protein mediated signal, resulting in neutrophil adhesion and subsequent transendothelial migration (Fig. 10-3). At this stage neutrophils are well activated and with expression of increased levels of FC receptors for antibody and receptors for complement bind more effectively to antibody or complement coated pathogens. In addition, the neutrophils also release granules (proteases, phospholipases, elastases, and collagenases), which result in phagocytosis marked by pus formation.

MIP-1a and MIP-1b- are the macrophage inflammatory proteins acting as mediators to the inflammatory response. Vascular changes (vasodilation, vascular permeability) and extravasation of leukocytes induce local masT cell degranulation with release of histamine. Vascular dilation and smooth muscle contraction are caused by histamine (a potent inflammatory mediator) and prostaglandins cause vascular permeability. The macrophage inflammatory proteins arrive about 5-6 hours after the onset of an inflammatory response at the site of inflammation, characterized by vasodilation and vascular permeability.

Chemokines- contribute to the influx of neutrophils by increasing their adhesion to vascular endothelial cells. IFN-g and TNF-a activate macrophages and neutrophils promoting increased phagocytic activity and increased release of lytic enzymes into the tissue spaces.

Cytokines- released at the site of inflammation and facilitate both the adherence of immune system cells to vascular endothelial cells and their migration through the vessel wall into the tissue spaces. This results in the influx of lymphocytes, neutrophils, monocytes, eosinophils, basophils and mast cells to the site of tissue damage, which further contribute to the killing of antigen and healing of the tissue. The duration and intensity of the phagocytic activities during an inflammatory response require careful regulation to control tissue damage and tissue repair. In this phase of phagocytosis, failure of proper leukocyte adhesion can result in disease known as leukocyte adhesion deficiency (LAD) in humans and cattle. The LAD was first understood through a report of a rare autosomal recessive disease in which there occurs a swelling and reddening around the stalk of the umbilical cord in a pus free state. In LAD patients, there are antibiotic-resistant strains of many pathogenic bacteria, where cytotoxic chemicals fail, but bone marrow transplantation ensues. The latter ensures proper leukocyte adhesion to the damaged tissues.

IL-1, TNF-α, IL-6 are effective to induce a fever response by acting on the hypothalamus within 12-24 hours of the onset of acute inflammatory response. TNF-α secrets colony-stimulating factors (CSF, M-CSF, G-CSF, GM-CSF) stimulating hematopoiesis (increase in WBC to fight the infection). Increased levels of cytokines form a systemic response to inflammatory process.

Mediators of Chronic Inflammatory Response

Among the mediators of chronic inflammatory response are macrophages activated by cytokines. In humans, rheumatoid arthritis, Crohn's disease, ulcerative colitis, Graves disease, Hashimotos thyroiditis, and diabetes mellitus, characterize an influx of leukocytes contributing to chronic inflammation. These diseases are caused by development of HEV (high-endothelial venules) like regions. In other words, HEV-like vasculature in the affected organs causes chronic inflammatory response. The accumulation and activation of macrophages is the central characteristic of chronic inflammation. In chronic inflammation, a type of sear tissue develops at sites of chronic inflammation by a process called fibrosis (a wound healing reaction), which can intervene with normal tissue functioning. In chronic inflammation a tumor-like mass consisting of a central area of activated macrophages surrounded by activated lymphocytes called granuloma is formed. In the center of granuloma, multinucleated giant cells resembling epithelial cells are found. These cells are also called epitheloid cells. Macrophages are activated by cytokines, especially IFN-γ. Accumulation of large number of macrophages at the site of infected cells can cause damage to the surrounding tissues, besides intensive killing of intracellular microbial pathogens. This action of IFN-γ is balanced by secretion of cytokine TNF-α (through macrophages), which exerts a cytotoxic effect on tumor cells only. This is a very interesting aspect of immune system, which will be discussed in detail in a subsequent chapter with regard to treatment of cancer.

Anti-inflammatory Agents

Long inflammatory responses induced by body's defense mechanism (production of macrophages, cytokines, etc.) can also be complicated; hence there arises a need to reduce the period of inflammatory response. Reduction in inflammation can be achieved by medicines, which either block the activity of various adhesion molecules with antibodies, decrease lymphocyte circulation patterns or lympholysis, stabilize the lysosomal membrane and inhibit the production of prostaglandins and thromboxanes.

Antibodies- impede the process of leukocyte extravasation, by blocking the activity of various adhesion molecules with antibodies. A combination of antibodies (anti-ICAM-1 and anti LFA-1) i.e. integrins binding on cell adhesion molecules has been found effective in reducing the likelihood of rejection of human kidney transplant.

Corticosteroids- are cholesterol derivative (prednisone, prednisolone, methyl prednisolone) acting upon lymphocytes. Corticosteroids decrease lymphocyte circulation pattern or induce lysis of lymphocytes (lympholysis). Corticosteroids also induce increased transcription of the NF-kB inhibitor i.e., I-κB, reduce both the phagocytic and killing ability of macrophages and neutrophils, and this effect may contribute to their anti-inflammatory action, reduction in T_H cell activation lysosomal membrane is also stabilized and released at the site of inflammation.

Pain and inflammation are also treated by non steroidal anti-inflammatory drugs (NSAIDs), which inhibit the cyclooxygenase pathway by which prostaglandins and thromboxanes are produced from arachidonic acid.

CHAPTER – 11

CELL-MEDIATED EFFECTOR RESPONSES

The intracellular pathogens are the major challenges to immunity. This challenge is well encountered by effector responses consisting of cells, so called cell-mediated immunity. Both antigen specific and antigen-nonspecific cells contribute to the cell-mediated immune response. $CD4^+$ T_H cells, $CD8^+$ cytotoxic T lymphocytes are antigen specific cells. Natural killers (NK) cells, non-lymphoid cells such as macrophages, neutrophils, and eosinophils are nonspecific cells. The cell-mediated immunity is a function of both the above types of cells, so they are not independent of each other. This chapter deals with initiation of cell death carrying intracellular pathogens by CTLs (cytotoxic T-lymphocytes) and NK cells together with their mechanisms and a measuring device for measuring cell-mediated cytotoxicity i.e. antibody-dependent cell mediated cytotoxicity (ADCC), besides delayed-type hypersensitivity (DTH) i.e. responses mediated by effector T_H cells.

Effector Responses

Lack of T cell components, causes repeated infections with viruses, intracellular bacteria and fungi. This lack is termed as cell-mediated immunodeficiency. Children born without a thymus are more susceptible to life threatening infections even with the attenuated virus present in a vaccine than normal individuals, because in normal individuals, these viruses are not multiplied. This state of lack of T cells is called cell-mediated immunodeficiency. Cell-mediated immune response affects in two ways- one the effector cells directly kill foreign cells and attract self-effector cells by exerting a cytotoxic reaction, whereby these cells lyse the foreign cells. The lysing is carried on by antigen-specific cytotoxic lymphocytes (CTLs) and NK cells together with secondary CD4⁺ T cells (T_H1, T_H2 and $CD8^-$), mediated delayed-type hypersensitivity reactions.

General Properties of Effector T Cells

The effector cells are more easily activated than naive T_H and T_C cells. Effector cells express higher levels of cell adhesion molecules, exhibit different trafficking patterns and produce both soluble and membrane effector molecules (see Tables 11-1).

Table 11-1: Effector molecules produced by T cells

Cell type	*Effector molecules*	*Membrane-bound effectors*
CT	Cytotoxins (perforins and granzymes), IFN- γ TNF-β	Fas ligand (Fas L)
TH1	IL-2, IL-3, IFN- γ TNF-β, GM-CSF (high)	TNF-β
TH2	IL-3, IL-4, IL-5, IL-6, IL-10, IL-13,	CD40 ligand

Many populations of naïve and effector T cells express different isoforms of CD44, designated CD45 RA and CD45 RO, which are produced by alternative splicing of the RNA transcript of the CD45 gene. This membrane molecule has phosphatase activity and mediates TCR signal transduction by catalyzing dephosphorylation of a tyrosine residue on the protein tyrosine kinases Lck and Fyn; this change activates these kinases and triggers the subsequent steps in T cell activation. The CD45 RO isoforms, which are expressed on the effector T cell, associates with the TCR complex and its co-receptors CD4 and CD8. This association is much better than that of CD45 RA isoform expressed by naive T cells.

Effector T cells have higher level of LFA-1 and CD2 so they bind more effectively to antigen-presenting cells and to various target cells than naïve T cells. Thus, effector T cells show more cell association. Recognition of a peptide-MHC complex by the TCR produces a signal, which increases the affinity of LFA-1. The T_H1 effector cells remain bound to macrophages that display peptide-class II MHC complex. The T_H2 effector cells remain bound to B cells that display peptide-class II MHC complexes and CTL effector cells bind to virus-infected target cells displaying peptide class I MHC complexes.

Effector T cells also express certain effector molecules, which may be membrane bound or soluble. The effector molecules, which are membrane bound, include Fas ligand (Fas L) and CD40 ligand (CD40 L). Effector T cells also secrete cytotoxins (perforins and granzymes) and cytokines (IFN-γ and TNF-β). The Fas ligand, perforins and granzymes mediate target cell distribution by the CTL and TNF-β, and IFN-γ promotes macrophage activation.

Cytotoxic activity is absent in naïve T cells, so they are non-functional in killing target cells. Naïve T cells are therefore termed as CTL precursors (CTL-Ps). They are activated or generated as CTLs (cytotoxic T lymphocytes) by certain signals, which are - i) transmitted by the TCR complex, ii) transmitted by the interaction of the CTL-P and the antigen-presenting cell, and iii) transmitted by the interaction of IL-2 with the high affinity IL-2 receptor.

IL-2 is the principal cytokine required for the proliferation and differentiation of activated CTL-Ps into effector CTLs. The IL-2 receptor is not expressed until after a CTL-P has been activated by antigen plus a class I MHC molecule (Fig. 11-1). In the absence of IL-2, CTL mediated cytotoxicity is not accountable for the proliferation and differentiation of both antigen activated T_H1 cells and CTL-Ps. The level of IL-2 declines after clearance of antigen and this induces programmed cell death of T_H1 cells and CTLS by apoptosis. Thus, the likelihood of nonspecific tissue damage from the inflammatory response is lessened. This is the picture of cell mediated immune response. It must be borne into mind that though the generation of naïve cells (CTL-PS) into effector cells (CTLS) presents a complex scenario in terms of generation, however, IL-2 and co-stimulation play a major role in the transformation of CTL-PS into CTLS.

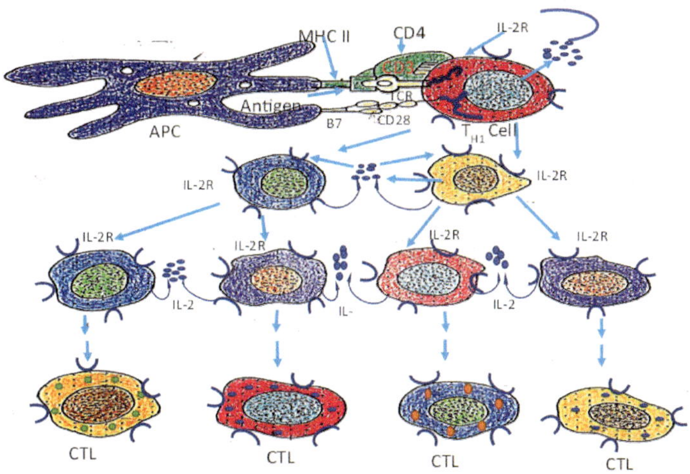

Fig.-11-1.Mechanism of generation of effector cytotoxic T cell lymphocytes (CTL) under the effect of IL-2 released upon activation.

Destination of Target Cells by CTLs

A number of membrane molecules and membrane events are involved in the termination of target cells as observed in long-term cultures of CTL clones (studies with mouse strains) (Figs. 11-2). In CTL-mediated death the first event are conjugate formation, membrane attack, CTL dissociation and target cell annihilation. When antigen-specific CTLs are incubated with appropriate target cells, the two cell types interact and conjugate. This is called CTL target cell conjugation. Within several minutes of conjugation, an energy immediate stop- $Ca2^+$ dependent-programs of the termination of Cell. At this juncture, the CTL dissociates from the target cell and binds to another target cell causing the death of the target cell by apoptosis.

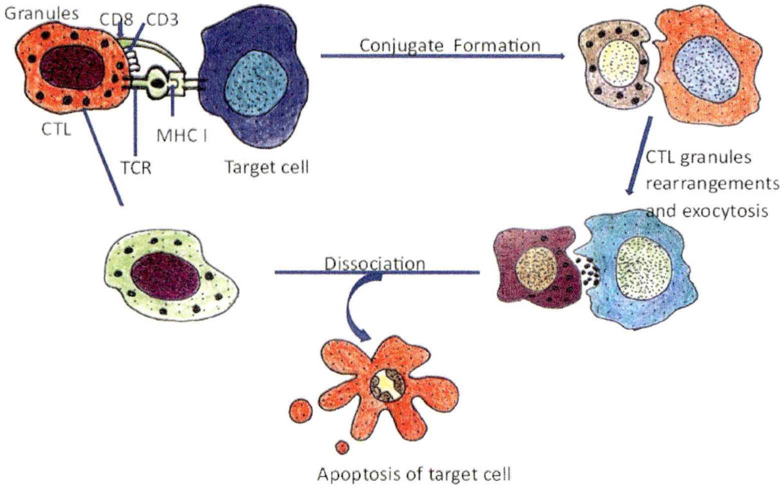

Fig.-11-2. CTL-mediated killing of target cells. The CTL-cell after injecting granzyme dissociates and goes to recycling, and target cells undergoes the process of apoptosis (Adapted from P. Henkart. 1985. *Annu. Rev. Immunol.* 3:31; with permission).

As regards membrane molecules, the TCR-CD3 membrane complex on a CTL recognizes antigen in association with class I MHC molecules on a target cell. This is particularly a stage of antigen-specific recognition, LFA-1 (the integrin receptor) on the CTL membrane binds to intracellular cell-adhesion molecules (ICAMs) on the targeT cell membrane. As a result, conjugate is formed and LFA-1 is converted from a low-avidity state to a high-avidity stage. At this juncture, CTLs adhere to proper target cells (conjugation), that display antigen peptides associated with class I MHC molecules. Thus, the antigen activates molecules, and LFA-1 operates in a high-avidity state for a couple of minutes, then shifts to low-avidity state. This shifting of LFA-1 from high-avidity to low-avidity state causes dissociation of the CTL from the target cell. At this stage, the CTL is recycled and the target cell dies by apoptosis. In CTL mediated termination of target cells, there is a major role of perforin. CTL contain 65 kDA monomers of a pore-forming protein called perforin and several serine proteases called granzymes.

After CTL-target cell conjugation, perforin monomers and the granzyme proteases are released. Perforin polymerizes on the cells membrane of target cells and forms perforin pore through which granzymes enters the target cells. Thereafter, granzymes activate an apoptotic pathway within the target cell. In this CTL-mediated killing, even the viral DNA in the target cells are also killed and viral replication prevented at all. Besides target cell death by perforins and granzymes, Fas L (a transmembrane protein found on the membrane of CTLs) also deliver a death signal by binding to its receptor Fas on the target cells. Thus, in all the CTL-mediated killing-perforins, granzymes and Fas play active role in combination or independently. Substantially speaking, based on experiments, target cell death occurs by two mechanisms, one release of cytotoxic proteins (perforins and granzymes) from CTLs by entering target cells, and the other by interaction of the membrane-bound Fas ligand on CTLs with the Fas receptor on the surface of target cells (Fig. 11-3). Actually, cell death (of target cell) by apoptosis (CTL-mediated) involves caspase family of cysteine proteases, which cleave after an aspartic acid residue. Caspase (cysteine aspartate proteases) are present in all the CTLs as inactive proenzyme, procaspases that require proteolytic cleavage for conversion into the active forms having their own specificity. Cleavage of procaspases produces an active initiator caspase, which cleaves other procaspases, thereby activation of their proteolytic activity. This all results in systemic and orderly death of target cell (apoptosis). Increasingly, CTLs force the targets cells to die by themselves, in the midst of initiation or activation of caspase cascades.

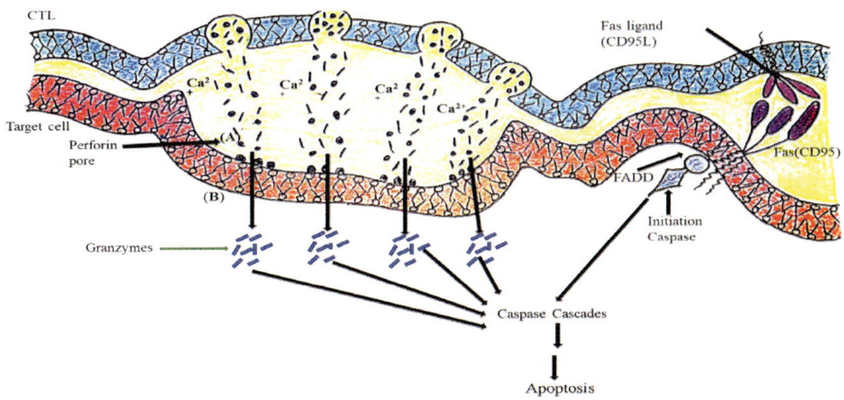

Fig.-11-3. Diagrammatic representation of target cell apoptosis generated by CTL. The cascade of CTL response involve in two distinct steps (a) perforin polymerization is independent on Ca^{2+}, and (b) Granzyme enters into the target cells by perforin pores.

NK-Cell Mediated Killing of Target Cells

A population of about 5-10% granular lymphocytes is called natural killers (NKs) as they secrete cytokines (IFN-γ) which act on target cells. Thus, NK cells play an important role in immune regulation. NK cells respond primarily to certain viruses and intracellular bacteria, by exerting cytokines e.g. IFN-α, IFN-β, IL-12. During the time of viral replication (activation, proliferation and differentiation), NK cells are the first line of defense against virus infection.

Different NK cells express different sets of membrane molecules- CD2, 75-kDa b subunit of the IL-2 receptors and CD-16 (a receptor for the Fc region of IgC). Unlike T cells and B cells, NK cells do not undergo rearrangement of receptors genes.

NK cells contain perforin and granzymes like CTLs, and are constitutively cytotoxic. When NK cells come in contact with an infected cell, perforin and granzymes interact with it, and kill the viruses and intracellular bacteria by process of apoptosis. Thus, the mechanisms of killing by NK cells is similar to that of CTLs. NK cells recognition of target cells is not MHC restricted i.e., no immunologic memory. However, recent observation suggests that they can recognize lipoid antigens through CD1 surface molecules, therefore, NK-cells are said to be CD1 restricted unlike to MHC restriction of T-lymphocytes. This is unique and much interesting feature of NK cells, which enables it to recognize lipid antigen whereas T lymphocytes only recognize protein antigen through MHC. This feature of NK cells clearly differentiates them from CTLs, which have memory cells, and do not express antigen-specific receptors (however, it can be said that NK-cells have memory too for lipoid antigens only). But the question arises as to how NK cells distinguish altered self-cells and normal body cells.

Experiments have been carried out to find a solution to the problem, as to how the normal body cells and altered self cells is distinguished by NK cells in the process of apoptosis. That is to say that how healthy cells are distinguished from infected or cancerous cells, by NK cells. This implies that NK cells must have receptors, may-be CD1. NK cell killing is regulated by the balance between positive signals generated by the engagement of activating receptors (NKR-P1 and others) and negative signals from inhibitory receptors (CD94/NKG2 and the KIR family (Fig. 11-4).

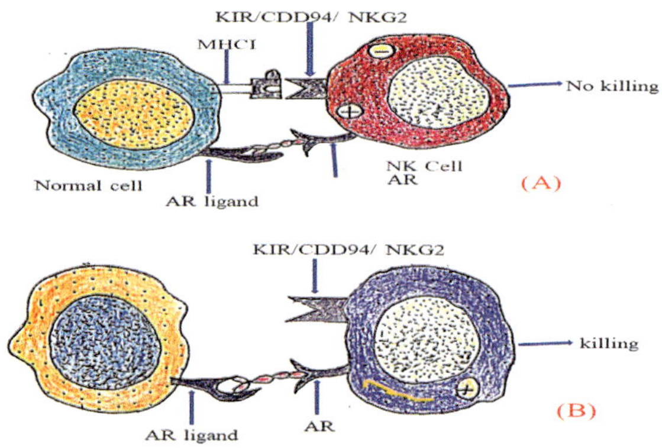

Fig.-11-4. Opposing signal model of cytotoxic activity of NK-cells.

Carbohydrate-binding proteins known as C-type lectins (having calcium dependent carbohydrate-recognition domains, NKR-P1), have activating properties in NK cells. Other activating receptors include CD28 (receptor for B-7 family of co-stimulatory molecules), CD2 (receptor for the adhesion molecule LFA-3) and the FcγIII receptor CD16. These receptors help NK cells to recognize target cells. The Killing of the target cells (infected or cancerous cells) involves these activating receptors.

Healthy cells are protected from killing by NK cell inhibitory receptors. In humans, the C-type lectin inhibitory receptors are CD94/NKG2 (a disulfide-bonded heterodimer made up of two glycoproteins). The CD94/NKG2 receptors are not specific for a particular HLA allele and send inhibitory signals (negative signals) to the NK-cells resulting in the inhibition of killing potential target cells, expressing adequate levels of class I. That is to say that CD-14/NKG2 inhibits the lysis of target cells. In other words, cells expressing normal levels of unaltered MHC class I molecule are escaped from the NK cell killing mechanism.

Antibody-Dependent Cell-Mediated Cytotoxicity

NK cells, macrophages, monocytes, neutrophils, and eosinophils are nonspecific cytotoxic cells, which damage the target cell membrane by a binding to the Fc region of antibody on target cells. The damage is caused by release of lytic enzymes, perforin or TNF. The process in which the above cells having cytotoxic potential express membrane receptors and bind to the antibody Fc region on target cells, resulting in lysis of target cells is called antibody-dependent cell-mediated cytotoxicity (ADCC). Various substances e.g. lytic enzymes, TNF, perforin, and granzymes secreted by the above nonspecific cytotoxic cells mediate target cell destruction. When these nonspecific cytotoxic cells bind to a target cell by way of the Fc receptor metabolically they become more active, and increase the lytic enzymes in their cytoplasmic lysosomes or granules. Lytic enzymes in contact with the target cells cause death to them. The mechanism of cell death is similar to that of CTL-mediated apoptosis (Fig. 11-5).

Experimental Assay of Cytotoxicity

The activation and effector phases of cell-mediated cytotoxic responses can be measured by the following experimental devices:

- in vitro assaying of T_H cells proliferation in a cell-mediated response- Mixed Lymphocytic Reaction (MLR).
- in vitro assaying of effector cytotoxic function- Cell-Mediated lympholysis (CML), and
- in vivo assaying of graft-versus-host reaction (GVH).

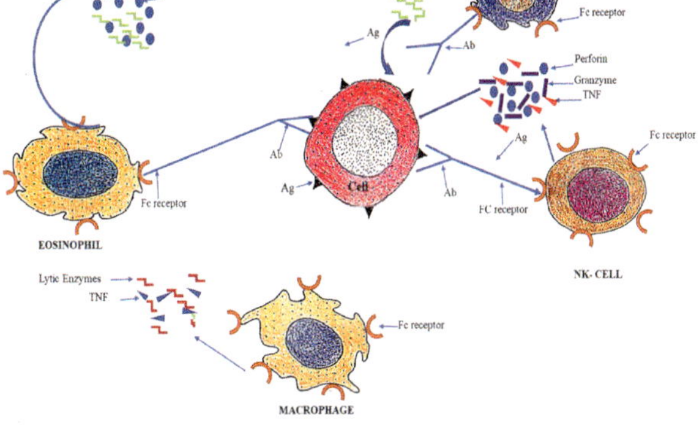

Fig.-11-5. Antibody dependent cell-mediated cytotoxicity (ADCC) by different immune cells.

MLR: In this assay, the function of macrophages is observed in vitro, in activating the class II MHC restricted T_H cells. That is the proliferation of T_H cells in terms of function of macrophages is measured in this experimental device. The degree of proliferation of T-lymphocytes by co-culturing allogeneic spleen cells in the MLR system can be measured by adding 5H thymidine to the culture medium and monitoring uptake of label into DNA in the course of repeated cell divisions. Allogeneic T-lymphocytes may be rendered unresponsive by addition of mitomycin C or lethal x-irradiation. The unresponsive population of T lymphocytes expresses alloantigens foreign to the responder cells. In this way, functional CTLs can be generated and their activity assessed with various effector assays. Actually in the MLR assay of T_H cell proliferation, responder T_H cells recognize allogeneic class II MHC molecules on the stimulator cells (unresponsive population of T-lymphocytes) and proliferate in response to these differences. Removal of the $CD4^+$ T_H cells from the responder cells with anti CD4 plus complement abolishes the proliferation and averts generation of CTLs.

CML: This assay provides a better understanding of the mechanism of target cell killing by CTLs. In CML assay, suitable target cells are labeled intracellularly with chromium-51 (51Cr) by incubating the target cells with Na251CrO4. After the 51Cr diffuses into a cell, it binds to cytoplasmic proteins, reducing passive diffusion of the label out of the cell. When specifically activated, CTLS are incubated for one to four hours with such labeled target cells, the cells lyse and the 51Cr is released. The amount of Cr released corresponds to the number of lysed cells (target cells) as a function of CTLS for allogeneic cells, tumor cells, virus infected cells and other modified cells by chemical treatment.

GVH: This assay provides an in vivo indication of cell-mediated cytotoxicity. In this method, there are two bodies; one the host (who has immunocompetent lymphocytes) and the other recipient (graft). Since the two bodies are not genetically identical, the donor lymphocytes (grafted lymphocytes) begin to attack the host (recipient). As a result, in patients having had radiation exposure or undergone transplantation of bone marrow, or AIDS, or leukemia, or autoimmune diseases, GVH reactions like diarrhoea, skin lesions, jaundice, spleen enlargement develop, which sometimes result in death of the host. The grafted lymphocytes generally are carried to a number of organs, including the spleen, where they proliferate in response to the allogeneic MHC antigens of the host, an influx of host cells is induced. This results in splenomegaly (enlargement of the spleen). Splenomegaly results from proliferation of both $CD4^+$ and $CD8^+$ T cell populations. GVH reaction is also induced by NK cells resulting in skin lesions and intestinal wall damage.

Responses Mediated by Effector T_H Cells: Delayed type Hypersensitivity

Activated macrophages are more effective killers of intracellular pathogens, as effector cells. Activated T_H cells on encounter with certain types of antigens secrete cytokines, resulting in large influxes of nonspecific inflammatory cells.

This reaction of macrophages with antigens is called delayed type hypersensitivity (DTH). This reaction was first observed by Robert Koch in individuals infected with Mycobacterium tuberculosis. He developed a localized inflammatory response by injecting intradermally a filtrate derived from a mycobacterial culture. As the inflammatory response was observed developing as a reaction with other type of antigens also, the phenomenon was termed as DTH in reference to the delayed onset of the reaction and to the extensive tissue damage (hypersensitivity). Tissue damage may be extensive or limited. However, the response is in defense against intracellular pathogens and contact antigens (synthetic or naturally occurring substances such as poisoning- in contact with skin or mucous membrane.

Mechanism and Impact of DTH Response

DTH response develops in phases known as:

- Sensitization phase- one to two weeks of primary contact with the antigen.
- Effector phase- in which macrophages and other non-specific inflammatory cells are activated (on an average 24 hours of secondary contact culminating within 48-72 hours).
- Activation phase- granulomatus reaction.

Sensitization phase: Within 10-15 days of primary contact with the antigen, T_H cells are activated and clonally expanded by antigen-presented together with the requisite class II MHC molecule on an appropriate antigen-presenting cell. The cells that activate T_H cells may be Langerhans dendritic cells found in the epidermis and macrophages. Antigens entering through the skin (epidermis level) are picked up by these cells and transported to regional lymph nodes, where T cells ($CD4^+$, $CD8^+$) are activated. $CD4^+$ and $CD8^+$ are subsets of T cells inducing DTH response, so these cells are called T_{DTH} cells.

Effector phase: T_{DTH} cells with subsequent exposure to the antigens secrete cytokines which activate macrophages and other nonspecific inflammatory cells by an average of 24 hours after secondary contact, and reach to the peak by 48-72 hours. This delay in activation of macrophages and nonspecific cells denotes DTH response. Cytokines secreted by T_{DTH} cells induce blood monocytes to adhere to vascular endothelial cells and migrate from the blood into the surrounding tissues, resulting in activation of macrophages. Activated macrophages express increased levels of class II MHC molecules and cell adhesion molecules to kill microorganisms.

Active phase: The pathogens are killed during effector phase of DTH response by macrophages. A prolonged DTH response by itself turns lethal to pathogens through the inflammation of T cells visibly as granules (Fig. 11-6). This is called granulomatous reaction. In this process, macrophages adhere closely to the cells forming multinucleated giant cells. These giant cells displace the normal tissue cells and high concentration of lytic enzymes destroys the surrounding tissue. Though this lysis clears the microorganisms, there is always the risk of tissue necrosis. Thus, the DTH response involving activation of macrophages produces and secretes

IL-3 and GM-CSF, which induce hematopoiesis generating a granulocyte-monocyte lineage.

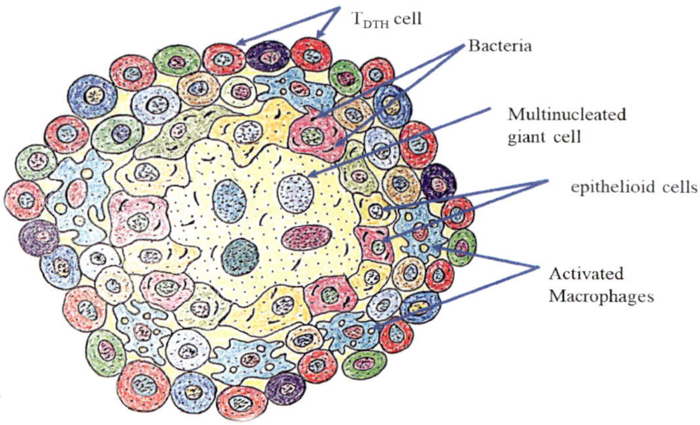

Fig.-11-6. Formation of granuloma by prolonged Delay type hypersensitivity (DTH) response.

Impact of DTH Response: DTH reaction involves the release of cytokines (IFN-γ, TNF-β, IL-1) by T_{DTH} cells. This reaction can be measured experimentally by observing the development of characteristic skin lesions following intradermal injection of antigens. Developed skin lesions indicate a positive DTH reaction i.e. the individual has been exposed to specific antigens. Since, macrophages accumulate at the site of infection; tissue damage is a necessary outcome, besides killing of specific pathogens through lysis. However, there is a double-nature pattern of DTH response. That macrophage, wall off pathogen from normal cells while killing and concentrating bacteria in the granulometric region called tubercle, thus protecting the surrounding tissues from damage.

CHAPTER – 12

HYPERSENSITIVE REACTIONS

In the process of eliminating antigens by effector molecules, the inflammatory response may either cause damage to the host's tissues. In this case, the response is known as hypersensitive or allergic being inappropriate in nature. This nature of inflammatory response dates back to the attempts of Portier and Richet, who were engaged by the Prince of Monaco to counteract the reaction of bathers in the Mediterranean Sea stung by jellyfish. The French scientists concluded that the localized reaction of bathers to the stings of the jellyfish was the result of toxins. The French scientists Portier and Richet isolated toxins from the jellyfish and used them as vaccines for dogs. They found that instead of producing antibodies against the purified toxins, the dogs instantly reacted with a lighted response as such as vomiting, diarrhoea, asphyxia and even death. This implied that dogs overreacted to the antigen. Portier and Richet coined this overreaction as 'anaphylaxis' (opposite of prophylaxis) or hypersensitive reaction.

The instant reactions or anaphylactic reactions are termed as instant hypersensitivity because the symptoms develop within minutes or hours of encounter with antigens induced by a humoral branch of the immune system. Symptoms or reactions developing until days after exposure to the antigens are also hypersensitive, but delayed in development are induced by cell-mediated branch of the immune system. Thus, there are two types of hypersensitive reactions- instant hypersensitivity and delayed hypersensitivity. Hypersensitive reactions may develop in the course of either humoral or cell-mediated responses. There are various types of hypersensitive reactions. The hypersensitive reactions as occur within a few hours are of three types:

- Type I - Individuals allergic to pollens, mold spores, certain foods,h or some drugs- IgE mediated.

- Type II - Unmatched blood transfusion- leading to the destruction of the transferred blood cells- antibody-mediated.
- Type III - Some autoimmune diseases and infectious diseases triggered by antigen-antibody complexes- immune complex-mediated.

A fourth type of hypersensitivity depends on reactions within the cell-mediated branch. These are initiated by T_{DTH} cells and are referred to as delayed-type hypersensitivity, since symptoms develop until days after exposure. Thus, another type of hypersensitive reaction is

- Type IV - Delayed type hypersensitivity (DTH). It is a response to intracellular pathogens and contact antigens (e.g. poison, oak, poison ivy, certain organic compounds) that takes 2 to 3 days to develop.

This chapter deals with these four types of hypersensitive reactions induced by different mechanisms and specific manifestations (consequences). Depending upon involvement of different immune mechanisms given rise to hypersensitive reactions. P.G.H. Gell and R.R.A. Coombs have divided the above four types of hypersensitive reactions (Fig. 12-1):

- Type I - IgE-mediated hypersensitivity.
- Type II - IgG-mediated cytotoxic hypersensitivity.
- Type III - Immune complex-mediated hypersensitivity.
- Type IV - Cell-Mediated hypersensitivity.

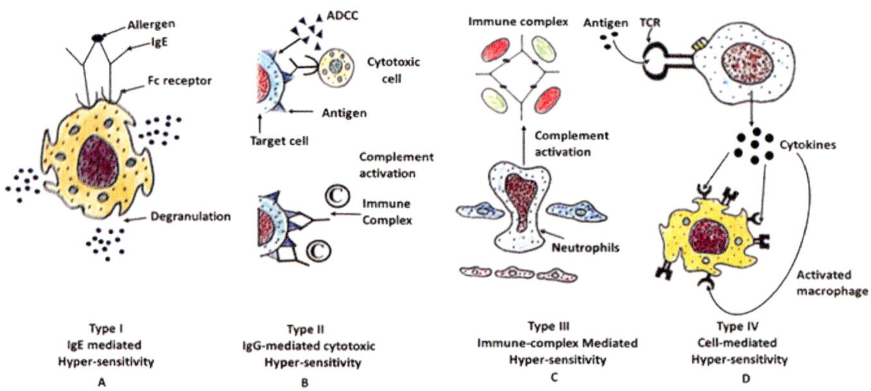

Fig.-12-1. Schematic representation of four types of hypersensitivity.

IgE-Mediated Hypersensitivity (Type – I)

Certain types of antigens are referred to as allergens. Like other soluble antigens allergens generate antibody-secreting plasma cells and memory cells. On encounter with allergens plasma cells secrete IgE. The IgE binds with high affinity to Fc receptors on the surface of tissue mast cells and blood basophils, resulting in sensitization of mast cells and blood basophils. A further exposure to the same allergen cross-links the membrane-bound IgE on sensitized mast cells and basophils, causing degranulation of these cells. Now, some pharmacologically active mediators are released from the granules which act on the surrounding tissues. As a result, depending on the extent of the mediator released, systemic or localized vasodilation and smooth-muscle convection ensues.

Components of Type I Reaction

There are two types of humans with respect to allergens reactivity. Some individuals react to a parasite by increasing levels of serum IgE until the parasite is cleared from the body. Some individuals may not immediately react to common environmental antigens. Individuals, who do not immediately react to common environmental antigens, lack a hereditary capacity (predisposition) and are called atopic, which means an IgE regulatory defect. As a result of this abnormality, IgE production is inappropriate causing tissue damage. Atopic individuals are thus allergic and susceptible to allergies like hay fever, eczema and asthma, running in families. Thus, atopy is inherited. Allergens are non parasitic antigens capable of stimulating type I hypersensitive response, were serum IgE level does not mount to the clearance of antigens that enter the body through inhalation or ingestion. Common allergens associated with Type I hypersensitivity have been listed in Table 12-1. Rye grass pollen, ragweed pollen, codfish, birch pollen, trimothy grass pollen and bee venom are sources of allergens i.e. allergenic agents. Some of these allergenic agents possess potent antigens (foreign serum and egg albumin) and plant pollens possess weak antigens. The effects of these allergens correspond to the genetic constitution of the exposed individuals.

Human serum factor released/reacting to allergens has been identified as P-K antibodies after K. Pramsmitz and Kushner who demonstrated them in 1921. A further characterization of these antibodies was experimentally conducted by K. and T. Ishizaka in mid-1960s. However, extensive chemical analysis of serum IgE levels was conducted later (Johansson and H. Bennich, 1967) to show that IgE is composed of two heavy and two light chains with a combined molecular weight of 190,000 (higher than the weight of IgC-150,000) which enables it to bind to glycoprotein receptors on the surface of mast cells and blood basophils, remaining stable for a number of weeks. Mast cells are abundantly found near blood and lymphatic vessels containing active mediators. Mast cells also secrete a variety of cytokines and serve as broad spectrum agents in the physiologic, immunologic and pathologic processes.

The intensity of the IgE response is regulated by two receptor–high-affinity receptor (FCeRI) and low affinity receptor (FCeRII). Allergen-mediated cross linkage of the bound IgE results in anaphylaxis and activation of B cells, alveolar macrophages and eosinophils. Atopic individuals have higher levels of FceRII (CD23) than non-atopic individuals. Allergen cross linkage of bound IgE induces many biochemical events. IgE-allergen complexes in ratio of 2:1 or greater may induce degranulation of mast cells when the ratio 1:2 may fail to induce any degranulation. Fig. 12-2 may be referred to visualize the biochemical events in mast cells activation and degranulation. Actually IgE-mediated degranulation begins with an allergen cross links IgE that is bound to the FC receptor on the surface of a mast cell or basophils.

Table 12-1: Common allergens Associated with Type I Hypersensitivity

Allergens	*Nature*
Foreign serum	Protein
Vaccines	Protein
Rye grass	Plant pollens
Ragweed	Plant pollens
Timothy grass	Plant pollens
Birch trees	Plant pollens
Penicillin	Drugs
Sulfonamides	Drugs
Local anesthetics	Drugs
Salicylate	Drugs
Nuts	Food
Seafood	Food
Eggs	Food
Peas	Food
Beans	Food
Milk	Food
Bee venom	Insect product
Ant venom	Insect product
Cockroach calyx	Insect product
Dust mites	Insect product
Animal hair and dander	Anima product

Besides IgE allergen complexes, intracellular events also lead to the degranulation of mast cells. From intracellular stores in the endoplasmic reticulum $Ca2^+$ is increased, resulting in the formation of arachidonic acid, which is converted into two potent mediators of type I hypersensitive reactions- the leukotrienes and the prostaglandins. The increase of $Ca2^+$ is also obtained by the use of drugs such as cromolyn sodium, which block this influx and allergy. The increase in $Ca2^+$ is followed by a transient increase in membrane-bound adenylate cyclase (cAMP) activity. Simultaneously there is a rapid peak of cAMP, which phosphorylates the granule-membrane proteins, thereby changing the permeability of the granules to water and $Ca2^+$. There is also a sequential rise and fall in cAMP levels, which can be regulated by the use of drugs.

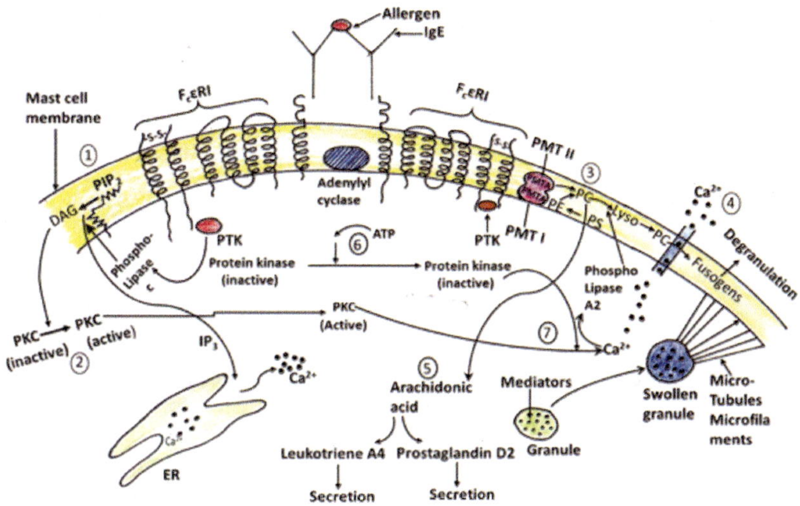

Fig.-12-2. Schematic overview of the biochemical events that take place in mast cell activation and deregulation.

Mediators of Type I Hypersensitivity Reactions

Type I hypersensitive reactions are related to two types of mediators- primary and secondary, released during mast cell or basophil degranulation. These mediators act on local tissues, eosinophils and neutrophils, thereby amplifying antigen-antibody interaction, bringing an influx of plasma and inflammatory cells to attack the pathogen (Table 12-2).

Primary Mediators

Heparin, histamine, proteases, neutrophil chemotactic factor, and eosinophil chemotactic factor are primary mediators, of which histamine is the most potent one, formed by decarboxylation of the amino acid histidine. Histamine is a major constituent of mast cell granules and binds to specific receptors on various target cells. The receptors of histamine exert varied biological effects in allergic reactions. Binding of histamine to H1 receptors induces contraction of intestinal and bronchial smooth muscles, increased permeability of venules, and increased mucous secretion by goblet cells. The binding of histamine to H2 receptor increases vasopermeability and dilation and stimulates exocrine glands. Degranulation is suppressed when histamine binds to H3 receptors on mast cells and basophils.

Secondary Mediators

Leukotrienes, prostaglandins and cytokines are the secondary mediators of type I reactions. More potent than histamines, these secondary mediators act on local tissues, eosinophils and neutrophils. These are more pronounced and long-lasting than histamines and other primary mediators. Secondary mediators are formed after degranulation of mast cell and enzymatic breakdown of phospholipids in the plasma membrane. The enzymatic cascade generates the leukotrienes and the prostaglandins taking longer time than primary mediation. The leukotrienes mediate broncho-constriction, increased vascular permeability and mucous production. Prostaglandin causes broncho-constriction. Leukotrienes are active at nano mole levels, so they are 1000 times more potent mediators than histamines, which are also mediators in human bronchial and tracheal smooth muscle contraction. That is to say that leukotrienes and prostaglandins further constrict the bronchial and tracheal smooth muscles. Thus, they are more potent stimulators of vascular permeability and mucous secretion as found in asthma patients.

Table 12-2: Some important mediators in type I hypersensitivity

Mediators	*Effects*
Histamine	Increased vascular permeability; smooth muscle contraction
Serotonin	Increased vascular permeability; smooth muscle contraction
Eosinophil chemotactic factor (ECF-A)	Eosinophil chemotaxis
Neutrophil chemotaxis factor (NCF-A)	Neutrophil chemotaxis
Proteases	Bronchial mucus secretion, degradation of blood vessel basement Membrane, generation of complement split products
Platelet-activating factor	Platelet aggregation and degranulation, contraction of pulmonary smooth muscles
Leukotrienes (slow reactive substance of anaphylaxis SRS-A)	Increased vascular permeability, contraction of pulmonary, smooth muscles
Prostaglandins	Vasodilation, contraction of pulmonary smooth muscles, platelets aggregation
Bradykinin	Increased vascular permeability, smooth muscles contraction
Cytokines (IL-1, IL-2, TNF-α, TGF-β and GM-CSF	Systemic anaphylaxis, increased expression of CAMs on vascular endothelial cells.

Cytokines released from mast cells and eosinophils manifest in type I reactions by altering the local microenvironment and recruiting neutrophils and eosinophils. Consequently, clinical manifestations as systemic anaphylaxis, asthma, hay fever, and eczema may set in.

Clinical Manifestations of Type I Reaction

The following are the prominent clinical manifestations of type I reactions:

- Systemic Anaphylaxis
- Localized Anaphylaxis

Systemic anaphylaxis is a resultant of type I hypersensitivity reaction characterized by a shock like fatal state manifesting massive edema, shock and bronchial constriction leading to death by asphyxiation within minutes of onset of type I reaction. In animals, anaphylaxis has been induced by injection of a foreign protein such as egg albumin. In humans, it has been induced by the a wide range of antigens-venom from bee, wasp, hornet, besides the drugs such as penicillin, insulin, antitoxins, seafood. The asphyxiation in type I reactions is caused by systemic vasodilation and smooth-muscle contraction brought on by mediators released in the course of the reaction. The predominant mediators are histamines and leukotrienes; the drug Epinephrine has been found of counteracting the effect of mediators by relaxing the smooth muscles and reducing vascular permeability. Epinephrine improves cardiac output, increases cAMP levels and blocks further degranulation.

Localized anaphylaxis is inherited and called atopy. Atopy includes a wide range of IgE mediated disorders such as-

- Asthma (tissue injury)
- Atopic dermatitis (eczema)
- Allergic rhinitis (hay fever)
- Food allergies (Wheal and flare reaction)

Asthma is a localized anaphylaxis in which airborne or blood borne allergens such as pollens, dust fumes, insect products or viral antigens cause an asthmatic allergy. In some other cases, asthma can be induced by exercise or cold. That asthma can develop independent of allergen stimulation (internal). Asthma is triggered by degranulation of mast cells with release of mediators affecting the lower respiratory tract. The contraction of bronchial smooth muscles causes broncho-constriction, developing into airway edema, mucous secretion and inflammation. Airway edema, mucous secretion and inflammation cause bronchial constriction and airway obstruction. That is why asthmatic patients gasp very badly for respiration. The principal mediators in asthmatic atopy are histamines, and leukotrienes (LTC4 and prostaglandin D2; PGD2), which cause broncho-constriction, vasodilation and some buildup of mucous. The secondary mediators in asthmatic atopy are IL-4, IL-5, IL-6, TNFα, eosinophil chemotactic factor (ECF) and platelet-activating factor (PAF). The sum action of these mediators is to augment endothelial cell adhesion, as well as the recruit inflammatory cells including eosinophils and neutrophils into the bronchial tissue. Eosinophils and neutrophils release toxic enzymes, oxygen radicals and cytokines, which cause significant tissue injury, fluid buildup (edema) and hypertrophy of the bronchial smooth muscles.

In asthma, both the genetic and environmental factors are involved as a complex anaphylaxis. Among the ethnic groups in the United States, studies of genes have suggested it to be an inherited disease (predisposing). The environmental linkage has been confirmed in a study of African-American inner city children infested by cockroaches. Since, IgE is the principal mediator in type I reactions, genetic factor causing asthmatic characteristics is well confirmed.

Atopic dermatitis (allergic eczema) is a hereditary disease often developing during infancy with serum IgE levels increased. The skin eruptions in atopic dermatitis are erythematous and filled with pus. It has increased the number of eosinophils and TH2 cells.

Allergic rhinitis commonly known as hay fever, the most common atopic disorder, results from the reaction of airborne allergens with sensitized mast cells in the conjunctivae and nasal mucosa. Pharmacologically active mediators are released from mast cells, which cause localized vasodilation and increased capillary permeability. Allergic rhinitis is characterized by watery exudation of conjunctivae, nasal mucosa, and upper respiratory tract, sneezing, and coughing.

Food Allergies is caused by various foods inducing localized anaphylaxis in allergic individuals by allergen cross linking of IgE on mast cells along the upper or lower gastrointestinal tract. When a food allergen is carried to sensitized mast cells in the skin there occurs smooth-muscle contraction and vasodilation manifesting in symptoms such as vomiting or diarrhoea. The symptoms ensue depending on where the allergen is deposited. That is, if the allergens attack by cross linking IgE on mast cells, the upper and lower gastrointestinal tract is affected resulting in vomiting or diarrhoea. If the allergens are carried to the blood stream by mast cell degranulation symptoms like asthmatic reaction develop or red, swollen eruptions appear on the skin. The later response is called a wheal and flare reaction.

Types and Regulation of the Type I Reaction

Bronchial hyperresponsiveness is a late phase reaction in asthma, characterized by inflammation of the airway. Likewise all type I reactions are grouped in acute and late-phase reactions. Late-phase reactions follow as type I hypersensitive reaction begins to subside, characterized by infiltration of neutrophils, eosinophils, macrophages, lymphocytes and basophils including cytokines released by mast cells. The expression of cell adhesion molecules on venular endothelial cells is increased by TNF-α and IL-1 facilitating the buildup of neutrophils, eosinophils and monocytes, which marks the late-phase response.Extensive tissue damage occurs in late-phase reaction. The influx of eosinophils is causative to the chronic inflammation of the bronchial mucosa as found in asthmatic response. Besides, eosinophils, neutrophils are another causative factor to the late-phase reactions in type I reactions, involving neutrophil chemotactic factor, released from degranulating mast cells and activating cytokines (IL-6, IL-8).

Type I reaction can be regulated by the determination of the susceptibility or sensitivity of an organism (animal/human) to an antigen/allergen. Some of the major determinants of type I reactions are as follows:

- Antigen dosages (Mounting of IgE levels) - low doses of antigen induce a transient IgE production shifting to IgG with high doses.
- Cytokines secreted by T_H2 cells (IL-3, IL-4, IL-5 and IL-10) increase IgE production, whereas IFN-γ secreted by T_H1 cells inhibits IgE response. Thus, a balance of IL-4 and INF-γ determines the amount of IgE production. IFN-γ

is secreted by T_H1 subset and IL-4 is secreted by TH2 subset. The production of IgE is related to T_H2 subset, and found in atopic individuals. Non-atopic individuals involve T_H1 subset and produce IgM or IgG. IL-4 is down-regulated in allergic individuals.

Assessment of Type I Reaction

There are three methods of assessment of type I hypersensitivity:

- Skin testing
- Radioimmunosorbent test (RIST)
- Radioallergosorbent test (RAST)

Skin testing is conducted by intradermal injection or superficial scratching of specific skin sites (on the fore-arm or back of an individual). Under skin test, a person allergic to the allergen expresses degranulation of local mast cells and produces a wheal and flare reaction within half an hour of such introduction of allergens. In skin test, the disadvantageous aspect is sometimes seen as late phase reaction taking 4-6 hours to induce tissue damage; otherwise this test being inexpensive is beneficial in allowing screening of a large number of allergens on a single sitting.

Radioimmunosorbent Test (RIST) is a highly sensitive technique based on radioimmunoassay to detect and assess the serum level of total IgE antibody (microgram levels of total IgE). In this test, the patient's serum is reacted with agarose beads or paper disks coated with rabbit anti-IgE. After washing the beads or disks, 125I labeled rabbit anti-IgE was added. The radioactivity of the beads or disks is measured with a gamma counter; it is proportional to the level of IgE in the patient's serum (Fig. 12-3. a).

Radioallergosorbent Test (RAST) detects the serum level of IgE specific for a given allergen. In this test, the allergen is coupled to disks or beads; the patient's serum is added. Unbound antibody is washed away. The amount of specific IgE bound to the solid phase allergen is measured by adding 125I-labeled rabbit anti-IgE and counting the bound radioactivity (Fig. 12-3. a).

Therapy for Type I Hypersensitivities

The severity of type I reactions may be reduced or completely eliminated by a number of controlling or curing devices:

- **Identification of offending allergens:** Contact and inhalation cause the allergic reactions as the first possible effectors. So removal of home pets, dust control, avoidance of foods and elimination of inhalant allergens may be adopted as the first line of control of type I reaction.
- **Introduction of allergen by subcutaneous injections:** is another potent measure for controlling allergic reactions, especially allergic rhinitis. By injection of allergens, there happens a shift toward IgG production, which competes for the allergen bound to it and forms a complex that can

be removed by phagocytosis; as a result, the allergen is not variable to crosslink the fixed IgE on the mast cell membrane. Consequently, it follows a decrease in allergic symptoms.

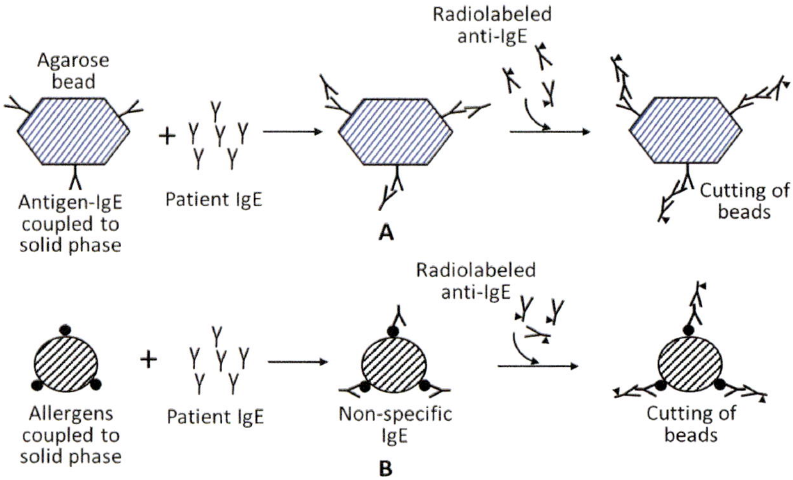

Fig.-12-3. Procedures for assessing type I hypersensitivity.

- **Use of drugs:** Anti histamines have been found the most effective way of controlling type I reactions. The Antihistaminic drugs act by binding to the histamine receptors on target cells and block the binding of histamine. Similarly, drugs the block the release of allergic mediators by interfering with various biochemical steps in masT cell activation and degranulation also serve as potent controller of allergic reactions. Theophylline in asthmatics, epinephrine in anaphylactic shock, cortisone and various other anti-inflammatory drugs are very useful in controlling type I hypersensitivity.

IgG-Mediated Cytotoxic Hypersensitivity (Type II Hpersensitivity)

Transfusion reactions resulting from unmatched blood from the donor and recipient host and hemolytic disease of the newborn are two clinical manifestations of type II reactions. These manifestations are autoimmune diseases involving auto-antibody-mediated cellular destruction by type II mechanism. The blood transfusion reactions and hemolytic disease of the newborn (erythroblastosis fetalis) follow in detail.

Blood Transfusion Reactions

Blood transfusion reactions occur when host antibodies react with foreign antigens on the unmatched transfused blood cells and mediate destruction of these cells by activating the complement system. A large number of proteins and glycoproteins on the membrane of red blood cells are encoded by different genes, each of which has a number of alternative alleles. An individual possessing one allelic form of a blood-group antigen can react to other allelic forms of transfused blood recognizing them as foreign. This causes production of IgG class antibodies resulting in cell destruction by antibody-dependent cell-mediated cytotoxicity (ADCC). Thus, transfusion of blood into a recipient possessing antibodies to one of the blood-group antigens can result in a transfusion reaction from massive intravascular haemolysis of the transfused red blood cells by antibody plus complement. The manifestations may be immediate or delayed depending upon the blood-group unmatching to the recipient. As a result of transfusion of unmatched blood, some of the hemoglobin of the recipient gets converted into bilirubin which is toxic at high levels. This unmatched blood transfusion often gets reacted with ABO blood-group. The symptoms associated with unmatched blood-group are fever, chills, nausea, clotting within blood vessels, pain in the lower back, and hemoglobin in the urine. Hemoglobin in the kidney can cause acute tubular necrosis. The treatment of unmatched blood transfusion is prompt termination of the transfusion and maintenance of urine flow with a diuretic. The above symptoms of unmatched blood transfusion occur within minutes of transfusion.

The delayed reaction of transfusion occurs within 2-6 days after transfusion. The transfused blood induces clonal selection and production of IgG against a variety of blood-group membrane antigens viz. Rh, Kidd, Kell and Duffy. With transfusion of these red blood cells, many of them are destroyed at extravascular sites by agglutination, opsonization, and subsequent phagocytosis by macrophages. The symptoms associated with delayed reaction of transfusion are fever, low hemoglobin, increased bilirubin mild jaundice and anemia. Since the RBC destruction occurs at extravascular sites, free hemoglobin is usually not detected in the plasma or urine in delayed reactions. Blood transfusion reactions can be prevented by cross-matching the presence of the antibodies in donor or recipient sera responsible for such reactions.

Erythroblastosis fetalis

With an Rh– fetus expresing antigen on its blood cells that the Rh– mother does not express, there develops severe hemolytic disease of the newborn. That is when an Rh– mother carries on Rh^+ foetus, IgG anti-Rh antibodies are formed which cross the placenta and damage the fetal red blood cells. As a result, the fetus may develop severe anemia or fatal results may follow. Thus, Rh incompatibility may result in hemolytic disease of the newborn or even death due to severe anemia. This factor of Rh incompatibility resulting in hemolytic disease of the newborn can be almost entirely prevented, in a subsequent pregnancy by administering antibodies called Rhogam, against the Rh antigen to the mother within 24-48 hours after the first delivery. This is because the mother having received Rhogam is unlikely to produce IgG anti-Rh antibodies and the fetus is protected from the damage that occurs when these antibodies cross the placenta. The need to administer Rhogam is determined by testing natural serum at intervals during pregnancy for antibodies to the Rh antigen. If a rise in the titer of these antibodies is detected as pregnancy progresses, it indicates that the mother has been exposed to Rh antigens and is producing increasing amounts of antibody. Another test for determining the administration of Rhogam is the Coombs test. In this test, if maternal IgG is bound to the fetal red blood cells, agglutinate with the Coombs reagent.

On diagnosis of hemolytic disease in the fetus either due to Rh incompatibility or by ABO blood-group incompatibility between the mother and fetus, treatment may be given corresponding to the specific cause, as above, during pregnancy.

- For a severe reaction caused by Rh incompatibility, the fetus can be given an intrauterine blood-exchange transfusion to replace fetal Rh^+ red blood cells with Rh– cells, every 10-21 days until delivery.
- For a comparatively less severe reaction caused by Rh incompatibility, the mother can be treated during pregnancy by plasmapheresis- a procedure in which the mother's blood is separated into two fractions cells and plasma. Plasma containing the anti-Rh antibody is discarded and the cells are reinfused into the mother in an albumin or fresh plasma solution. The plasmapheresis is conducted by a separation machine.
- ABO blood group incompatibility between the mother and fetus usually results in fetal anemia or slight elevation in bilirubin creating jaundice-like situation. Therapy for such complications is blood exchange transfusion. Bilirubin can be broken down by exposure of the infant to UV light. The exposure can also be useful in managing any threatened cerebral damage.

Hemolytic anemia can also be induced by administration of certain antibiotics viz. penicillin, cephalosporin, streptomycin which can adsorb nonspecifically to proteins on RBC membranes, forming a complex similar to hapten-carrier complex i.e. formation of antibodies. Antibodies bind to the adsorbed drug on RBC, inducing complement mediated lysis. This results in the development of anemia. Actually, the production of IgM and IgG is the cause of development of hemolytic anemia.

Immune Complex-Mediated Hypersensitivity (Type III Hypersensitivity)

The type III hypersensitivity reaction is mediated by the formation of immune complexes (reaction of antibody with antigens), and the activation of complement. The complexing of antigen with antibodies facilitates the clearance of antigen by phagocytic cells. Of course, large amounts of antibodies cause damage to tissues. Thus, the quantity of immune complexes as also their distribution within the body determines the quantum of reaction and resulting destruction of tissues at or near the site of antigen entry. The damage of tissues is always local- at the site of accumulation of antibodies (immune complexes). The sites are commonly blood vessel walls, in the synovial membrane of joints, on the glomerular basement membrane of the kidney and on the choroid plexus of the brain. The reaction is initiated by the recruitment of neutrophils at the site of antigen entry. The tissue damage is caused by granular release from the neutrophils. Neutrophils attempt to phagocyte immune complexes by release of lytic enzymes and much of tissue damage results there-from. Tissue damage can also be caused by activation of membrane attack mechanism of the complement system. Thus, complement split products serve as immune effector molecules that elicit localized vasodilation and chemotactically attract neutrophils.

Besides, the localized reaction known as Arthus Reaction, which is manifested in mild swelling and reduces to tissue necrosis (Arthus type-reactions are induced by bacterial spores, fungi, or dried faecal proteins), generalized type III reactions occur when circulating immune complexes are deposited at various sites. It was observed that after administration of antitoxins containing foreign serum (horse anti tetanus or anti diphtheria serum), the recipient developed antibodies specific for the foreign serum proteins within days or weeks after exposure. Symptoms such as fever, weakness, rashes (vasculitis) with edema and erythema, arthritis, glomerulonephritis (complex deposition in the kidney) and lymphadenopathy develop after exposure to foreign serum antigens. These symptoms often occur in combination and generally called serum sickness. Since antibodies specific for the foreign serum antigens form circulating immune complex, tissue damages occur at various sites. However, it must be borne in mind that is localized type III reactions; the amounts of antigen enter the blood-stream in smaller quantities than in generalized type III reactions. Fig. 12-4 can be referred to have a view of localized reaction.

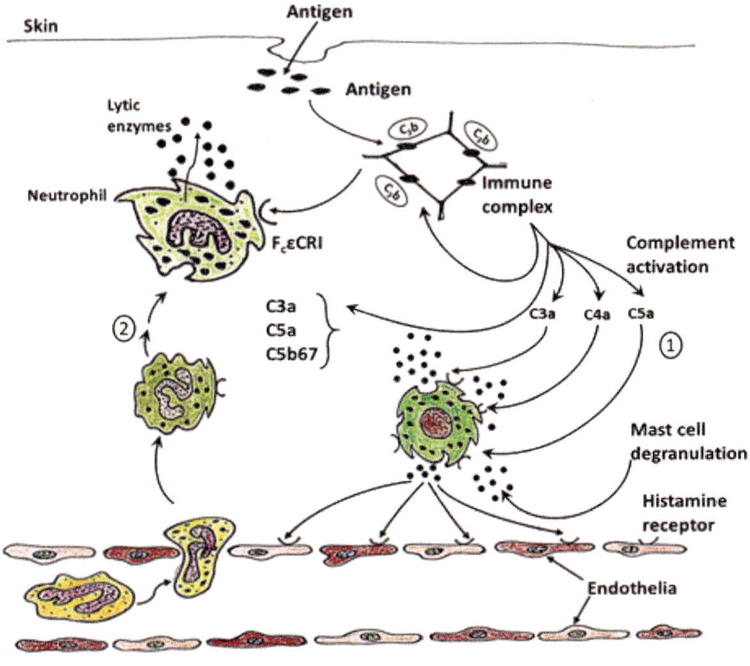

Fig.-12-4. Schematic representation of Arthus reaction or type III hypersensitivity.

Cell-Mediated Hypersensitivity (Type IV)

The type IV hypersensitive reaction involves the cell-mediated immune response. The main point of difference between type III and type IV reactions is recruitment of neutrophils and macrophages respectively. Type IV reaction is a delayed type hypersensitivity, which manifests within 48-72 hours after activation of T_{DTH} cells by antigens on appropriate antigen-presenting cells resulting in secretion of cytokines (IL-2, IFN-γ, macrophage-inhibiting factor MIF, and tumor necrosis factor b; TNF-β). As a result of secretion of these cytokines, macrophages are drawn into the required area and lytic enzymes carry out the postphagocytic activity. Tissue damage is a concomitant phase, as such in type IV reaction also. Tissue damage is amply manifested in granulomatous skin lesions seen with Mycobacterium leprae and lung cavitations seen with Mycobacterium tuberculosis.

Besides activation of T_{DTH} cells by antigens, contact with formaldehyde, trinitrophenol, nickel, turpentine, various cosmetics and hair dyes, poison oak and poison ivy also induce type IV reactions by complexing with skin proteins. Substantially these substances are small molecules and activate T_{DTH} cells by contact with the skin, so called contact dermatitis. In contact dermatitis, T_{DTH} cells induce cytokine production, which cause the accumulation of macrophages at the contact site. The characteristic symptoms of contact dermatitis are redness and pustules, which mark the release of lytic enzymes from macrophages to clear the site intradermally.

CHAPTER – 13

IMMUNE RESPONSE TO INFECTIONS DISEASES

Infectious diseases pose a challenging threat to human populations despite the development of vaccines and drug therapy. The challenge of infectious diseases is more visible in Third World, in comparison to developed countries where therapeutic techniques have reduced the mortality arising out of infectious diseases. However, some drug-resistant strains of pathogens have also emerged challenging the advanced techniques to combat them. This chapter deals with the major types of pathogens, the numerous diseases caused by them and some vital emerging infectious diseases.

Pathogens invading a susceptible host are counteracted upon by innate (nonspecific) and adaptive (specific) immune defense mechanisms involving:

- Epithelial surfaces of the skin (epithelial cells),
- Production of cytokines, and
- NK cells

Epithelial cells and the lining of the gut provide the first barrier to the entering pathogen as a result of which most pathogens never gain productive entry into the host and instead pass through it without establishing a productive infection. Some chemicals are also produced by epithelial cells to prevent infection- viz. gastric enzymes, which lower the pH of the stomach and upper gastrointestinal tract. Other specialized cells produce antibacterial peptides in the gut. The presence of normal gut flora is a major feature of innate immunity, which inhibits the binding of pathogens to gut epithelial cells. Some bacteria produce endotoxins such as LPS which stimulate the production of cytokines such as TNF, IL-1 and IL-6 by macrophages or endothelial cells. Macrophages and other phagocytic cells cause phagocytosis by bacteria in addition to NK cells.

Pathogens use a variety of strategies to escape adaptive immune mechanisms. Some pathogens reduce their own antigenicity either by growing within host cells or by shedding their membrane antigens. Some pathogens express molecules with amino acid sequences like those of host cell membrane molecules. Some pathogens adopt continual variation or antigenic variation to elude the immune system.

Infectious diseases are caused by four main types of pathogens-viruses, bacteria, protozoa, and helminths (parasitic worms).

Immune Response to Viral Infections

The immune response to viral infections involves both humoral and cell-mediated components (Table 13-1). A great resistance is posed by viruses to subvert the humoral and cell-mediated responses. The life of viruses in attempt of establishing within the host body depends upon the effectivity of latter to counteract the replication of viruses. The innate immune response against the viruses comprises IFN-α and IFN-β production by macrophages, monocytes, fibroblasts and NK cells.

Table 13-1: Mechanism of humoral and cell-mediated immune responses

Response	*Effector molecules*	*Functions*
	IgA	Blocks binding of virus to host cells, thus
	IgG, IgM, and IgA	preventing infection or reinfection
Humoral immune response	IgG and IgM	Blocks fusion of viral envelope with hosT cell plasma membrane
	IgM	Enhances phagocytosis of viral particles
	Complement activated by IgG or IgM	Agglutinates viral particles
		Mediates opsonization by C3b and lysis of enveloped viral particles by membrane-attack complex
Cell mediated Immune response	IFN- secreted by TH or TC cells	Has direct antiviral activity
	Cytotoxic T-lymphocytes	Kill virus infected cells
	NK cells and macrophages	Kill virus infected cells by antibody-dependent cell mediated cytotoxicity

Several genes are also synthesized which activate a ribonuclease (RNAse) that degrades viral RNA. IFN-a and IFN-b bind to the IFN receptor, which in turn induces the synthesis of 2-5(A) (an enzyme known as 2'-5'-oligo-adenylate synthetase) and protein kinase (PKR). The action of 2-5(A) synthetase results in the activation of RNAse I, which can degrade mRNA. PKR inactivates the translation initiation factor eIF-2 by phosphorylating it. Thus, both pathways result in the inhibition of protein synthesis and effectively block the viral replication in infected cells. In response to viral infection, the cytokine IL-12 is produced and the activity of NK cells is also increased.

A virus is blocked in its very first stage of attachment/penetration to membrane cells by antibodies and CTL precursors after occurrence of infection. In other words, antibodies play an important role in the host's protection against the virus and the CTL precursors or cell-mediated antiviral mechanism gets into action after the infection is established.

Humoral and Cell-Mediated Response against Viruses

Most viruses express surface receptor molecules that enable them to initiate infection by binding to specific hosT cell membrane molecules. If antibodies to the viral receptors is produced, it can block infection altogether by preventing the binding of viral particles to host cells. The use of oral polio vaccine is an example of this antibody activity against poliovirus. The attenuated oral polio vaccine induces production of secretory IgA, which blocks viral attachment to mucosal epithelial cells. Thus, antibodies specific for viral surface antigens are often crucial in containing the spread of a virus during acute infection and in protecting against reinfection or subsequent infection. However, viruses are neutralized after viral attachment to host cells. Antibodies have again block viral penetration by binding to epitopes that are necessary to mediate fusion of the viral envelope with the plasma membrane. Antibodies also act as opsonizing agent to facilitate Fc or C_3b-receptor mediated-phagocytosis of the viral particles. Thus, preventing or blocking the infection by viruses is the humoral response of host body.

After the infection has set inside the host as bound to membrane molecules or the DNA of virus is integrated into host chromosomal DNA. This state of virus is known as its established state. At this stage cell-mediated immune mechanism comes to the aid of host in defense against virus. The main components of cell-mediated immune mechanism are $CD8^+$ T_C cells and $CD4^+$ T_H1 cells and in some cases $CD4^+$ T_C cells are also implicated. A number of cytokines are produced by activated TH1 cells. Cytokines IL-2, IFN-g and TNF defend against viruses either directly or indirectly. NK cells are also activated by the actions of above cytokines, bringing CTL response into play. CTLs specific for the virus eliminate virus-infected cells causing elimination of potential sources of new viruses.

Immunosuppression by Viruses and Viral Evasion

Host responses to viruses are evaded by viruses in a number of ways. That is to say that some viruses develop strategies to evade the action of IFN-α/β such as hepatitis C virus, herpes simplex viruses (HSV), adenoviruses, cytomegalovirus (CMV), which reduce the surface expression of class I MHC molecules. HIV reduces level of class II MHC molecules on the cell surface thus, blocking the function of antigen-specific anti-viral helper T cells.

A number of viruses have strategies for evading complement-mediated destruction such as vaccinia virus. Vaccinia virus secretes a protein that binds to the C4b complement component inhibiting the classical complement pathway and HSV (herpes simplex virus), which have a glycoprotein component that binds to the C3b complement component inhibiting both the classical and alternative pathways.

Paramyxovirus (causes mumps), measle virus, Epstein-Barr virus (EBR), cytomegalovirus and HIV are among the viruses that either directly destroy the immune cells by cytolytic mechanism or alter their function thus, develops a cytokine imbalance means decreased levels of IL-2, TNF and INF-α/β.

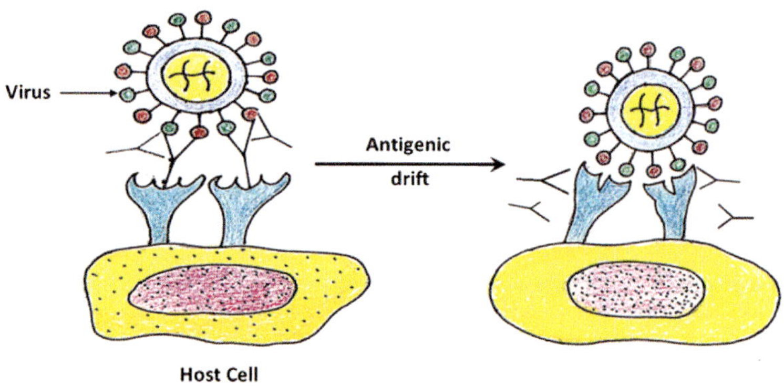

Fig.-13-1. A Distinct mechanism generates variations in influenza surface antigen, which shows the antigenic drift.

Influenza virus infecting the upper respiratory tract and major central airways is one of the dreaded viruses distinguished by differences in their nucleoprotein and matrix proteins. There are three basic types of influenza virus- A, B and C. Type A influenza viruses contain 13 different hemagglutinins and 9 neuraminidases. Influenza contains eight separate strands of ssRNA responsible for the reassortment of the RNA strands of human and animal virions within a single cell infected with viruses. This fact is reason for antigenic shift of influenza virus, wherein the population is unprepared for new types of viruses. The antigenic shift provides answer for outbreaks of influenza as epidemic.

Actually by antigenic shifts, the influenza viruses undergo antigenic drift generating minor antigenic variations, which account for strain differences within a subtype. Though a given influenza strain is eliminated by an effective immune response, the accumulation of point mutations sufficiently alters the antigenicity of some variants so they are able to escape immune elimination (Fig. 13-1). Continual changes in amino acid residues also allow the influenza virus to evade the antibody response. Serum antibody levels play an important role in resistance to infection by influenza virus, when high for a particular hemagglutinin (HA) molecule. CTLs also play an important role in defense against influenza viruses.

Immune Response to Bacterial Infections

There are two routes through which bacteria enter the body-natural (the respiratory tract, the gastrointestinal tract and the genitourinary tract) and unnatural (opened by breaks in mucous membranes or skin). Lower inoculums or organisms are met by localized tissue phagocytosis and larger inoculums are met by specific (adaptive) defense mechanisms.

Innate and Adaptive Responses to Bacteria

Under innate response, the humoral antibodies secreted by plasma cells in regional lymph nodes and the sub mucosa of the respiratory and GITS (gastrointestinal tracts) provide the main protective response against extracellular bacteria, by inactivation of bacterial toxins (Fig. 13-2). antibodies to a bacteria toxin may bind to the toxin and neutralize it. The antibody-toxin complexes are then cleared by phagocytic cells in the way adopted by other A-B complexes.

Table 13-2: Immune responses against different modes of bacterial infections

Mode of infection	*Immune response*	*Mechanism of bacterial invasion*
Attachment to host cells	Blocking of attachment by secretory IgA secretion	Secretion of proteases that cleave IgA dimers, antigenic variation in attachment structure
Proliferation	Phagocytosis (Ab-C3b-mediated opsonization)	Production of different surface structure hat prevent phagocytic cells to work, mechanism for surviving within phagocytic cells, induction of apoptosis in macrophages
	Complement-mediated Lysis and localized Inflammatory response	Generalized resistance of gram-positive to complement mediated lysis, insertion of membrane-attack complex prevented by long side chain in cell-wall LPS
Invasion of host	Ab-mediated agglutination	Secretion of elastase that inactivates C3a and C5a
Tissues		
Toxin induced		
Damage to host cells	Neutralization of toxin by antibody	Secretion of hyaluronidase, which enhances bacterial invasiveness

Under adaptive response, the intracellular bacteria activate NK cells besides a cell-mediated immune response (DTH) - IFN-γ which activates macrophages.

Macrophages kill the ingested pathogens. For a comprehensive appreciation of host immune response to bacterial infection and bacterial evasion mechanism Table 13-2 may be referred to.

The following steps are involved in bacterial infection:

- Step 1 – Bacteria attach to the host cell
- Step 2 – Bacteria proliferate in the host body
- Step 3 – Host tissues are invaded by bacteria
- Step 4 – Toxins secreted by bacteria damage host cells

Some bacteria have pili (long hair-like projections), which enable them to attach to host cells-specially to the membrane of the intestinal or genitourinary tract. Adhesion molecules in bacteria also help them to attach to host cells-especially ciliated epithelial cells of the upper respiratory tract. Bacteria having pili are inhibited to attach to mucosal epithelial cells by secretory IgA antibodies. However, IgA secretion is evaded by some bacteria (such as *Neisseria gonorrhoea, Haemophilus influenza,* and *Neisseria meningitis*) by secreting proteases that cleave secretory IgA resulting in an inability to agglutinate microorganisms. The protein component of the pili has a highly variable structure. This variation is caused by gene rearrangements of bacteria known as mini-cassetes; the pathogenicity of bacteria is increased, which help in evading IgA.

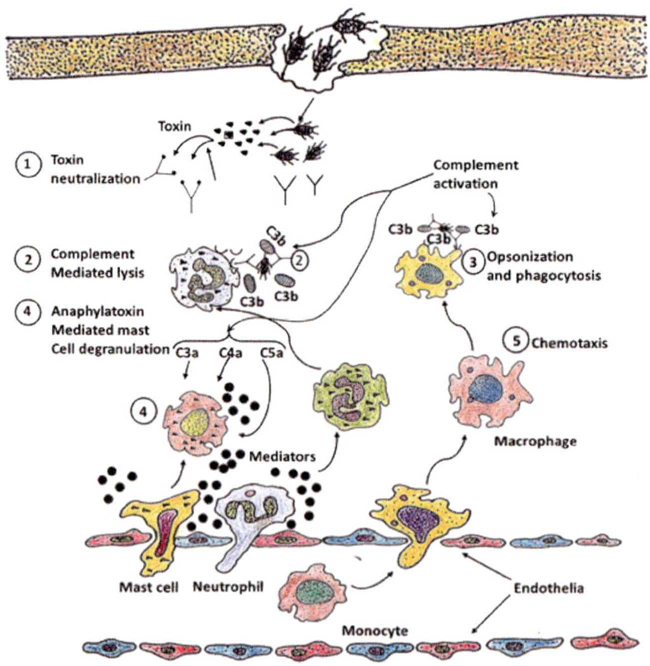

Fig.-13-2. Visualization concept of antibody-mediated mechanisms for neutralizing extra cellular bacterial infection.

The polysaccharide capsule in some bacteria also helps them to evade phagocytosis. The polysaccharide capsules are "serotypes" numbering differently in different bacteria. During infection the antibody produced in host provides defense against the infecting serotype. If the host is infected by a different serotype, the earlier antibody fails to provide defense against this different serotype, as is the case with *Streptococcus pneumoniae*. That is why pneumonia is caused many times.

Phagocytosis is also evaded by some bacteria, which assemble a protective coat from host proteins. These bacteria secrete a coagulase enzyme that precipitates a fibrin coat around them, which serves as shield against phagocytic cells while some bacteria survive within phagocytic cells by escaping from the phagolysosome to the cytoplasm since the latter provides them a more favorable environment for growth.

Pattern-Recognition Receptor (PRRs)

Unlike adaptive immunity, innate immunity does not recognize every possible antigen. Instead, it is designed to recognize molecules shared by groups of related microbes that are essential for the survival of those organisms and are not found associated with mammalian cells. These unique microbial molecules are called pathogen-associated molecular patterns or PAMPs, which include LPS (gram-negative cell wall), peptidoglycan and lipoteichoic acids (gram-positive cell wall), mannose, bacterial and viral unmethylated CpG DNA, bacterial flagellin, the amino acid N-formylmethionine (bacterial proteins), double-stranded and single-stranded RNA from viruses, and glucans from fungal cell walls. In addition, unique molecules displayed on stressed, injured, infected, or transformed human cells, so called danger molecules like heat shock proteins (HSPs), also act as PAMPs.

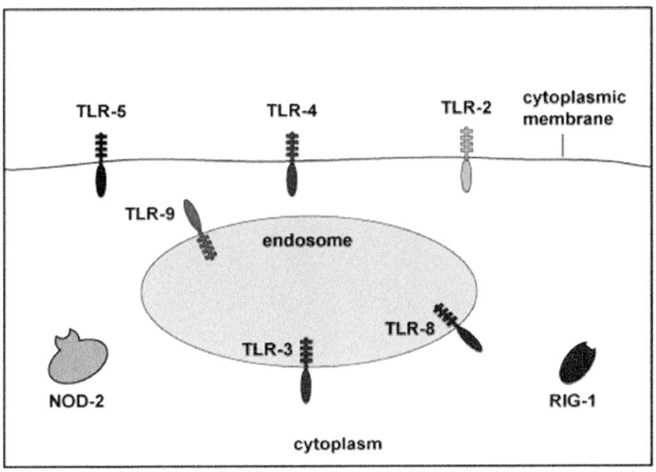

Fig.-13-3. Pattern-Recognition Receptors (PRRs) present on the immune cells

These PAMPs are recognized by specialized receptor found on the surface of the cell like macrophages, dendritic cells, endothelial cells, mucosal epithelial cells, and lymphocytes and in the cytosol and/or phagolysosomes, called the pattern-recognition receptors or PRRs. A number of PRRs are found on the cell surface of innate immune cells and play an indispensable role in the induction of innate and adaptive immune responses against microbes. Some PRRs are found in the cytosol of the cell (Fig.-13-3 & Table 13-3). Therefore, the innate immune system is thought to recognize approximately 103 types of microbial molecular patterns.

There are two functionally different classes of pattern-recognition receptors (PRRs): endocytic pattern-recognition receptors and signaling pattern-recognition receptors.

A. Endocytic Pattern-Recognition Receptors: Endocytic pattern-recognition receptors are found on the surface of phagocytes and promote the attachment of microorganisms to phagocytes leading to their subsequent engulfment and destruction. It includes mannose receptor, scavenger receptors, opsonin receptor and N-formyl Met receptors.

- **Mannose receptors**: Mannose receptors are found on the surface of phagocytes. It binds with mannose-rich glycans, the short carbohydrate chains having mannose or fructose as the terminal sugar. It is commonly found in microbial glycoproteins and glycolipids but rare in humans (human glycoproteins and glycolipids typically have terminal N-acetylglucosamine and sialic acid groups). C-type lectins found on the surface of phagocytes are mannose type receptors that bind to mannose-rich glycans in order to attach microbes to phagocytes (Fig.-13-4). Mannose- binding lectin (MBL), also known as mannan-binding protein, is soluble pattern recognition TLR-toll-like receptor; NOD-nucleotide binding oligomerization domain; MDA-melanoma differentiation-association gene.

Table 13-3: Various pattern-recognition receptors and their corresponding pathogen-associated molecular patterns and their location in the cells.

PRRs	*PAMPs*	*Location*
TLR-2	Peptidoglycan, lipoproteins, lipoteichoic acid and porins	Cell surface
TLR-3	Viral double-stranded DNA	Endosomes
TLR-4	Lipopolysaccharide (LPS) from gram-negative cells wall, fungal mannans, viral envelope proteins, parasitic phospholipids, heat shock proteins	Cell surface
TLR-5	Bacterial flagellin	Cell surface
TLR-8	Viral single stranded RNA	Endosomes
TLR-9	Viral and bacterial unmethylated CpG sequences	Endosomes
NOD-1	Muramyl dipeptide NAG-NAM gamma-D-glutamyl-meso diaminopimelic acid in peptidoglycan monomer in common gram-negative bacteria	Cytoplasm
NOD-2	Muramyl dipeptide (NAG-NAM-L-alanyl-isoglutamin) from bacterial peptidoglycan	Cytoplasm
RIG-1	Viral RNA	Cytoplasm
MDA-5	Viral RNA	Cytoplasm

Receptor present in plasma and tissue fluid of a cell that binds to mannose-rich glycans present on microbes in order to activate the lectin complement pathway.

- **Scavenger receptors:** Scavenger receptors found on the surface of phagocytic cells bind to bacterial cell wall components such as LPS, peptidoglycan and teichoic acids (Fig.-13-5). There are several other scavenger receptors are also present on the cell surface to recognize certain components of other types of microorganisms, as well as for stressed, infected, or injured cells. Scavenger receptors include CD-36, CD-68, and SRB-1.

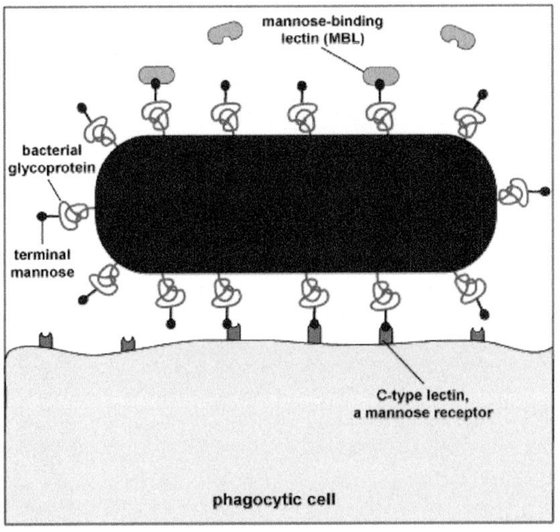

Fig.-13-4. Terminal mannose of bacterial glycoprotein is recognized by mannose receptors, which is C-type lectin present on the cell surface of phagocytic cells.

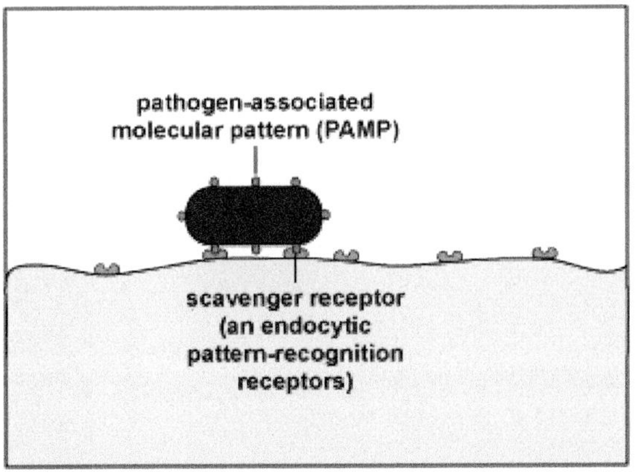

Fig.-13-5. Scavenger receptors such as CD-36, CD-68, and SRB-1 binds to the certain bacterial components and, endogenous danger molecular released due to stress and injury and induces immune cell reactivity.

- **Opsonin receptors:** Opsonins are soluble molecules produced by a cell that provide immune defense and help in the binding of microbes to phagocytes. One portion of the opsonin binds to a PAMP on the microbial surface and another portion binds to a specific receptor present on the phagocytic cell (Fig.-13-6). The acute phase proteins (mannose binding proteins and C-reactive protein), complement pathway proteins (C3b and C4b), surfactant proteins in the alveoli of the lungs (SP-A and SP-D) and immunoglobulin IgG are the example of opsonin that bind to their receptor present on the phagocytic cell.

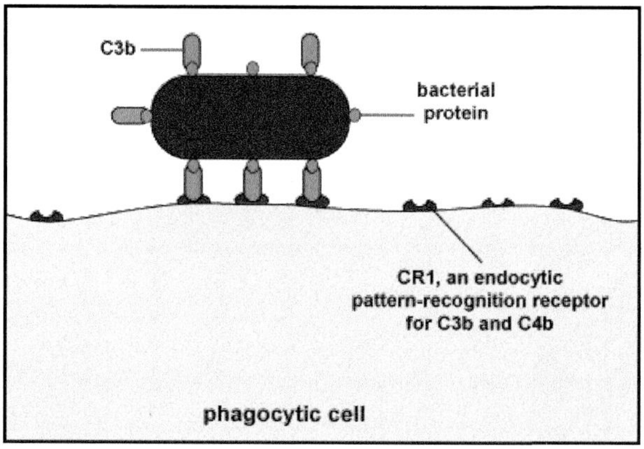

Fig.-13-6. Opsonin binds to the opsonin receptor and results in opsonization and subsequent immune response. Mannose-binding lectin (MBP) binds on one hand with sugar mannose or fructose (terminal sugar) and on the other side binds to mannose receptors. Likewise, C-reactive protein (CRP) binds to phosphorylcholine in bacterial membranes and phosphatidylethanolamine in fungal membranes at one side and opsonin receptor on the other side. One portion of C3b binds to bacterial surface proteins and another portion binds to C3b receptors such as CD-1 on phagocytes. IgG binds to an antigen present on the bacterial surface and also binds to its Fc-γ receptor and thus functions as an opsonin.

4. *N-formyl* Met receptors: N-formyl methionine is the first amino acid produced in bacterial proteins. This form of the amino acid is typically not seen in mammalian proteins. FPR and FPRL1 are N-formyl receptors present on the surface of neutrophils and macrophages. The Binding of N-formyl Met to its receptor promotes the motility and chemotaxis of these phagocytes. It also promotes phagocytosis.

B. Signaling Pattern-Recognition Receptors: Signaling pattern-recognition receptors bind a number of microbial molecules like LPS, peptidoglycan, teichoic acids, flagellin, pilin, unmethylated cytosine-guanine dinucleotide or CpG sequences from bacterial and viral genomes; lipoteichoic acid, glycolipids, and zymosan from fungi; double-stranded viral RNA, and certain single-stranded viral RNAs. Binding of microbial PAMPs to their signaling PRRs promotes synthesis and secretion of intracellular regulatory molecules such as cytokines that play a crucial role in induction of innate and adaptive immunity.

Signaling cell surface PRRs present on the cell surface binds to bacterial and fungal cell wall components and stimulate the transcription and translation of genes responsible for the formation of cytokines such as interleukin-1 (IL-1), tumor necrosis factor-alpha (TNF), interleukin-8 (IL-8), and interleukin-12 (IL-12) through the activation of a transcription factor NF-κB (nuclear factor-kappa B) with the help of adaptor protein MyD88 (Fig.-13-7). These cytokines trigger innate immune defenses such as inflammation, fever, and phagocytosis in order to provide an immediate response against the invading microorganism. Because cytokines such as IL-I, TNF IL-8, and IL-12 trigger an inflammatory response, they are often referred to as inflammatory cytokines.

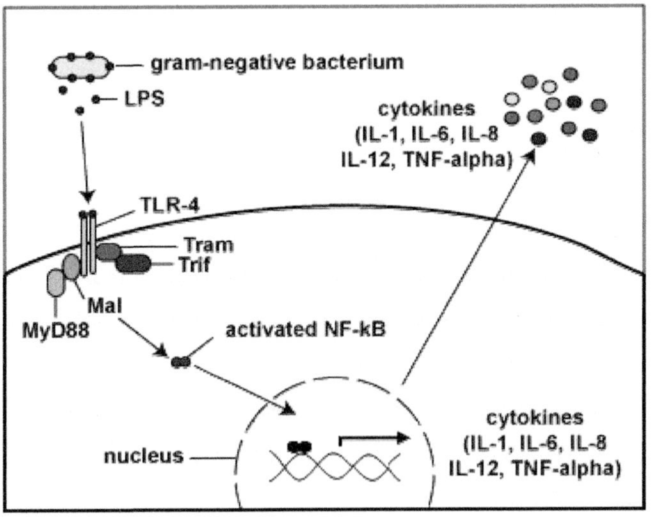

Fig.-13-7. Molecular mechanism of triggering signaling events through PRRs by PAMPs. The details is given in chapter 8 with respect to the signaling mechanism induced by heat shock proteins through the TLR-4 and CD14 (a type of PRRs present on the cell surface of phagocytic cell population).

However, those PRRs, which are found in the membranes of phagolysosomes such as TLR-3, TLR-7, TLR-8 and TLR-9 or soluble PRRs such as NODs (NOD-1 and NOD-2), CARD (caspase activating and recruitment domain) containing proteins (RIG-1 and MDA-5) recognize their respective PAMPs (Table 13-2) and induces cascade of biological reactions. Signaling PRRs found in the membranes of the endosomes (phagolysosomes) help in the degradation of pathogens. Most of the TLRs that bind to viral components trigger the synthesis of cytokines called interferons that block viral replication within infected host cells. Likewise, binding of the muramyl dipetides to NOD-1 or NOD-2 leads to the activation of genes coding for inflammatory cytokines such as IL-1, TNF-α, IL-8, and IL-12 in a manner similar to the cell surface TLRs. CARD (caspase activating and recruitment domain)-containing proteins, such as RIG-1 (retinoic acid-inducible gene-1) and MDA-5 (melanoma differentiation-associated gene-5) bind to viral RNA molecules produced in viral-infected cells and trigger the synthesis of cytokines called interferons that block viral replication within infected host cells in a manner similar to the endosomal TLRs.

Another cell surface PRR is CD14 which is found on monocytes, macrophages, and neutrophils and promotes the binding of TLR-4 to LPS. Interaction of LPS with CD14 and TLR-4 leads to an elevated synthesis and secretion of proinflammatory cytokines such as IL-1, IL-6, IL-8, TNF-α, and platelet-activating factor (PAF). These cytokines then bind to cytokine receptors on target cells and initiate inflammation and thus activate both the complement pathways resulting into the lysis of bacterial cell (Fig.-13-8).

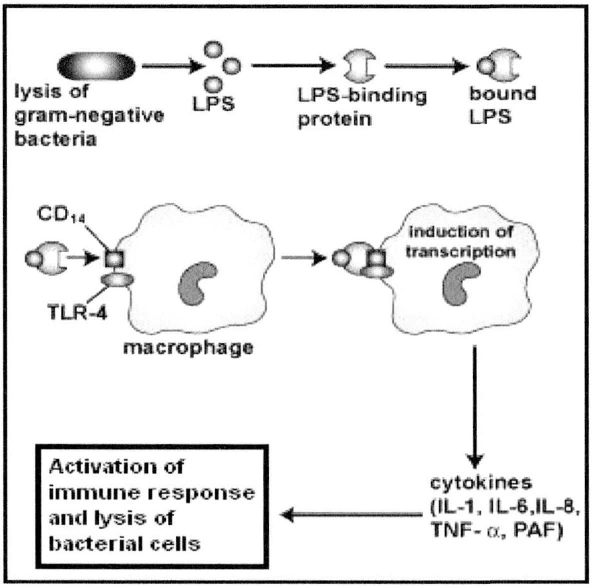

Fig.-13-8. Signaling through the PRRs (TLR4 and CD14) and its biological consequences. The lysis of gram-negative bacteria causes them to release lipopolysaccharide (LPS; endotoxin) from the outer membrane of their cell wall. The LPS binds to a LPS-binding protein circulating in the blood and this complex, in turn, binds to a receptor molecule (CD14) found on the surface macrophages and stimulates the macrophage to release various defense regulatory molecules called cytokines, including IL-1, IL-6, IL-8, TNF, and PAF. The cytokines then bind to cytokine receptors on target cells and initiate inflammation and immune cell activation which results into bacterial cell lysis.

In addition to the PRRs found on or within cells, there are also secreted PRRs. These PRRs bind to microbial cell walls and enable them to activate the complement pathways, as well as by phagocytes. For example, mannan-binding lectin (also known as mannan-binding protein) is synthesized by the liver and released into the bloodstream as part of the acute phase response. Here it can bind to the carbohydrates present on bacteria, yeast, some viruses, and some parasites and activates the lectin complement pathway. As a result the production of a variety of activated complement proteins that are able to trigger inflammation, chemotactically attract phagocytes to the infection site are induced. It also promote the attachment of antigens to phagocytes via enhanced attachment or opsonization, and causes lysis of gram-negative bacteria and infected or transformed human cells.

CD4^{+} T cell Mediated Responses against Bacterial Pathogens

CD4^{+} T cell mediated responses act upon the following bacterial infections:

- Diphtheria,
- Tuberculosis,
- Lyme Disease

The above infections lead to massive tissue destruction and require a cell mediated DTH type response.

Diphtheria is caused by a gram-positive rod-like organism called Corynebacterium diphtheriae. This organism was found to causing diphtheria in guinea pigs and rabbits, as observed by Klebs and Loeffler in 1883-84. In this disease, autopsies on the infected animals revealed that, while bacterial growth was confined to the site of inoculation, there was widespread damage to a variety of organs including the heart, liver, and kidney. A toxic substance was recognized to cause the neurologic and cardiologic manifestations. The antigenicity of the toxoid produced by this organism (C. diphtheriae) was treated with heat and formalin exposure resulting in the destruction of antigenicity.

In humans, *C. diphtheriae* causes little tissue damage except for a mild inflammatory reaction in the nasopharyngeal tract as the organism remains in the superficial layers of the respiratory mucosa. The virulence of the organism is due completely to its potent exotoxin, which causes the destruction of the underlying tissue resulting in the formation of a tough fibrinous membrane (pseudo membrane) composed of fibrin, while blood cell and dead respiratory epithelial cells. The exotoxin may cause congestive heart failure (due to local inflammation) and mild weakness to complete paralysis (neurologic damage). The exotoxin produced by C. diphtheriae is extremely potent, as a single molecule can kill a cell. There are disulfide-linked chains in diphtheria exotoxin- a binding chain and a toxin chain. Ganglioside receptors on susceptible cells are interacted with binding chain facilitating internalization of the exotoxin. The toxin chain exerts an inhibitory effect on protein synthesis. The exotoxin can be turned into nontoxic by removing the binding chain. An immunotoxin can be prepared by replacing the binding chain with a monoclonal antibody specific for a tumor cell. The toxin chain can thus be targeted to local infection as to tumor cells. A diphtheria toxoid prepared by treating diphtheria toxin with formaldehyde, tetanus toxoid and inactivated *Bordetella pertussis.* A combined vaccine prepared of above, renders immunization by binding to the toxin and neutralizing its activity- given to children beginning at 6-8 weeks of age followed by booster doss at 10 year intervals to maintain antitoxin levels within the protective range.

Tuberculosis is caused by *Mycobacterium tuberculosis,* an organism which spreads easily and pulmonary infection usually results from inhalation of small droplets of respiratory secretions containing a few bacilli. The inhaled bacilli are ingested by alveolar macrophages and are able to survive and multiply intracellularly by inhibiting formation of *phagolysosomes*. When the infected macrophages lyse, as they eventually do, large numbers of bacilli are released. Within 2-6 weeks after infection $CD4^+$ T cells are activated inducing the infiltration of large numbers of activated macrophages. These cells wall off the organism inside a granulomatus lesion called a "tubercle". As a result of this formation of tubercles, macrophages are activated resulting in concentrated release of lytic enzymes. The nearby healthy cells are destroyed by these lytic enzymes, resulting in circular regions of necrotic tissue, which finally form a lesion with a cheese like consistency. Cytokines produced by $CD4^+$ T cells (T_H1 subset) activate macrophages, which either kill the bacilli or inhibit their multiplication. The cheese like (caseous) lesions on healing appears as calcified and readily visible on x-rays. Thus, $CD4^+$ T cells mount control and later protect the individuals infected with M. tuberculosis.

However, in some individuals a different pattern of tuberculosis is seen, in which years after primary infection, accumulation of large concentrations of mycobacterial antigens within "tubercles" leads to extensive and continual chronic $CD4^+$ T cell activation and ensuing macrophage activation. As a result of large concentrations of mycobacterial antigens and subsequent release of lytic enzymes, the necrotic lesions are liquefied creating a rich medium that enables the tubercle bacilli to proliferate extracellularly. The lesions finally rupture and the bacilli disseminate in the lung and/or are spread through the blood and lymphatic vessels to the pleural cavity, bone, urogenital system, meninges, peritoneum and skin. The disease is referred to in reference to the site infected- such as bone TB, pleural effusion, urogenital TB and so on.

Treatment consists of chemotherapy by drugs- isoniasid, rifampin, pyrazinamide, ethambutol and streptomycin in combination (of two or more drugs) for a specified period. Failure to continue therapy long before the recommended period may not eliminate organisms and a multidrug-resistant strain of organisms may develop. Such individuals are potentially dangerous to population as carriers of tuberculosis and compromise efforts to contain the spread of the disease. However, efforts are on to develop vaccine against multidrug resistant strains of M. *tuberculosis*. BCG vaccine is in extensive use as immunization against tuberculosis.

Lyme disease is caused by a new species of gram-negative spirochete called *Borrelia burgdorferi* inhabiting heavily wooded areas and a close geographic clustering of infected individuals, invading during summer months between June and September. The incidence of bite by a tick at the site of rash characteristic in appearance was noticed first in 1977. The tick bite was examined by Willy Burgdorfer, a noted authority on tick-borne diseases, and the bacterium was named after him as Borrelia burgdorferi.

The tick takes a blood meal, so it enters the blood stream. At the site of bite, a rash appear indicating intrusion of the spirochete and then proceeding to the kidneys, spleen, liver, cerebrospinal fluid and brain. The rash appears as a bulls-eye 10-15 cm in diameter. This is the superficial symptom of Lyme disease. Some 80% of individuals with Lyme disease develop arthritic symptoms, ranging from joint pain to chronic joint destruction. Some 60% individuals develop neurologic symptoms- headaches, meningitis and encephalitis. Therapy includes broad spectrum antibiotics like penicillin and tetracycline. Antibodies do not have any fruitful effect on Lyme disease. A vaccine based on outer-surface protein (OspA) is an approved protector against Lyme disease. The vaccine was developed in studies of SCID mice which unlike normal mice are susceptible to the disease. Monoclonal antibodies made in normal mice, to the outer-surface proteins protected SCID mice from disease, whereas monoclonal antibodies to the flagellar antigen did not.

Protozoan Diseases

Protozoans are unicellular eukaryotic organisms causing infection. Protozoans have life-cycle stages in which they are free within the bloodstream and infect human beings. Several serious diseases caused by these bloodstream organisms are amoebiasis, Chagas disease, African sleeping sickness, malaria, Leishmaniasis and toxoplasmosis. These diseases are called blood-borne diseases, since caused by protozoans which have their life-cycle stages in the bloodstream. Blood is the favorable environment for the intracellular growth of protozoans.

Malaria is caused by various species of the genus Plasmodium of which P. falciparum is the most virulent and prevalent. Female Anopheles mosquitoes which feed on blood meals serve as the vector for Plasmodium and part of the parasite life cycle takes place within the mosquito. Male Anopheles mosquitoes, since feed on plants do not serve as vector for Plasmodium. That is male Anopheles do not transmit Plasmodium. When an infected mosquito takes a blood meal on human being sporozoites (one of the Plasmodium stages) are introduced into the bloodstream of that human being. It is from thence that the malarial (the name given to infection caused by *P. species*) infection sets inside the human being. Within half an hour of introduction of P. sporosites into the bloodstream, they disappear from the blood and migrate to the liver, where they infect hepatocytes. The long slender cells covered by a 45 kDa protein (called circumsporosoite antigen; CS) helps adhesion of sporozoites to hepatocytes.

In about a week, the sporosites multiply within the liver and through a complex series of transformation release merozoites. The rate of merozoites released by a single sporozoites ranges from 5,000 to 10,000 in a liver hepatocyte. The released merozoites then infect red blood cells, initiating the symptoms of malaria-recurrent chills, fever and sweating. The symptoms attain it peak about every 48 hours in coherence with the replication and differentiation of sporosites undertaking their life-cycle in the red blood cells, which rupture and produce more merozoites. Some of the merozoites differentiate into male and female gametocytes, which may be ingested by a female Anopheles mosquito during a blood meal. This is how the mosquito turns to be an infected mosquito. The male and female gametocytes differentiate into gametes that fuse to form a zygote, which multiplies and differentiates into sporosites within the salivary gland of the mosquito. The mosquito is now ready to infect human beings by blood meal on them. In an infected individual the large number of merozoites formed can block capillaries causing intense headaches, renal failure, heart failure, or cerebral damage.

Malaria affects individuals in endemic regions, for the immune responsiveness in these areas is poor. The intracellular phases leading to the maturation of *Plasmodium* i.e. from sporosites to merozoites to gametocytes manifest continual changes to weaken the immune response and being shielded from antibodies attack. However, drug therapy and vaccines have aided to combat against this menacing disease.

African Sleeping Sickness is caused by the bite of the tsetse fly carrying Trypanosoma species in African origin. This debilitating disease is also a blood-borne disease, since in the bloodstream a *Trypanosome* differentiates into a long slender form that continues to divide every 4-6 hours proceeding infects the CNS causing meningoencephalitis and eventually the loss of consciousness.

This parasite is met by humoral antibody response after entering into the bloodstream, whereby the trypanosomal surface is covered by variant surface glycoprotein (VSG) causing death to the parasites. However, a very small number of trypansomes bearing antigenically different surface escapes the VSG mediated lysis. These surviving trypanosomes begin to proliferate in the bloodstream challenging the humoral antibody response by devising antigenic shift.

This is the stage of parasitemia. Trypanosomes having different surface sites are peculiar as containing more than 1000 VSG genes in its genome clustered at multiple chromosomal sites. This is peculiar to Trypansomes to express a new gene (VSG) displacing the previous gene in the process of activation and thus challenging the immune response to a known VSG. This continual shift in the VSG of this parasite poses a threat to the host gripping him in the sickness for a long duration. Interestingly the continual shifts in epitopes displayed by the VSG make the development of a vaccine possibly difficult.

Leishmaniasis is caused by a unicellular protozoan that lives in the phagosomes of the macrophages and survives quite emphatically the IFN-g response. TH1 and TH2 type responses, however withstand the resistance of *Leishmania* by developing high levels of IL-4 early in combat to the parasite. Effective treatments and successful vaccines are being extensively tried against this parasite, with the exception that Leishmaniasis has a feeble record of occurrence.

Diseases Caused by Parasitic Worms (Helminthes)

Helminthes are large multicellular organisms the live in humans and are carried from individual to individual. However, all the helminths are not carried from infected to non-infected individuals. Although, helminthes are parasitic in nature, the immune response is often very poor because of the mode of infection by these parasites.

The common helminthes and their expressions are:

Parasitic worm	*Disease symptoms*	*Prognosis*
Ascaris	(round debilitating disease worm in small intestine)	responsive to drugs (albendazole etc.)
Schistosoma	(round debilitating to fatal and tape worm of (schistosomiasis) animals)	Infectious larvae are released from snails. These larvae called cercariae, in contact to human skin, shed their tails and are transformed into schistosomules, and affect lungs, liver, urinary bladder. Number of these parasites is increased by production of egg (mating of male and female worms). By enveloping themselves in glycolipid and glycoprotein coat derived from the host, these Schistosoma species escape the host defense mechanism. These are eliminated by ADCC, aided by drugs. Vaccines are underway.

Emerging and Re-emerging Pathogens Causing Diseases

Diseases which once established to have been well controlled and contained are making their reappearance. These diseases are called re-emerging diseases caused by same species of pathogens viz. tuberculosis, and diphtheria. Tuberculosis has made its appearance besides general spread through HIV patients. As HIV is on an alarming increase, the susceptibility of AIDS patients to tuberculosis cannot be underestimated. WHO has predicted that one billion more people will become infected beginning from 1998 to 2020 by tuberculosis and over 70 million will die from this disease if proven preventive measures are not adopted. Similarly immunization against diphtheria has been immensely augmented to bring the mortality graph down.

Included in the list of emerging diseases are some new pathogens viz.

- *Legionella pneumophilia* causing bacterial infection called Legionnaires' disease.
- Ebola virus causing a viral disease.

The characteristic symptom caused by Ebola is hemorrhagic fever claiming more than 50% of those infected. Interestingly the disease is controllable by isolation of infected person. Virulent pneumonia is another disease caused by Legionella pneumophilia. This bacterium is associated with cool and dump areas.

The large air-conditioning systems producing aerosols are the major sources of this bacterium. With the devising systems to control the spread of aerosols, success has been achieved upon the incidence of aerosols produced L. pneumophilic disease. Among the factors leading to the emergence of pathogens of emerging diseases are over the increased ability to surmount mounting barriers by hop steps and the density of population, the latter of which is a devastating health hazard.

CHAPTER - 14

VACCINES

Vaccines provide immunity against life-threatening diseases to a maximum degree unless the individual's vital system is disturbed by any acquired disease or physiologic deficiency. Eradication of naturally acquired smallpox, since 1977, is an irrevocable fact to prove that vaccines do play a major role in the prevention of diseases. The noticeable decline in diseases like diphtheria, measles, mumps, pertussis (whooping cough), rubella (German measles), poliomyelitis, and tetanus etc. is enough to confirm the role of prevention of diseases by vaccines. However besides above, there are diseases for which vaccines have yet to be made, because of the mortality and morbidity rates claimed by them worldwide.The most important aspect latent in vaccines, so far as they prevent diseases, is their cost and delivery to the entire needed areas. Another important aspect of vaccines is their efficacy for human usage, if prepared by a vaccine candidate that was successful in laboratory or animal studies. This is a negative aspect of vaccines.

A further negative aspect of vaccines is the nature of some vaccines, which cause side effects or worsen the disease, they were meant to prevent. This fact holds an accessible impact on vaccines prepared by live viruses, which pose a special threat to those human beings with primary or acquired immunodeficiency. This aspect requiresa stringent taste of vaccines meant for healthy persons at large. Keeping in view the above aspects of vaccine preparation, a new and promising branch called genetic engineering has evolved to maximize the immune response, to sustain the optimum quality of vaccines for the prevention of various existing and emerging diseases.

Induction of Immunity

Immunity against infectious microorganisms can be obtained by two methods-natural and artificial. The methods imply processes. Immunity obtained by natural processes is also called passive immunity and immunity obtained by artificial processes is called active immunity. In obtaining immunity by natural processes (passive immunity), antibodies from mother humans or animals are transferred to fetus. In obtaining immunity by artificial processes (active immunity), microbial pathogens are introduced into humans, which do not cause diseases but rather prevent the development of disease against which the humans are inoculated.

Passive Immunization

When preformed antibodies are transferred from mother to fetus, immunity is developed in the fetus through natural processes. Maternal antibodies to diphtheria, tetanus, streptococci, rubeola, rubella, mumps, and poliovirus all provide passive (natural) immunity to the developing fetus. Antibodies present in the colostrum (first milk of a woman who gives birth) and milk of the mother provide natural immunity to the developing fetus. Since maternal antibodies are transferred across the placenta to the developing fetus, there are certain conditions wherein immunity cannot be achieved and induced in the fetus/infant,

- In case the mother is suffering from diphtheria, tetanus or botulism, antibodies may ameliorate or help suppress the effects of toxin present in the mother.
- Provided to mother is having congenital or acquired B cell defects alone or together with other immunodeficiencies, synthesis of antibody will be deficient in her and as a result, passive immunization will be totally lacking in the infant.
- In case of a child having leukemia, if exposed to varicella or measles, passive immunization (antibodies transferred across the placenta) fails to defend him against the complications arising out of such exposures.

Passive immunization is also induced by previous infection by the organism. Table 14-1 shows common agents used for passive immunization. As stated above, passive immunization or immunization induced by natural processes always implies the transfer of antibodies across the placenta to the fetus and previous infection by the organism. The main point with passive immunization to note is that it does not provide any permanent protection for which memory cells should be developed i.e. activation of the immune system.

Table 14-1: Common agents used for passive immunization

Immunizing Agent	*Disease*
Horse antivenin	Black widow spider bite
Horse antitoxin	Botulism
Horse antitoxin	Diphtheria
Pooled human immune - γ globulin	Hepatitis A and B
Pooled human immune - γ globulin	Measles
Pooled human immune - γ globulin	Rabies
Horse antivenin	Snake bites
Pooled human immune -γ-globulin or horse antitoxin	Tetanus

Passive immunization, despite its shortcomings has yet been found a recommended mode of treatment in certain diseases e.g. respiratory failure of children caused by respiratory syncytial virus (RSV). In treating respiratory failure of children, injection of complete antibody derived from mouse has been found most effective mode of treatment. Antibody is prepared in the body of mice, and then humanized by splicing the constant regions of human IgG to the mouse variable regions. The Preparation of monoclonal antibodies is the latest developed technique in passive immunization. However, passive immunization induced by preformed antibody i.e. produced in another species is not always free of hazardous effect as it generates antibody in the recipient. For example, antibodies produced in a horse on injection in human, can develop isotype response in the recipient- IgE antibody. Immune complexes of this IgE bound to the passively administered antibody can mediate systemic mast cell degranulation, resulting in systemic anaphylaxis. Transient anti-allotype response to the human immunoglobulin may develop with the use of human gamma globulin. Substantially speaking, protective immunity and immunologic memory cannot be induced by passive immunization.

Active Immunization

Immunity capable of developing immunologic memory in the recipient is also known as active immunizations or immunity developed through artificial processes- natural infection or by inoculation with microbial agents or components derived from them. Active immunization induces clonal selection of lymphocytes and results in the formation of memory cells. This type of immunity since latent in the recipient provides permanent immunity against diseases, without causing the disease specific to the microbial agents or pathogens. In this sense, immunity achieved by induction of microbial pathogens is active immunization, as it plays an active role in proliferation of antigen-reactive T and B cells followed by the development of memory cells. Since memory cells are involved in active immunization, this is highly effective in prevention of diseases. Since immunity acquired by artificial processes (natural infection with a microorganism) is highly effective in the prevention of diseases, vaccines have come along way. So far as prevention of diseases is concerned, vaccination starts in childhood, from the age of 2 months. The vaccines immunize a child against the following diseases:

Name of the Vaccine	*Name of the disease prevented by the Vaccine*
DTaP	Diphtheria, Tetanus, and Pertussis
OPV and IPV	Polio
MMR	Measles-mumps-rubella
Hib	Haemophilus influenza
Var (Varicella zoster)	Chickenpox
Rv (Rotavirus)	Infant diarrhoea

In order to prevent the occurrence of diseases or adequate immunity, multiple boosters (repeated inoculations) are administered at appropriately timed intervals.

Some vaccines e.g. poliovirus is administered at a frequency of time to ensure immunization with all the strains of a particular virus, corresponding to the most susceptible ages of occurrence.

Vaccines have been recommended for various risk groups of human beings e.g., military personnel are immunized against anthrax spores widely used by terrorists in biological warfare; international travelers are immunized by vaccines to prevent the spread of endemic diseases viz. cholera, yellow fever, plague, typhoid, hepatitis, typhus and polio. However in a given span of time, the entire population is never likely to be immunized. And in this case, if the majority of population is immunized against an infectious agent, the chance of contacting infection by a susceptible unimmunized individual is very low or not likely to be infected by a specific infectious agent. This phenomenon is known as herd immunity. Thus, in case the majority of population is not vaccinated against an infectious agent, the herd immunity population will increase, resulting in appearance of that particular infectious disease against which the entire population or majority of population could not be immunized. Probably this leads to the immunization at regular intervals of time viz. measles is administered one at 12-15 months of age and the other at early school (nursery / kindergarten). And inoculation is recommended for combined measles-mumps-rubella vaccine. Herd immunity is always corresponding to the quantum of population immunized. If one third of all infants are unvaccinated and about half of all children under the age of 2 are not vaccinated, herd immunity will decrease means the population of susceptible children to infection will increase. This state of affairs will be responsible for a rise in the endemic in various regions of a country. Therefore, vaccination is imperative at an extensive level to yield a lower herd immunity factor.

Development of Vaccines

In developing a vaccine, the branch of immune system, humoral or cell mediated- to be activated should be the prime consideration. Secondly, in developing a vaccine, development of immunologic memory should be well considered. This is because immunization with a vaccine fails to provide protection to the recipient after its primary response to an infectious agent subsides.

Effective protection against infection is always relevant to the development of memory cells. Repeated immunization provides the development of memory cells, so that memory cells remain activated to combat the infection during and after the incubation period of an infection. Diseases of longer incubation period, for example, poliovirus (incubation period 3 days to affect the central nervous system) require memory response to maximize the production of high levels of serum antibody. That is why the vaccine for polio is designed to induce high levels of immunologic memory. Polio vaccine reaches its peak level within 2 weeks and produces maximum memory cells at 6 months and persisting for years. Thus, with repeated immunization against poliovirus render the recipient to effectively defend against this dreadful infection.

Now a discussion will follow about the types of vaccines currently used in humans in reference to their ability to activate humoral or cell mediated response to infection besides the development of immunologic memory.

Types of Vaccines

Three types of vaccines are currently used in humans:

- Live but attenuated (avirulent) microorganisms,
- Inactivated (killed) microorganisms, and
- Purified macromolecules.

Vaccination by live but attenuated (avirulent) microorganisms: By growing a pathogenic bacterium or virus for prolonged period under abnormal culture conditions, microorganisms can be attenuated for vaccination. In the attenuated state, the microorganisms lose their ability to cause significant disease but retain their capacity for transient growth within an inoculated host. The attenuated microorganisms are less capable of growth in the natural host. BCG vaccine (Mycobacterium bovis) developed on a medium containing increasing concentrations of bile, Sabin polio vaccine attenuated by growth in monkey kidney epithelial cells, and rubella vaccine attenuated by growth in duck embryo cells and later in human cell lines is examples of attenuated viral or bacterial vaccines.

Advantages of attenuated vaccines: Attenuated vaccines because of their capacity for transient growth within the host provide prolonged immune-system exposure with only a single immunization and induce a cell-mediated response (memory cells). The advantage of attenuated vaccines is more noticeable in case of the Sabin polio vaccine. The three strains of poliovirus usually administered orally in sugar liquid, colonize the intestines and induce production of secretory IgA, which serves as an important defense against naturally acquired poliovirus, besides induction of IgM and IgG classes of antibody. However, for an optimum immunization against poliovirus at least three immunizations are needed to achieve defense against all three strains, so boosters of this vaccine is recommended.

Disadvantages of attenuated vaccines: Where advantage of avirulent vaccination is attractive, the disadvantages are considerable. The following are the major disadvantages of attenuated vaccines.

Reversion to virulent form: The development of a paralytic disease, though one case in four million doses of vaccine, portends reversion of the disease in a virulent form. The reversion implies that pathogenic forms of the virus are being passed by a few vaccinated individuals and can find their way into the supply in areas with rigorous sanitation standards or faulty wastewater treatment (recycling).

Fusion of contaminants in attenuated vaccines: The risk of contamination of vaccines during their attenuation is another disadvantage with attenuated vaccines. In 1960, it was observed that the oncogenic virus SV40 had contaminated some monkey kidney cultures used in the production of the Sabin vaccine. This probability requires stringent testing of the polio vaccine to eliminate SV40 from the oral polio vaccine.

Postvaccine complications: The development of post vaccine encephalitis in a small percentage of the recipients of measles vaccine, and rejection of measles vaccine are also considerable. Encephalitis when acquired naturally is different in manifestations than as a post-vaccine manifestation in an individual vaccinated for measles. The percentage of such a development is also very low.

Further, due to passive immunization (by natural antibodies), children given the measles vaccine at 4-6 months may develop common endemic disorders such as diarrhoea, pneumonia, and parasitic diseases. This is because of rejection of measles vaccine being an immunosuppressant at this very early age. The Measles vaccine after 9 months of age does not show any such rejection by host. This implies that measles vaccine, an attenuated strain of the virus called the Edmonston-Zagreb strain, in the presence of maternal antibodies get associated with potent immunosuppressive effects placing vaccinated infants at risk for infection by other endemic pathogens that unvaccinated infants can resist. Infant diarrhoea is a serious cause of mortality accounting for about 2 million deaths per annum, caused by an infection with rotavirus, an intestinal pathogen. In a series of experiments, it was observed that maternally acquired antibodies were highly effective in preventing diarrhoea in infants.

As there are four serotypes of rotavirus, the quadrivalent rhesus rotavirus vaccine has been recommended for infant diarrhoea (WHO). This vaccine is backed by research in genetic engineering techniques wherein to counteract the complications arising in inoculations and providing immunization, implies that the safest way to attenuate a virus irreversibly is to selectively remove genes that are necessary for virulence.

Vaccination by inactivated (killed) microorganisms: Unlike attenuated vaccines wherein the microorganism is capable of transient growth (replication) within the host, in producing vaccines by an inactivated microorganism, the pathogen is inactivated (killed) by heat or by chemical means to render it incapable of replication within the host. During inactivation, the structure of epitopes on surface antigens is well maintained. Table 14-2 compares the properties of attenuated and inactivated vaccines.

In the process of inactivation by heat, the epitopes are altered significantly. In other words, the protein structure of epitopes is denatured. This is a negative side of the production of vaccine by heat treatment.

In the process of inactivation of a pathogen by chemical means formaldehyde or various alkylating agents are used. The Salk polio vaccine and the pertussis (whooping cough) vaccine are produced by inactivating the pathogens in formaldehyde. This process of chemical inactivation of microorganisms has been found satisfactory in repeated doses of immunization, whereby an effective humoral antibody response is generated within the host.

Table 14-2: Comparison of attenuated and inactivated vaccines

Attenuated vaccine	*Inactivated vaccine*
Virulent pathogen grown under adverse culture condition	Virulent pathogen is activated by chemicals or irradiation with γ-rays
Usually required a single booster dose	Require multiple booster dose
Less stable produces humoral as well as cell-mediated immunity	More stable Mostly humoral immunity is provoked
May revert to virulent form	Cannot revert to a virulent form

On the side of advantage, the vaccines are produced by inactivating the microorganisms, since repeated doses are employed, the immune status of the host is well maintained. On the side of disadvantage, the vaccines produced by the process of inactivation, memory cells are not formed and at times for want of killing in a chemical media paralytic polio and encephalitis type reactions have been reported.

Vaccines consisting of purified macromolecules derived from pathogen: Attenuated vaccines undergo transient growth in the host/recipient and so strengthen the immune response by generation of memory cells without needing repetition (additional booster doses), though pose a risk of reversion to a pathogenic state. Reversion to a pathogenic state is not possible with killed microorganisms (inactivated). Vaccines consisting of purified macromolecules are as improvement over the above two types of vaccines. The following are the three types of such vaccines in use;

Vaccines consisting of purified capsular polysaccharides (polysaccharide vaccines): Some pathogenic bacteria have a virulent hydrophilic polysaccharide capsule, exerting antiphagocytic reaction. Once this capsule is coated with antibodies and/ or complements, the macrophages and neutrophils become dominantly phagocytic and phagocytose such bacteria. This concept has led to the production of polysaccharide vaccines. The vaccines induce formation of opsonizing antibodies suitable for administration to high risk groups such as infants, immune-suppressed individuals and the elderly. Polysaccharide vaccines are available for *Streptococcus pneumoniae, Neisseria meningitides.*

Polysaccharide vaccines fail to activate T_H cells and form memory cells (IgA-specific memory B cells). However, if polysaccharide protein is conjugated to a protein carrier, it becomes more immunogenic than the polysaccharide alone activates T_H cells, but it cannot induce memory T cells specific for the pathogen.

Vaccines consisting of inactivated exotoxin (toxoid vaccines): Bacterial pathogens causing diphtheria and tetanus produce exotoxins and require toxins to be neutralized. Toxoid vaccines are produced by purifying the bacterial exotoxin and then inactivating the toxin with formaldehyde to form a toxoid. Administration of such a toxoid vaccine induces antitoxin antibodies by binding to the toxin and neutralizing its effects. Care is taken to achieve detoxification without excessive modification of the epitope structure. Large quantities of exotoxin are obtained by cloning the exotoxin genes and then expressing them in easily grown host cells. This is how exotoxin is purified and subsequently inactivated for administration.

Recombinant surface antigens are produced by cloning the genes encoding surface antigens from viral, bacterial and protozoan pathogens into bacterial, yeast, insect or mammalian expression systems. The expressed antigens are used for vaccine development. The hepatitis B vaccine for human use is an example of such recombinant antigen vaccine. The hepatitis B virus (HbsAg) antigen is first cloned in yeast cells (grown in huge fermenters) resulting in intracellular accumulation in the cells. The yeast cells are harvested and disrupted by high pressure, releasing the recombinant HBsAg, which is then purified by conventional biochemical techniques to induce the production of antibodies.

Vaccines consisting of recombinant vectors (Recombinant Vector Vaccines): Recombinant Vector vaccines consist of attenuated organism serving as vector, growing within the recipient and expressing the gene product of the pathogen. Thus, organisms are used for vector vaccines as attenuated strains of viruses (of canarypox virus, attenuated poliovirus, adenoviruses, Salmonella, M. bovis, vaccinia virus). These vector vaccines carry enormous genes to serve as a potent immunogen in an inoculated host. The success of vector vaccines depends on increased production of secondary IgA at mucous membrane surface with the expression of desired epitopes. Recombinant vectors can be engineered to carry multiple genes from infectious microorganisms. Figure 14-1 shows the production of vaccinia vector vaccine (used to eradicate smallpox).

Vaccines that allow selective activation of either the humoral or the cell-mediated branch of the immune system have come up recently in vaccine strategy.

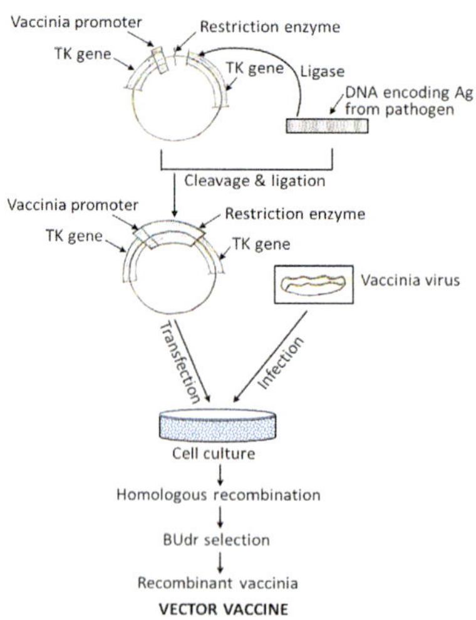

Fig.-14-1. Procedure for generating vaccinia vector vaccines.

Specific Immunity Dependent Vaccines

Specific immunity dependent vaccines are those vaccines, which may permit selective activation of either the humoral or the cell-mediated branch of the immune system. These vaccines are:

- DNA vaccines
- Synthetic Peptide vaccines
- Multivalent Subunit vaccines

DNA vaccines: Plasmid DNA encoding antigenic proteins is injected directly into the muscle of the recipient. The DNA is taken up by muscle cells and the encoded protein antigen is expressed, leading to both humoral antibody response and a cell-mediated response. The DNA taken up by the muscle cells either integrates into the chromosomal DNA maintained for long period in an episomal form. The viral antigen is expressed in the area that takes up the plasmid DNA by muscle cells and dendritic cells. Dendritic cells play an important role to the development of antigenic responses to DNA vaccines. Fig. 14-2 is illustrative to the mode of immune responses induced by DNA vaccines.

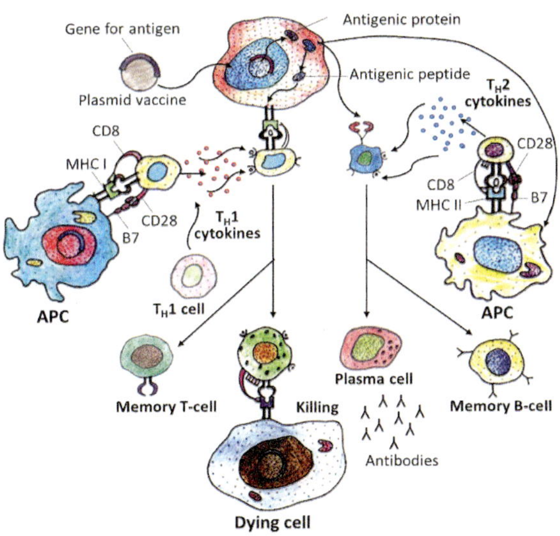

Fig.-14-2. Overview of the effect of DNA vaccines. DNA vaccines can induce both arm of immunity; humoral and cell mediated immune responses.

DNA vaccines for their following plus points have come over existing vaccines in laboratories.

- Immune response induced by DNA vaccines is directed to the antigen exactly as it is expressed by the pathogen. This means that encoded protein is expressed in the host in its natural form i.e., neither denatured nor modified. Because of prolonged expression of the antigen, significant immunological memory is intensively generated.

- The technique of delivery/administration of DNA vaccines are to coat microscopic gold beads with the plasmid DNA and deliver the coated particles through the skin into the underlying muscle with a gene gun. Thus, the complexity of delivery of DNA vaccines is eased. Besides, refrigeration is not required in the case of DNA vaccines nor needles and syringes.
- In manufacturing DNA vaccines, the same plasmid vector can be custom tailored to make a variety of proteins each encoding an antigen from a different pathogen.

However, DNA vaccines are very useful and shortly they are to be used for human immunization, some shortcomings have yet to be removed for example pneumococcal and meningococcal infections cannot be handled by DNA vaccines, since protein antigens can be encoded. Pneumococcal and meningococcal infections require protective polysaccharide antigens. Further DNA vaccines cannot be given orally or as a nasal spray that are used for mucosal surfaces.

Synthetic Peptide Vaccines

Synthetic peptides that represent immunodominant T or B cell epitopes are being evaluated as vaccines for several diseases. Such epitopes can be identified by determining the dominant antibody in the sera of individuals who are convalescing from a disease and then testing various synthetic peptides for their ability to react with that antibody with a high affinity. Substantially speaking, synthetic peptides are extensively under experiment to determine their use in providing immunity from the standpoint of generation of memory cells.

Multivalent Subunit Vaccines

These types of vaccines present multiple copies of a given peptide or a mixture of peptides to the immune system. For a CTL response, the vaccine must be delivered intracellularly so that the peptides can be processed and presented together with class I MHC molecules. Multivalent submit vaccines are to come in to vogue provided experiments back their use for immunity (Fig. 14-3).

Recently, the potential to development of multivalent vaccines has been proposed for the treatment of cancer patients. By considering the molecular chaperone property of heat shock proteins (HSPs), it is possible to associate HSPs with many types of tumor-associated testis antigens (TSTAs) and other marker cell surface proteins specifically expressed by tumor cells to form HSPs-antigens complex. After administration, the HSP-antigens complex is taken up by APCs by endocytosis, the tumor antigens are processed and presented to the cell surface of APCs through MHC I, thereby inducing profound anti-tumor immunity in the host making CTLs able to recognize multiple epitopes present on tumor cells, and therefore, increasing the possibility of anti-tumor CTL response. However, this concept is not practicised yet but it paves a way to design and develop a vaccine with multivalent potential.

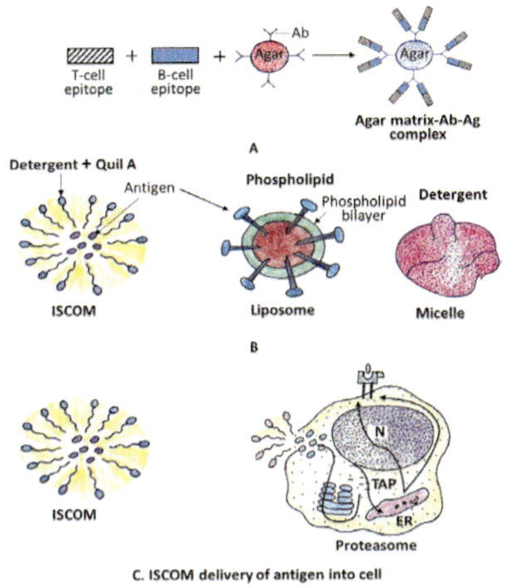

Fig.-14-3. Visualizing concept of the multivalent sub unit vaccines: (a) Agar matrix-Ag-Ag complex, (b) use of detergent with antigens and (c) ISCOM.

CHAPTER – 15

IMMUNODEFICIENCY DISEASES

An immune system that fails to protect the host from disease-causing agents or from malignant cells is called immunodeficiency. Immunodeficiency is of two types- primary and acquired. Acquired immunodeficien is also called secondary immunodeficiency. A genetic or developmental defect present at birth is called primary immunodeficiency. That is the deficiency in the immune system is present at birth. The immune system which acquires deficiency, as a result of exposure to disease-causing agents or malignant cells (a developmental factor) is called immunodeficient state of the host. A person with deficient immune system, acquired later in life is always at risk of death, on exposure to microorganisms, other than those which caused deficiency in the immune system. An understanding of cellular and molecular structure of host's immune system renders chances to induce improvement or treat the immunodeficiency disorders.

Since, deficiency in the immune system is caused by latent or innate genetic defects or developmental defects or acquired as a result of exposure to microorganisms (causing immune deficiency) this chapter deals with congenital defects and acquired immunodeficiency together with their treatment.

Primary Immunodeficiencies

A genetic or developmental defect present at birth is called congenital defect to cause immunodeficiency. A congenital defect causes failure of the host's immune system to adapt to the infection in a defensive way. The chief components of adaptive immunity are T and B cells (lymphoid and myeloid). Congenital defects are inherited hence present at birth. Primary immunodeficiencies are caused by congenital or inherited defects in the immune system. The number and type of immune system components involved, determine the consequences of primary immunodeficiency. A defect in the hematopoitic developmental scheme affects the entire immune system. In this type of defect, the leukocytes are not matured resulting in general failure of immunity to withstand infection by a variety of microorganisms. Defect in phagocytic function results in the increased susceptibility of the host to bacterial infection.

Defects in more highly differentiated compartments of the immune system results in the increased susceptibility of the host to respiratory and genitourinary infections. Table 15-1 shows diseases associated with genetic defects, causing impaired functions in the host's immune system. Figure 15-1 shows the overall cellular development in the immune system indicating sites of defects that give rise to congenital deficiencies or primary immunodeficiencies.

The chief congenital defects that cause immunodeficiencies are:

Lymphoid Dysfunction

- Lymphoid development (SCID)
- X-Linked defects – (WAS) – (XLA) - (XHM)
- Defect in the receptor for IFN-g
- Hypogammaglobulinemia / CVI
- Immunoglobulin deficiencies
- Thymic malfunction

Myeloid Dysfunction

- Reduced production of neutrophils during hematopoiesis
- Defective cytochrome (CGD)
- Chediak-Higashi Syndrome
- Leukocyte Adhesion Deficiency (LAD)
- Defects in X-chromosome

A. Lymphoid dysfunction related to primary immunodeficiency involves B cells, T cells or both of these lineages. The complications related with both of these cells arise from defects in developmental pathways within the first few years of life and may turn lethal.

For want of mature recirculating B cells, plasma cells and immunoglobulin, patients run the risk of recurrent bacterial infection, caused by Staphylococci, Streptococci and Pneumococci.

This happens because antibodies are deficient in production. For want of T cell deficiency, DTH responses and cell-mediated cytotoxicity both are affected, resulting in decrease in antibody levels.

Thus, defects in lymphoid lineage may result in defects in the mechanisms of gene rearrangement and other functions preventing normal B or T cell responses against specific agents. The specific defects and the inability of the host to respond to various agents are described below.

Table 15-1: Some immunodeficiency disease of humans

Immunodeficiency disease	*Specific defect*	*Impaired function*
Severe combined immunodeficiency	RAG1/RAG2 deficiency ADA and PNP deficiency JAK-3 and IL-2Rγ deficiency ZAP-70 deficiency	No TCR or Ig gene rearrangement Toxic metabolite in T and B cells Defective signal from IL-2, IL-4, IL-7, IL-9, IL-15 Defective signal from TCR
Bare lymphocyte syndrome	Defect in MHC class II gene promoter	No class II MHC molecules
Wiskott-Aldrich syndrome (WAS)	Cytoskeleton protein (CD43)	Defective T cells and platelets
Interferon gamma receptor	IFN-γ receptor defect	Impaired immunity to mycobacteria
DiGeorge syndrome	Thymic aplasia	T and B cell development
Ataxia telangiectasia	Defective cell-cycle kinase	Low IgA and IgE

Lymphoid development: The defects, which arise from defects in lymphoid development are called severe combined immunodeficiency (SCID) affecting either T cells or both T and B cells. The disorders caused by defective lymphoid development are:

- Lymphopenic immunologic deficiency, in which there is no immunoglobulin in the circulation, so the patients fail to mount immunity and suffer various acute infections characterized by diarrhoea, pneumonia, skin, mouth and throat infections.
- Failure to respond to stimulation by mitogens. With defective lymphoid development, the thymus does not develop, so the few circulating T cells in such patients do not respond to stimulation by mitogens as they cannot proliferate in response to antigens.
- Impaired function of IL-2 receptor. A depleted functioning of this receptor causes deficiency in receptors for all cytokines.
- Deficiency of the enzyme adenosine deaminase (ADA). An especially high activity of lymphocytes is due to ADA and its deficiency causes autosomal recessive disorders, in which the gene responsible is not encoded on the X or Y chromosome and only homozygote are affected.
- Defects in ZAP-70 tyrosine kinase (an important element in T cell signal transduction). This defect causes $CD4^+$ T cells nonfunctional.
- Bare-Lymphocyte syndrome. Without participation of genes that encode class II MHC molecules, the patient's lymphocytes cannot interact with T helper cells. A defective interaction between a 5' promoter sequence of the gene for the class II MHC molecule and a DNA binding protein necessary for gene transcription is called bare-lymphocyte syndrome. Viral infection is ascribed to this deficiency.

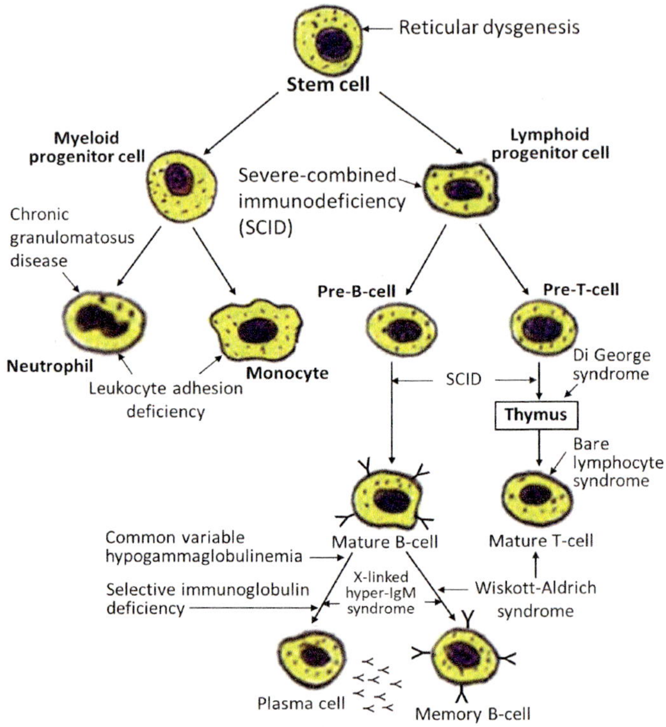

Fig.-15-1. Overview of congenital defect of impaired functioning for the immune system.

X-Linked Defects: Congenital immunodeficiency (primary immunodeficiency) related with X-linked defects are as follows:

Wiskoft-Aldrich Syndrome (WAS)

A Fatal bleeding (thrombocytopenia) and eczema usually appearing at a very early age in life are symptoms associated with this X-chromosome related defect leading to immunodeficiency due to CD43. A WAS patient also manifests defective responses to bacterial polysaccharides with growth in age.

Agammaglobulinemia (XLA)

This is another X chromosome related congenital defect which fails to mount specific antigens- especially bacteria. The deficiency is since related to the X chromosome called X-linked agammaglobulinemia (XLA). The defect is characterized by extremely low levels of IgG, and other IgS.

X-linked Hyper-IgM syndrome (XHM)

This is yet another X-linked immunoglobulin deficiency resulting from a defect in a T cell surface molecule manifesting deficiency of IgG, IgA and IgE and elevated level of IgM (normal IgM concentration is 1.5 mg/ml)- 10 mg/ml. The severity of deficiency in XHM patients is caused mainly by high counts of IgM-secreting plasma cells in peripheral blood and lymphoid tissue. Besides, XHM patients often have high levels of auto-antibodies to neutrophils, platelets and red blood cells, causing recurrent respiratory infections (carried by *Pneumocystis carinii*). XHM results from defects in CD40L that preclude normal maturation of B cells **(Fig. 15-2).**

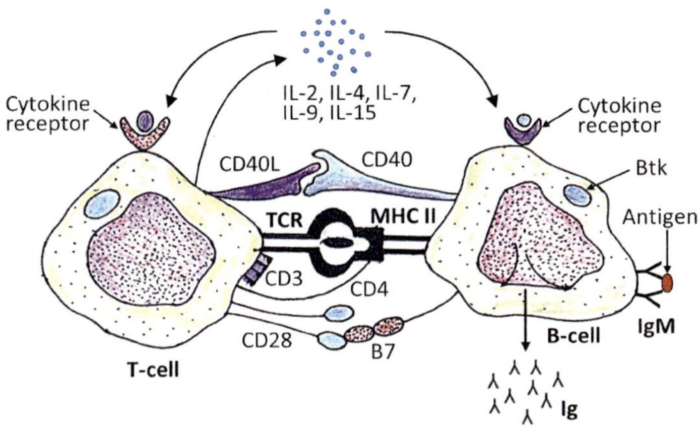

Fig.-15-2. Defect in T cell activation and signaling pathways leads to severe immunodeficiency.

Defect in the receptor for IFN-g

This is a mixed-cell category immunodeficiency involving a defect in the receptor for interferon gamma (IFN-γ). The deficiency was observed in patients of tuberculosis and leprosy. The defect is categorically inherited and the susceptibility to infection with intracellular bacteria (mycobacteria) is selective and those who survive the defect are not usually susceptible to other agents including other intracellular bacteria. Here the role of IFN-g mounts to dominance in protecting the host from infection with mycobacteria.

Hypogammaglobulinemia/CVI

This defect is characterized by low levels of most immunoglobulin isotypes (hypogammaglobulinemia) and recurrent infections. Common Variable Immunodeficiency (CVI) characterized by a profound decrease in the number of antibody producing plasma cells, is a defect often termed similar to hypogammaglobulinemia.

Immunoglobulin deficiencies

There are two specific defects related to IgA deficiency and IgE deficiency which group under this Ig deficienc.

Lack of secreted IgA on mucosal surfaces results in recurrent respiratory and genitourinary tract infections. In addition, intestinal malabsorption, allergic diseases and autoimmune disorders (a state wherein the system loses its sense of self and begins to attack host cells and tissues) are also associated with low IgA levels) Lack of secreted IgE on mucosal surface, results in a deficiency in maintaining balance (ataxia) and apparent broken capillaries (telangiectasia). However, the relationship of these two IgS with innate immune deficiencies is extensively under research to reach a stage of establishment.

Thymic malfunction

This immunodeficiency is caused by failure of the thymus to undergo normal development resulting in profound effects on T cell function, all populations of T cells including helper, cytolytic and regulatory T cells. Thymic malfunction manifests itself in affections from viruses and fungi. Substantially speaking, this defect is a congenital developmental defect associated with the deletion in the embryo of a region on chromosome 22. The symptoms are characteristic dysplasia of ears and mouth and abnormally long distance between the eyes (facial abnormalities), hypoparathyroidism and heart disease. These symptoms are known as DiGeorge syndrome or thymic aplasia. In this syndrome, T cell defects are grossly involved, so T cell responses are nil. Though the chances of survival of a DiGeogre patient is very weak, thymic transplantation has been found to yield a transient span in survival.

DiGeorge syndrome is of embryonic origin and similar to it is another syndrome known as Nezelof syndrome (hypoplasia), a rare inherited disorder. This congenital defect has varied manifestations and is difficult to be diagnosed. T cells in Nezelof patients are non-functional and B cells are deficient as secondary to the T cell defects. The characteristic symptoms are chronic diarrhoea, viral and fungal infections and a general failure to thrive.

B. Myeloid Dysfunction related primary immunodeficiency involves impaired phagocytic processes manifested by recurrent microbial infection of greater or lesser severity. That those affected by myeloid dysfunction suffer from increased susceptibility to bacterial infection. The specific defects arising out of myeloid dysfunction and the inability of the host to respond to bacterial agents, characteristic symptoms are described below.

Reduced production of neutrophils during hematopoiesis

Peripheral blood neutrophils are circulating granulocytes. Granulocytes may either be totally absent (agranulocytosis) or below 1500/mm3 in concentration (granulocytopenia or neutropenia). These quantitative deficiencies may either be congenital or acquired through extrinsic factors. The latter is more common. Children born with innate neutropenia show a severe reduction in neutrophils count; less than 200neutrophils/mm3 and suffer from frequent bacterial infections right from the first month of life, when normal children are protected by maternal antibodies and normal neutrophil cont. The abnormal condition (reduced neutrophils count) stems from a failure of the myeloid stem cell to differentiate along the granulocyte lineage. Causes for permanent neutropenia are yet to be established through transient neutropenia that worsens from chemotherapeutic agents and radiation, in which cells are rapidly damaged aggravating neutropenia.

Defective Cytochrome

A defect in the oxidative pathway by which phagocytes generate hydrogen peroxide and this resulting reactive products such as hypochlorous acid that kill phagocytosed bacteria, is prominent in this abnormality/immunodeficiency. A missing or defective cytochrome (cyt b_{588}) that functions in an oxidative pathway and defects in proteins (phagocyte oxidases, or phox) that stabilize the cytochrome are manifested through various symptoms viz. bacterial and fungal infections. Replacement of defective cytochrome has been reported as a corrective measure in this immunodeficiency.

Chediak-Higashi Syndrome

In this immunodeficiency, the regulation of intracellular trafficking is severely affected by mutation in a protein (LYST). The mutation impairs the targeting of proteins to secretory lysosomes, resulting in inability of the patient to lyse bacteria. The characteristic symptoms are recurrent bacterial infections, lack of skin and eye pigment and nonmalignant infiltration of organs by lymphoid cells.

Leukocyte Adhesion Deficiency (LAD)

Cellular interaction is based on cell surface molecules belong to the integrin family of proteins that function as adhesion molecules. In this immunodeficiency, adhesion molecules are in a defective stage or deficient, so the disease is called leukocyte adhesion deficiency (LAD). With impaired adhesion of leukocytes to vascular endothelium recruitment of cells to the site of inflammation is considerably confined, resulting in increased susceptibility of the individual to gram-positive and gram negative bacteria and fungi, rendering a short life span.

Defect in X-chromosome

A defect in a gene located on X-chromosome is responsible for a deficiency in properdin, which stabilizes the C3 convertase in the alternative complement pathway. The defect renders immuno-complement deficiency increasing the susceptibility of the individual to bacterial infection or immune complex disorders.

Treatment of Immunodeficiency

Congenital defects causing immunodeficiencies are incurable, yet besides complete isolation of the individual to avoid exposure to any microbial agents, the following modes of therapy may be tried.

- Replacement of the defective or missing protein
- Administration of human immunoglobulin
- Bone marrow transplantation
- Replacement of a missing or defective gene (gene therapy)

Since antibody production is impaired in immunodeficiency caused by congenital defects, if the missing protein immunoglobulin is administered intravenously or subcutaneously, recurrent infection in many types of disorders caused by impaired antibody production can be successfully handled. Agammaglobulinemic patients are mostly benefited by this mode of treatment. The quantum of immunologically important proteins may be increased or replaced or their concentration increased, to cause production of antibodies to combat infection. The administration of recombinant IFN-g has been found most effective in impaired immunodeficiency.

As regards to bone marrow transplantation (cell replacement) in case of suspected immunodeficiency, replacement of stem cells with those from an immunocompetent donor may enable the immunodeficient/likely immunodeficient to develop a functional immune response. Injection of paternal CD34+ cells in utero to a mother likely to give birth to a SCID child, has also been reported to enable the child to withstand infection after birth. $CD34^+$ cells are also effective in chronic granulomatous disease (CGD) with defective $p67^{phox}$ as by this procedure, the defective gene is replaced. Cell replacement therapy is extensively under trial, for ADA (adenosine deaminase deficiency) and CGD deficiency.

The trials of gene therapy are being conducted on nude (hairless) and athymic (absence of thymus) mouse, whose mortality is 100% within 25 weeks and 50% within the first two weeks after birth. The SCID mouse has also been included in trials of gene therapy. The SCID mouse is unable to differentiate into mature B and T-lymphocytes in the locations; thymus, spleen, lymph nodes, and gut tissue.

Gene therapy strategies have been extended to study the effect of deletion of a gene contributing to immunodeficiency by creating mouse strains deficient in immunity caused by a particular gene.

Acquired Immunodeficiency Disease

Acquired immunodeficiency is not inborn, and has their roots in sources lying outside the cell lineage. The diseases caused by acquired immunodeficiency are:

Acquired Hypogammaglobulinemia: The origin of this acquired immunodeficiency disease is not known. Its major symptoms are recurrent infections in young adults having very low but detectable levels of total immunoglobulin. There are T cell defects, which may aggravate with recurrent infections complicating the disease and rendering its nature of acquisition. Hypogammaglobulinemia is reflective of maternal circulation in acquired hypogammaglobulinemic mothers initially delivering normal infants with deficient circulating immunoglobulin.

Drugs Induced Immunodeficiency: Immunodeficiency may be acquired by a number of chemical and biological agents used to combat diseases such as rheumatoid arthritis, and lupus erythematosus. Extensive use of corticosteroids for autoimmune diseases (when the system loses its sense of self and begins to assault host cells and tissues) in the process of relieving disease symptoms induce an immunodeficient state.

Immunosuppressant Induced Immunodeficiency: The immunosuppressant drug such as cyclosporine used to thwart the effect of immune system on the transplanted organs has been reported to induce immunodeficiency. The diseases caused as a result of this suppression of immunity is another acquired immunodeficiency. The efforts to circumvent the side effects (likely to cause immunodeficiency) of immunosuppressive drugs and increase the tolerance to the grafts, will be discussed in a subsequent chapter.

Cytotoxic Drugs Induced Immunodeficiency: A state of immunodeficiency is also induced by cytotoxic drugs and radiation, whereby cells in the body are badly damaged including those of the immune system.

AIDS: The most dreadful and life claiming disease, which has assumed an epidemic form since discovered first in the United States in 1981 is HIV-1 causes Acquired Immunodeficiency Syndrome (AIDS). It was reported that patients frequently contacting infections such as P. carinii pneumonia (PCP), Kaposi's sarcoma (a skin tumor) in common had a marked deficiency in cellular immune responses and a significant decrease in the subpopulation of T cells that carry the CD4 marker (T helper cells).

Epidemiologists examining the background of this new disease-AIDS found that patients with this syndrome were homosexual males, promiscuous heterosexual individuals of either sex and their partners, intravenous drug users, including persons who received blood or blood products prior to 1985 and infants born to HIV infected mothers. The disease has assumed considerable proportions since its discovery in 1981, in the range of 25-44 years old age. The epidemic has clamped some 20 lakh people in India up to 2005, and the number is on increase. Worldwide the syndrome has clamped an equal number of males and females with highest incidence in sub-Saharan Africa. WHO reported an average of 16,000 persons infected everyday during the year 1997. However, the syndrome has assumed a proportion of consciousness worldwide lowering the growth of mortality caused by it.

Transmission of HIV-1: The common means of transmission of HIV-1, a retrovirus is homosexual and heterosexual intercourse, receipt of infected blood or blood products and, passage of infection from mothers to infants (by blood transferred in the birth process and milk in the nursing period).

AIDS is transmitted through vaginal and anal (in homosexual) intercourse at the sites of lesions and open spores found in abundance in STDs (patients who are already infected with sexually transmitted diseases- gonorrhoea, syphilis, etc.). The infected blood passes through lesions and open spores carrying retrovirus. Besides, infection with the STD agents leads to immune stimulation and activated lymphocytes are more readily infected by HIV-1 than resting lymphocytes. Intravenous drug users, since mostly use and share hypodermic needles, get infected by exposure to infected blood, and similar is the case with blood transfusion. A higher incidence of such infections than sexual intercourse has been reported for spread of the AIDS.

Casual contact or touching an AIDS patient or airborne transmission has not yet been attributed to the spread of the epidemic. Blood, milk (breast milk), semen or vaginal fluid have been attributed to the spread of the epidemic from an infected person. Exposure of broken or abraded skin or any mucosal membrane (kissing an infected person may cause infection by HIV-1) to blood from a potentially infected person, hase been established in the transmission of AIDS. Significantly for want of clinical signs for long period after infection, during which the infected person comes or causes exposure of a healthy person to HIV-1 accounts for most of the incidents of transmission.

Nature of HIV-1

AIDS is caused by a type of virus called a retrovirus, a human retrovirus, human immunodeficiency virus 1 (HIV-1). Figures 15-3 show the structure of HIV and the mode of infection with it.

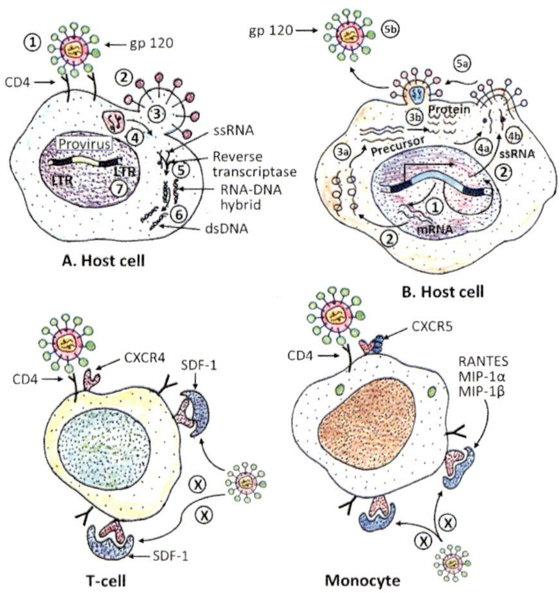

Fig.-15-3. Overview of HIV infection of target cells and activation of provirus.

The other human retrovirus is human T cell lymphotropic virus 1 (HTLV-1). This retrovirus is endemic in the southern part of Japan and in the Caribbean. T cell leukemia or a disabling progressive neurologic disorder called HTLV-1 associated myelopathy (tropical spastic paraparesis) is caused by HTLV-1.

Another human related virus called HIV-2 being less pathogenic in humans than HIV-1 infects certain non-human primates that are not infected by HIV-1. In India, three strains of AIDS virus have been reported; HIV1, HIV2 and HIV3. HTLV-1 and HIV-1 have no similar genomic sequences. HIV-2 is similar to viruses isolated from monkeys; it infects certain non-human primates that are not infected with HIV-1.

Simian immunodeficiency virus is related to HIV-1 causing immunodeficiency disease in certain infected monkeys, manifesting similar immunodeficiency as that of AIDS when injected to another species. African green monkeys have SIVagm in the wild, and cause severe and lethal immunodeficiency syndrome on infection.

Information about HIV-1 cannot be gained by infecting animals because it does not replicate in them. However, chimpanzee infected by HIV-1 develops sufficient symptoms, but for trials in vaccination. Feline, bovine immunodeficiency virus and the mouse leukemia virus have however, yielded some information about the general nature of retrovirus activation. Substantially, no animal model is available for study of factors needed for HIV-1 infection and its vaccine. The vaccine for HIV-1 is, therefore, hitherto non-existent.

In vitro studies of HIV-1 reveals that it infects T cells, monocytes and other cells that have CD4 on their surface due to a high-affinity interaction between a coat (envelope or env) protein of HIV-1 and cell surface CD4. Other cell surface molecules, co receptors present on T cells and monocytes ease infection by HIV-1. Recently, the chemokine receptors CXCR4 and CCR5 present on T cells and macrophages have been found associated with the binding of gp120 proteins of HIV coat. Binding leads to the insertion of its genomic materials together with enzymes such as reverse transcriptase, topoisomerase, proteases, etc. that is associated with viral RNA genomes. After the attachment of HIV-1 virus onto T cells, there occurs fusion of virus and T cells caused by the interaction of the viral envelope protein gp120 on the surface of infected cells with CD4. This is how the HIV-1 binds to cells and forms a characteristic fused ballooning membrane by the action of other cell adhesion molecules. After the virus has entered the cell, the RNA genome of the virus is reverse transcribed and a cDNA copy (provirus) integrate into the host genome. The integrated provirus is transcribed and the various viral RNA messages spliced and translated into proteins, which along with a completely new copy of the RNA genome are used to form new viral particles. The gag proteins of the virus are cleaved by the viral protease into the forms that make up the nuclear capsid in a mature infectious viral particle.

HIV-1 infection of T cells with certain strains of virus leads to the formation of giant cells or syncytia, formed by the fusion of a group of cells. Individual isolates of HIV-1 differ in their ability to induce syncytia formation. The ability of the virus to infect T cells or macrophages has led to the classification of the HIV-1 virus as syncytia inducing (SI) or non- syncytia inducing (NSI). T cell infecting strains are called T tropic and macrophages infecting strains are called M tropic. T tropic strains are SI and M tropic strains are NSI, or CXCR4 SI and CCR5 NSI.

Clinical and Immunologic Consequences of HIV-1 Infection

The time between the incubation period and death from the immunodeficiency caused by HIV-1 averages from nine to twelve years. Between the period of infection and the progression of disease, there appear some overt indications followed by persistent secondary symptoms such as tuberculosis, pneumonia, severe wasting diarrhoea or various malignancies to which generally the AIDS patients succumb. After infection primarily, there appear few symptoms such as fever, swollen lymph nodes (lymphadenopathy) and a rash which may be transient. These symptoms may not last for more than a few weeks. Then sores in the mouth (thrush) or in women, a vulvovaginal yeast infection appear which generally do not respond to conventional modes of treatment. The thrush and vulvovaginal yeast infection are the first overt indication of HIV-1 infection. Another early indication of infection may appear in the form of persistent lacking cough caused by *P. earinii* infection of the lungs.

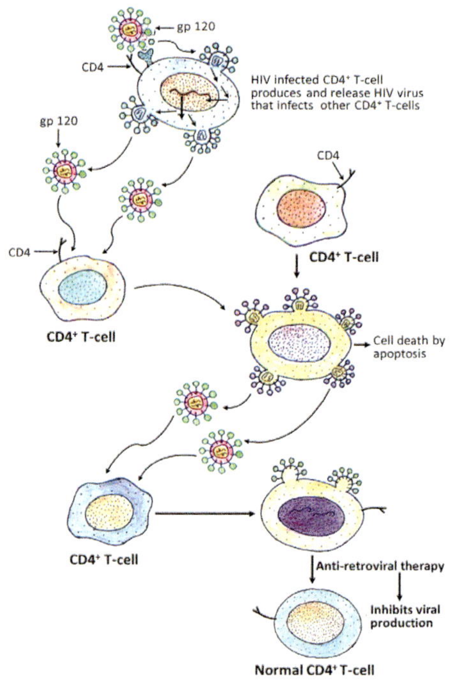

Fig.-15-4. Schematic diagram of production of virus by CD4⁺ T cells and treatment with anti-retroviral therapy.

The diagnosis of AIDS includes evidence for infection with HIV-1 (presence of antibodies or virus in blood), greatly diminished numbers of CD4⁺ T cells (<200 cells/mm), impaired or absent DHT reactions and the occurrence of opportunistic infections. A rise in the level of circulating HIV-1 in the plasma and concomitant drop in the number of CD4⁺ T cells are the established indication of infection by HIV-1. After the first few weeks of primary infection, manifesting the above symptoms, the disease enters into chronic stage in which the initial infection event causes dissemination of virus to lymphoid organs and a resultant strong immune response. Understanding how the immune response keeps the HIV-1 in check can lead to the design of effective therapy and preventive strategies. During the chronic stage of infection by HIV-1, when the immune response comes into play both antibody and cytotoxic CD8$^+$ T lymphocytes, keep viral replication in check. After the initial burst of viremia (high levels of virus in the circulation), the viral level in the circulation achieves a steady state.

This state can be detected by sensitive PCR assays for viral RNA, which measure viral load (the number of copies of viral genome in the plasma). This assay helps in the prognosis of the disease. Although the viral replication is kept in check by antibodies and cytotoxic CD8$^+$ T lymphocytes, throughout the chronic stage, yet large amounts of virus are produced in infected CD4$^+$ T cells (10^9 virions are released every day and continual infect and destroy additional host cells (Fig. 15-4). With this interplay of immune system and viral replication the host sustains the chronic stage of the disease expressing an increase in viral load and a decrease in CD4$^+$ T cell numbers. This marks the immunodeficiency associated with AIDS. As the patient passes through chronic stage succumbs to death followed by increased opportunistic infection.

Significantly, studies of the structure of the lymph nodes during the period of infection by HIV-1 entered into chronic phase, has revealed that they have been fully destroyed by virus long before plasma viral load reached the steady-state level. Chronic stage studies of AIDS have revealed depletion of CD4$^+$ T cells as killed by HIV, a decrease or absence of delayed hypersensitivity to antigens to which the individual normally reacts, a sharp increase in the serum levels of IgG and IgA immunoglobulins in the serum levels, decreased responses to specific antigens (influenza virus), alloantigens and mitogens (Table 15-2). Viral replication occurs in the central and peripheral nervous systems, with the progression of the disease. Abnormalities in cognition, motor performance and behavior add to the syndrome of AIDS in its progression, although noticed in a considerable percentage of patients are yet to be established if results of infection.

Table 15-2: Immunologic abnormalities associated with HIV infection

Type of cells/ response	*Stage of infection*	*Typical abnormalities observed*
Lymph node	Early	Infection and destruction of dendritic cells; some structural disruption
	Late	Extensive damage and tissue necrosis; loss of follicular dendritic cells and germinal centers; inability to trap antigens or support activation of T and B cells
T helper cells	Early	No in vitro proliferative response to specific antigen
	Late	Decrease in T h-cell numbers and corresponding helper activities; response to T cell mitogens or alloantigens

Type of cells/ response	Stage of infection	Typical abnormalities observed
Antibody production	Early	Enhanced non-specific IgG and IgA production but reduced IgM synthesis
	Late	No proliferation of B cells specific for HIV-1, undetectable secretion of serum antibody
Cytokine production	Early	Increased level of some cytokines in serum
	Late	Shift in cytokine production from TH1 to TH2 subset
Delay-type hypersensitivity	Early	Significant reduction in proliferative capacity of $_{TDTH}$ cells and reduction in skin test reactivity
T cytotoxic cells	Early	Normal reactivity
	Late	Reduction but not elimination of CTL activity due to impaired ability to generate CTLs from T cells

Therapeutic Agents for the Treatment of AIDS

The pharmaceutical agents used forthe treatment of AIDS/HIV-1 should be specific to HIV-1 retrovirus and minimally affect normal cell processes (Fig. 15-5). There are several considerations for the treatment of development of therapeutic agents for AIDS. The major pharmacological agents used in AIDS treatment are as follows:

- Zidovudine (AZT or Retrovir), a nucleoside analog causes termination of the cDNA chain of the retrovirus, but its long-term use causes the development of resistant viral mutants in treated patients. It is an inhibitor of viral reverse transcriptase. Besides, since precursors of red blood cells are especially sensitive to AZT, anemia is developed with several of side effects such as nausea, headache, anemia, neutropenia etc. Some other nucleoside analogs found to be useful are Zalcitabine (ddc or HIVID), Didanosine (Videx or ddl), Lamivudine (3TC or Epivir), Stavudine (d4T or Zerit), all have profound side effects to the AIDS patients. Lamivudine and zidovudine in combination named as Combivir has found to be more fruitful.
- Navirapine (Viramune) and Delavirdine (Rescriptor) inhibit the action of the reverse transcriptase enzyme of non-nucleoside origin, but is not free of side effects such as rash, hepatitis, etc.
- Ritonavir (Norvir) is a protease inhibitor, which is found to be effective in AIDS treatment. Some other protease inhibitors useful are Indinavir (Crixivan), Nelfinavir (Viracept) and Saquinavir (Invirase). All these drugs are also associated with profound side effects such as diarrhoea, kidney problem, nausea, headache, blurred vision, vomiting, abdominal pain etc.

♦ HAART (highly active antiretroviral therapy) is a combination of two nucleoside analogs and one protease inhibitor, is able to control drug-resistant mutants, lower viral load, but the recommended large number of pills to be taken every day and side effects vide put a barrier to its usage, in addition to cost ($ 15,000 per annum). It is also known as ART (Anti-Retroviral Therapy)

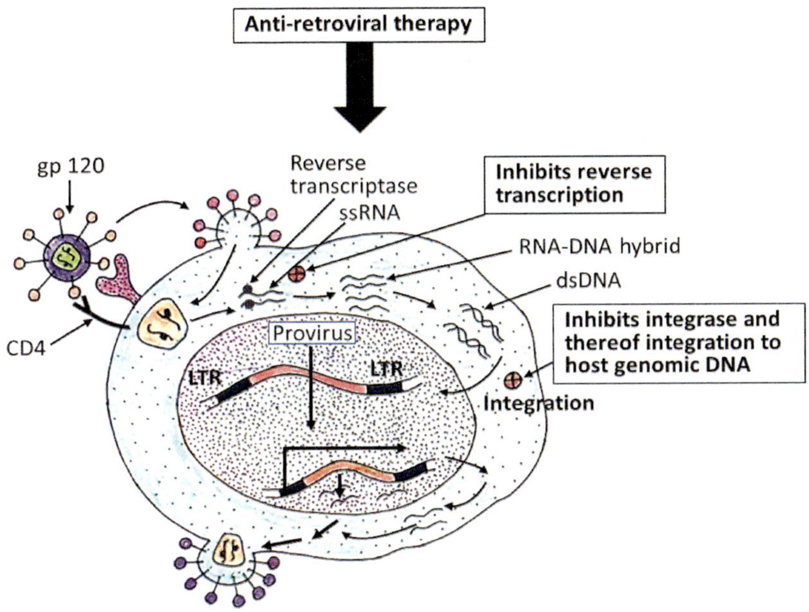

Fig.-15-5. Different stages in the viral replication cycle inside the cell that give rise to a target for therapeutic anti-viral treatment.

Antiretroviral drugs fail to kill the AIDS virus circulating in the brain, where they are not detectable. Immune modulators such recombinant IL-2 in combination with HAART are still in the experimental stage to ensure as a strategy to help reconstitute the immune system and restore normal function. However, the research to find drugs that meet the optimal advantages over the above considerations is in progress.

Prevention of AIDS by Vaccines

As the epidemic is at a stage of rage in developing and underdeveloped countries, effective cost-conscious (fairly low cost/inexpensive) and well tolerated drugs are under development, prevention seems to be a matter of prime consideration for the present.

As for the development of vaccines for prevention of AIDS is concerned, there are certain limitations in vaccine strategy.

♦ Antibodies directed against proteins of AIDS virus and active cellular immunity is not clear or consistent, so duplication of these immunopathogenic agents cannot be developed into a vaccine.

- Eradication of a retrovirus is not a simple matter, since it integrates into the host genome and may remain latent for long periods, duplication of every copy of the virus and every infected cell can be tried for eradication, which is not possible.
- HIV-1 evade the immune system by a rapid mutation of proteins targeted by antibody, vaccine to correspond to these mutations is difficult to be developed, as is the case with rhinoviruses against which no effective vaccine has yet been developed.
- While vaccines have been developed for infections through respiratory or gastrointestinal tracts, HIV-1 commonly enters through genital tract, for which even an animal model is not available for testing vaccine strategy, in the face of the nature of HIV-1.
- Inoculation with the gp^{120} molecules from different HIV-1 strains demonstrated the vaccinated individuals unprotected by the viral envelope glycoproteins.
- Behavioral and social issues pose a further barrier to the development of a vaccine against this menacing disease of our time since volunteers in a vaccine trial would practically be unavailable.

Thus, it is obvious that the development of a vaccine against AIDS infection is not a simple task like other conventional vaccines, yet if all the data are carefully analyzed and all the accessible means of creating immunity are extensively tested a safe and effective vaccine for AIDS may be developed.

CHAPTER – 16

AUTOIMMUNITY AND AUTOIMMUNE DISEASES

Normal healthy individuals have maturd, recirculating, self-reactive lymphocytes, whose activity must be regulated through clonal anergy or clonal suppression. A failure in this regulation can lead to the activation of self-reactive clones of T or B cells generating humoral or cell-mediated response causing damage to cells and organs. That is when the multi-component immune system fails in some or all of its parts, it loses sense of self and begins to attack host cells and tissues.

The diseases caused by this failure of immune system in part or full, are called, since generated by an individual's immune system, autoimmune diseases. Autoimmune hemolytic anemia (antigens on red blood cells are recognized by auto-antibodies and destroyed), rheumatoid arthritis (self-reactive T cells attack the tissue in joints causing inflammatory response-swelling- and tissue destruction), insulin-dependent diabetes mellitus, multiple sclerosis are some of the common autoimmune diseases, to which this chapter deals in addition to current treatment of autoimmune diseases and their prevention.

Human autoimmune diseases can be divided into organ specific (single organ or cell damaging) and systemic diseases. The organ specific diseases involve an autoimmune response directed primarily against a single organ or gland.

The systemic diseases are directed against a broad spectrum of tissues and have manifestations in a variety of organs resulting from cell-mediated responses and cellular damages caused by auto-antibodies or immune complexes. Table 16-1 shows some autoimmune diseases in humans.

Single Organ and Cell-Damaging Autoimmune Diseases

When the immune mechanism is directed to a target organ causing damage to its cells by way of lysis or an inflammatory response, the phenomenon is called an autoimmune disease. Examples of this type of autoimmune diseases are:

Diabetes Mellitus

The clinical name of this disease, since insulin dependent is insulin dependent diabetes mellitus (IDDM). Scattered throughout the pancreas, there are insulin producing cells called beta cells. The autoimmune attack is directed to these beta cells, which are destroyed by attack and the production of insulin, as a result is decreased to a high level, and concomitantly levels of blood glucose increases to very low level. As a result of destruction of beta cells there arise abnormalities in glucose metabolism characterized by ketoacidosis and increased urine production. As people, who suffer from diabetes mellitus go undetected for several years, irreparable loss of pancreatic tissue cause atherosclerotic vascular lesions- which in turn cause gangrene of the extremities due to impeded vascular flow- renal failure and blindness. Really diabetes is a lethal disease.

The treatment is aimed at providing insulin to the metabolism to keep the levels of blood glucose under control. However, this mode of treatment cannot compensate the metabolically regulated continuous and controlled release of the hormone, yet is the most common therapy. The chain of action of autoimmune attack includes CTL infiltration into the islets of beta cells, activation of macrophages, release of cytokines and auto-antibodies- (antibody-plus-complement lysis or antibody dependent cell mediated cytotoxicity).

Table 16-1: Some important autoimmune disease in humans

Autoimmune Disease	*Self-antigen*
Addison's disease	Adrenal cells
Autoimmune hemolytic anemia	RBC membrane proteins
Goodpasteure's syndrome	Renal and lung basement membranes
Graves' disease	Thyroid stimulating hormone receptor
Hashimoto's thyroiditis	Thyroid proteins and cells
Idiopathic thrombocytopenia mellitus	Platelet membrane proteins
Insulin-dependent diabetes mellitus	Pancreatic beta cells
Myasthenia gravis	Acetylcholine receptor
Myocardial infarction	Heart
Pernicious anemia	Gastric parietal cells, intrinsic factor
Poststreptococcal glomerulonephritis	Kidney
Spontaneous infertility	Sperm

Pernicious Anemia, Autoimmune Hemolytic Anemia, and Drug-induced Hemolytic Anemia

A pathological phenomenon in which the number of functional mature red blood cells decrease below normal for want of sufficient uptake of vitamin B_{12} from gastric parietal cells, is called anemia. When auto-antibodies are directed to gastric parietal cells, the uptake of vitamin B_{12} is circumvented in them, causing fall in the number of functional mature red blood cells. This autoimmune disease is called pernicious anemia treated by injections of vitamin B_{12}.

In hemolytic anemia auto-antibody is directed to RBC antigens, causing phagocytosis of RBC. This disease is diagnosed by the Coombs test. Administration of penicillin and methyldopa (anti-hypertensive drug) induce autoimmune anemia, as these drugs interact with RBC and become antigenic.

Hashimoto's Thyroiditis

In this autoimmune disease, there is intense infiltration of the lymphocytes, macrophages and plasma cells in the thyroid gland, causing decreased uptake of iodine from thyroglobulin and thyroid peroxidase, resulting in decreased production of thyroid hormones (hypothyroidism). The disease is after seen in middle-aged women.

Good Pasture's Syndrome

Damage to the glomerular and alveolar basement membrane is caused by specific auto-antibodies, characterized by kidney damage and pulmonary hemorrhage leading to death. Deposits of IgG and C3b along the basement membranes of kidney and lungs damage the related cells.

Some single-organ and cell-damaging autoimmune diseases are mediated by stimulating antibodies (Graves' disease) or blocking antibodies (Myasthenia gravis).

- **Graves' Disease:** Pituitary gland produces thyroid stimulating hormone (TSH). When auto antibodies bind to the receptor for TSH the normal function of TSH activating adenylate cyclase is mimicked and the thyroid is over stimulated. The stimulation of thyroid causes Graves' disease. A detailed mechanism of Graves's disease in humans has been shown in Figure 16-1.
- **Myasthenia Gravis**: When auto-antibodies block the normal binding of acetylcholine receptors on the motor end-plates of muscles, progressive weakening of the skeletal muscles ensues. As a result, drooping eyelids and inability to retract the corners of the mouth (snarling appearance) characterize a disease called myasthenia gravis.

Thus, the above autoimmune diseases are directed against a single organ or gland, in which auto reactive lymphocytes or auto antibodies bind to self antigens leading to direct cellular damage. And in some autoimmune diseases, either antibodies are stimulated or blocked as in Graves' disease and Myasthenia gravis. A detailed mechanism of Myasthenia gravis in human has been shown in Figure 16-2.

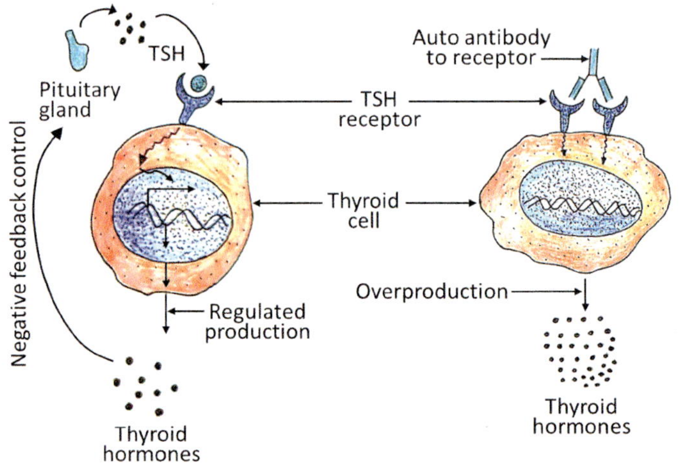

Fig.-16-1. Binding of auto-antibodies to the receptor for TSH in Graves disease.

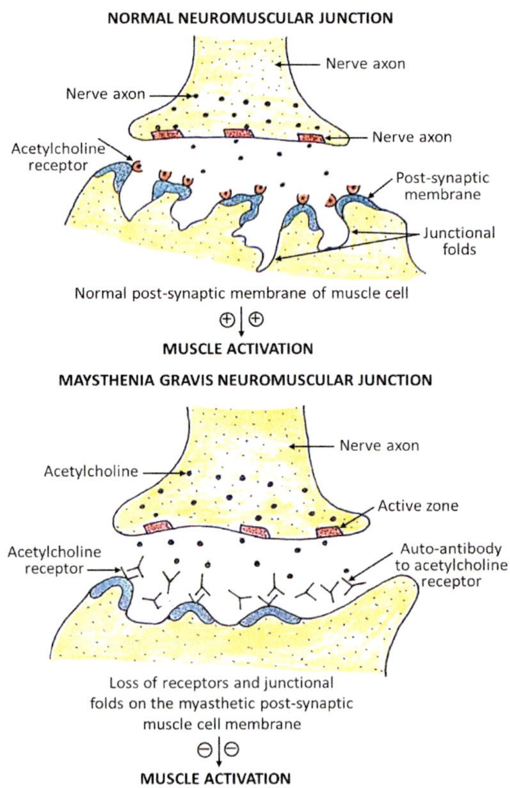

Fig.-16-2. The Binding of auto-antibody to acetylcholine receptors, blocks the normal binding of acetylcholine in *Myasthenia gravis* disease at neutron-muscular junctions.

Systemic Diseases

Unlike single organ or cell-damaging diseases in systemic autoimmune diseases, a number of organs and tissues are targeted antigens. The systemic diseases are directed against a broad spectrum of tissues and have manifestations in a variety of organs resulting from cell-mediated responses and cellular damage caused by auto-antibodies or immune complexes. Some systemic autoimmune diseases are:

Systemic Lupus Erythematosus (SLE)

This is a systemic disease mostly afflicting women in the range of 20 to 40 years of age, characterized by fever, weakness, arthritis, skin rashes, pleurisy and kidney dysfunction. In this disease, auto-antibodies are produced to a vast array of tissue antigens such as DNA, histones, RBCs, platelets, leukocytes and clotting factors. The characteristic symptoms of the disease are the result of interaction of these auto-antibodies corresponding to specific antigens. For example auto-antibody specific for RBCs can lead to complement-mediated lysis, resulting in hemolytic anemia.

Vasculitis and glomerulonephritis are produced by activation of immune complexes complicating in membrane-attack complexes and complement split products, which damage the wall of the blood vessels.

Multiple Sclerosis (MS)

Numbness in the limbs or paralysis or loss of vision are some of the mild to severe symptoms of multiple sclerosis, caused by auto reactive T cells that participate in the formation of inflammatory lesions along the myelin sheath of nerve fibers, induced by activated T lymphocytes. T lymphocytes infiltrate the brain tissue and cause characteristic inflammatory lesions, destroying the myelin. Numerous neurologic dysfunctions result from a breakdown in the myelin sheath functions to insulate the nerve fibers.

Environmental and genetic factors also contribute to the contraction of MS. Multiple sclerosis is more common in the northern hemisphere than the southern. Persons who move from southern hemisphere to the northern are susceptible to the contraction at a prevalence rate of 110-140 cases per 100,000. This proves the contribution of environmental factors. Similarly, as the disease is existent in the US, children at siblings of people with MS have been reported to contract the disease at a ratio of 1:100 or 1:50 indicating the prominence of genetic factors.

Rheumatoid Arthritis

This disease is characterized by chronic inflammation of the joints, afflicting women in the age group of 40-60 years. Rheumatoid arthritis has an IgM antibody, which binds to normal circulating IgG forming IgM-IgG complexes, which deposit in the joints and activate the complement cascade resulting in a type III hypersensitive reaction, which leads to chronic inflammation.

Scleroderma

Scleroderma is characterized by a gradual tightening of the skin usually in the extremities (hands, feet and face). The tightening of the skin is actually hardening of the skin. It is a chronic degenerative disease causing over production of collagen in connective tissue. Persons afflicted with this disease manifest a localized form of "CREST" development, whereby the symptoms complicate into a syndrome. This group of five symptoms is called CREST, where each letter stands for a symptom seen in Scleroderma viz.

C =Calsinosis, a painful deposit of calcium;

- in the skin

R =Reynaud's phenomenon;

- abnormal blood flow in response to cold or stress

E =Esophageal dysfunction;

- swallowing problems in esophagous.

S =Sclerodactyly;

- tightening of the skin

T =Telangiectasia;

- formation of red spots on the hands, palms, forearms, face and lips.

Scleroderma in its severe form attacks internal organs such as kidneys, lungs, heart, GI tract, joints and skin, and leads to premature death. This disease is again related to genetic factors causing over production of collagen in the connective tissue. Large numbers of fetal cells in the blood of women, who have scleroderma, point to the probability of foetal cells' immune response to the mother like graft versus host disease. The disease since is reported in childless mothers and men, the exact nature of contraction is under extensive study.

Spontaneous and Experimental Animal Models for Autoimmune Diseases

Animal models for autoimmune diseases provide insight into the development of autoimmunity arising spontaneously and arising by induction of specific antigenic auto-antibodies. Both of these approaches open a practical mechanism for alleviating autoimmune responses.

Animal models for spontaneous autoimmune diseases

Animals who develop autoimmune diseases serve as models for similar human/ autoimmune diseases. Such animals also serve as models for discovering ways to alleviate these diseases. Several mouse strains have been reported to develop and manifest symptoms on the line of human autoimmune diseases, such as;

- New Zealand Black and New Zealand White (NZB and NZW respectively) strains of mice spontaneously develop symptoms between 2-4 months, which closely resemble systemic lupus erythematosus (SLE) common in women. Antibodies in these mice include antibodies to nuclear proteins, DNA and T-lymphocytes.

- MRL/lpr/lpr a homozygous mouse strain having a defective fas gene develops SLE due to fas protein (TNF family), which when interacts with its ligand transduce a signal that leads to apoptotic death of the Fas-bearing cells and $CD4^+$ T cells. But, because of no link of fas expression and SLE in humans the lpr mouse cannot serve as a model for SLE in humans.
- NOD (nonobese diabetes) mouse also develops a form of diabetes resembling IDDM in humans. The mechanism of development of diabetes in NOD resembles that of IDDM in human's i.e., lymphocytic infiltration into the islets of pancreas. Recent experimental evidence implicate the role of T_H1 subset in disease development.

Experimental induction of autoimmunity in animals

The experimental models (animals) provide a very specific cause for development of an autoimmune disease, so are more reliable and effective in control and treatment of autoimmune diseases. Similar to humans, however, the autoimmune disease can be induced to an experimental system, so as to ease in study.

- EAMG (Experimental Autoimmune Myasthenia Gravis): Autoantibodies to the acetylcholine receptor are the cause of myasthenia gravis in humans. Rabbits immunized with acetylcholine receptors purified from electric cells soon developed nuclear weakness as seen in myasthenia gravis in humans. The antibodies to the acetylcholine receptor blocked muscle stimulation by acetylcholine in the synapse.
- EAE (Experimental Autoimmune Encephalomyelitis): Immunization with myelin basic protein (MBP) or proteolipid protein (PLP) in complete Freund's adjuvant may cause development of cellular infiltration of the myelin sheaths of the central nervous system, resulting in demyelination or paralysis. The study is intensively suggestive of development of MS in humans and treatment with orally administered MBP.
- EAT (Experimental Autoimmune Thyroiditis): Injection of thyroglobulin in complete Freund's adjuvant cause both humoral antibodies and T_{DTH} cells directed against the thyroglobulin, resulting in thyroid inflammation. The symptoms resemble Hashimoto's thyroiditis in humans.
- AA (Autoimmune Arthritis): Injection of *M. tuberculosis* in complete Freund's adjuvant induces characteristic arthritis in animals resembling that in humans.

CD4+ T cells, T cell receptor, MHC and Autoimmune Diseases

Transfer of autoimmune diseases from one animal into another by T cells from animals injected (in complete Freund's adjuvant) with either MBP or PLP or by cloned T cell lines from such animals has proved the role of CD4 in the development of autoimmune diseases. T cell recognition of antigen, of course involves a trimolecular complex of the T cell receptor, on MHC molecule and antigenic peptide. Thus, an individual susceptible to autoimmunity must possess MHC molecules and T cell receptors capable of binding self-antigens.

CD4+ T cells and T_H subsets Balance

T cell clones are obtained from the autoimmune animals in the presence of various T cell growth factors by culturing lymphocytes *in vitro* with MBP. When these MBP-specific T cell clones are injected IV into normal syngenic animals, the cells penetrate the blood-brain barrier and induce demyelination. EAE develops within 5 days (Fig. 16-3). Examination of such obtained T cells has revealed that they bear the CD4 membrane marker. The development of autoimmune diseases can be reversed by depletion of T cell population with antibody directed against CD4. Injection of anti-CD4 monoclonal antibody abolished the symptoms of autoimmune diseases in (NZB ´ NZW) mice and in mice with EAE.

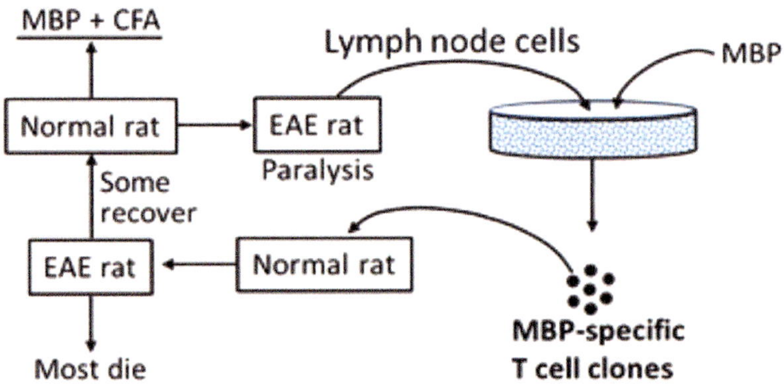

Fig.-16-3. Experimental demonstration in rat by injecting myelin the basic proteins in case of auto-immune encephalomyelitis.

Analysis of CD4+ T cells has revealed that autoimmune diseases develop as a result of imbalance of TH subsets or T_H1/T_H2 dichotomy. T_H1 cells contribute to the development of autoimmune diseases and TH2 cells protect against the development of autoimmune diseases. The protection is carried out by injection with MBP plus adjuvant, and development is carried out by TH1 cytokines (IL-2, TNF-β, and IFN-γ).

Role of MHC

The role of MHC as one of the trimolecular comples is related to particular MHC allele in developing autoimmune diseases. However, the phenomenon of development of autoimmune diseases is a complex one and is best seen in ankylosing spondylitis, an inflammatory disease of joints of vertebrae.

T cell Receptor (TCR)

In a study, T cells specific for various encephalitogenic peptides of MBP were cloned and their T cell receptors analyzed. Analysis of the T cell receptors on these clones revealed restricted TCR expression for MS and myasthenia gravis, indicating the strong association of TCR in these diseases.

Mechanisms for induction of Autoimmunity

A variety of mechanisms have been proposed for inducing autoimmunity. The results indicate that-

- autoimmunity does not develop from a single event but rather from a number of different events, and
- the autoimmune diseases affect selective sexes- male or female, called preferential susceptibility.

The proposed mechanisms for inducing autoimmunity (Fig. 16-4) are indicative of pathways to the development of diseases, as

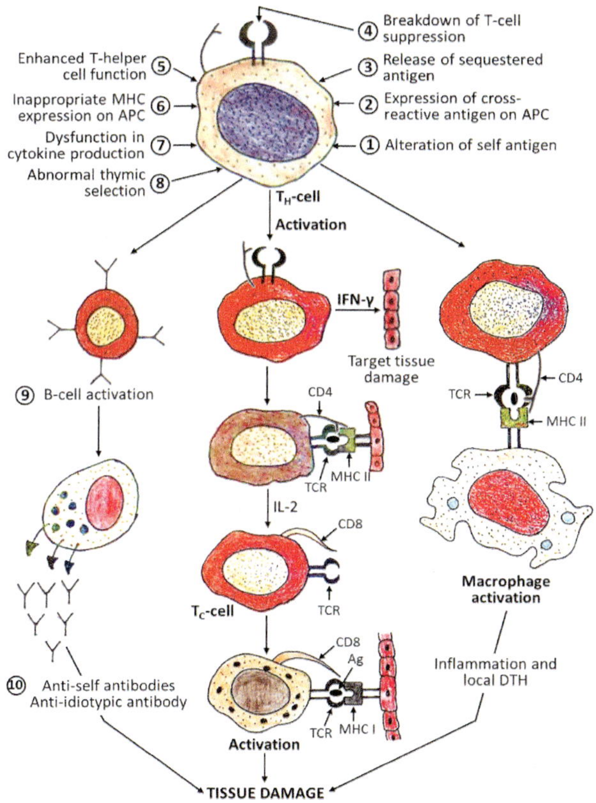

Fig.-16-4. Proposed and possible mechanism contributing factors in auto-immune disease.

Release of Sequestered Antigens

Any tissue antigens that are sequestered from the circulation, and therefore, are not seen by the developing T cells in the thymus will not induce self-tolerance. Exposure of mature T cells to such normally sequestered antigens at a later stage might activate them in the form of auto-antibodies. Trauma to tissues following either an accident or a viral or bacterial infection might also release sequestered antigens with the circulation. For example after a vasectomy some sperm antigens are released into the circulation, inducing auto-antibody formation in some men.

Molecular Mimicry

Post-rabies encephalitis is one of the best examples to understand this phenomenon. In the past rabies virus was grown in rabbit brain-cell cultures and preparations of the vaccine included antigens derived from the rabbit brain cells. In a vaccinated person, these rabbit brain cell antigens could induce the formation of antibodies and activated T cells which could cross-react with the recipient's own brain cells leading to encephalitis.

Inappropriate class II MHC expression on cells

Inappropriate expression of class II MHC molecules on cells may activate self-reactive T cells resulting in the impairment of the secretory function of human beta cells. IFN-γ also induces increases in class II MHC molecules on a wide variety of cells (pancreatic beta cells, intestinal epithelial cells, melanoma cells and thyroid acinar cells). IFN-γ is known to induce production of several other cytokines (IL-1 and TNF) involving antigen presentation by class II MHC molecules resulting in activation of self-reactive T cells.

Polyclonal Activators

A number of viruses and bacteria are nonspecific polyclonal B cell activators. These viruses and bacteria (Epstein Barr Virus, cytomegalovirus and gram positive and gram negative bacteria) induce the proliferation of numerous clones of B cells that express IgM in the absence of T_H cells. If B cells reactive to self-antigens are activated, auto-antibodies can appear. That polyclonal B cell activation leads to auto-antibody production.

Homology between MBP and Viral peptides

Infection with certain viruses expressing epitopes that mimic sequestered self-components, such as myelin based protein, can cause autoimmunity to those components. This suggests homology between MBP and viral peptides. In an experiment mice were infected with herpes simplex virus Type 1 which led to a disease known as herpes stromal keratinitis (HSK), a disease resembling autoimmune disease in which T cells specific for a particular viral peptide attack corneal tissue causing blindness. From all these studies, it comes out that a particular epitope of a virus is responsible for autoimmune disease like symptoms, because of the homologous factor.

Treatment of Autoimmune Diseases

The design of treatment of autoimmunity aims at reducing the symptoms irrespective of the side effects of drug administered. For example immunosuppressive drugs such as corticosteroids, azathioprine and cyclophosphamides are often given to slow proliferation of lymphocytes. Concomitantly, patients under this frame of treatment run the risk of developing cancer, because the general immune system is also greatly affected with immunosuppressive drugs.

Another mode of treatment currently in practice involves removal of organs (single organ) causing autoimmunity viz. thymectomy (for thymic abnormalities), plasmapheresis (plasma is removed from a patients' blood by continuous flow centrifugation) for diseases myasthenia gravis, Graves' disease, rheumatoid arthritis, systemic lupus erythematosus etc. By this mode of treatment only a short-term reduction in symptoms can be achieved. These therapies being relatively nonspecific may have significant side effects. Figure 16-5 shows some sites of immuno-intervention for the treatment of autoimmune diseases.

Therapies developed out of experiments on animals have, however, proved specific to the autoimmune disease leaving the least side effects. Experimental induction of autoimmunity in animals has paved the way for devising such a class of therapies. What specific treatment modules have been advised include:

T cell Vaccines

Rats injected with low doses of cloned T cells specific for MBP did not develop symptoms of EAE, instead they became resistant to the development of EAE with a lethal dose of activated MBP-specific T cells or MBP in adjuvant. The efficacy of these specific T cell clones could be enhanced by cross linking the cell-membrane components with formaldehyde or glutaraldehyde. Permanent remission of autoimmune symptoms was observed with these cross linked T cells, because these elicited regulatory T cells specific for TCR variable region determinants of the autoimmune clones.

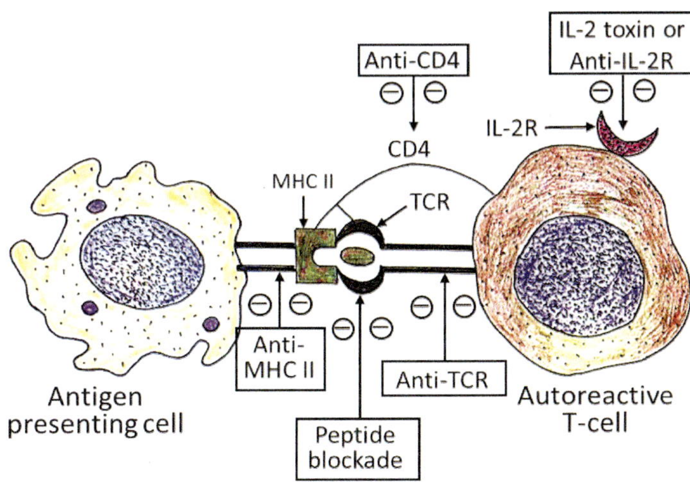

Fig.-16-5. Possible intervention at the different steps for the treatment of auto-immune disease.

Peptide Blockade of MHC Molecules

Sufficient amounts of encephalogenic peptides of MBP administered along with synthetic peptides of MBP administered along with synthetic peptides bind to MHC molecule and the clinical development of EAE is blocked. This experimental approach is of prime consideration in treatment of autoimmune diseases, specifically EAE.

Monoclonal Antibody Treatment

Lymphocytic infiltration, diabetic symptoms, and autoimmune lupus like symptoms have been treated by injections of anti-CD4 monoclonal antibodies, because anti-CD4 monoclonal antibodies block or deplete all T_H cells. The risk of depletion of all T_H cells has been, counter-acted by using monoclonal antibody directed against a submit of the high affinity IL-2 receptor, which is expressed only by antigen-activated T_H cells. By this strategy, the autoreactive T cells are blocked.

Similarly blocking of a MHC molecule that is associated with autoimmunity, and blockage of the preferred receptor with monoclonal antibody has also been found to reverse the symptoms of autoimmunity in animals. Exciting possibilities are awaited in treatment of autoimmunity with monoclonal antibodies.

Tolerance Induction of Oral Antigens

The state of immunologic unresponsiveness achieved by oral administration of antigens is called tolerance. For example, individuals with multiple sclerosis (MS) fed with 300 mg of bovine myelin every day for a year showed a reduction in the symptoms of MS, though the reduction was not of any statistical significance. Further trials of this approach, found promising on animals, are underway, to be concluded in the future.

CHAPTER – 17

IMMUNOLOGIC RESPONSE TO TRANSPLANTATION

The Transfer of cells, tissues, or organs from one site (one individual or one site of an individual) to another (other individual or other site of an individual) is called transplantation. The history of transplantation dates back to 1908, when Alexis Carrel who transplanted both kidneys in nine cats. As a result of this interchange (transplantation) of organs (kidneys), the cats who got the kidneys of other cats could not maintain normal renal function for more than 25 days and finally died. The experiment initially established the fact that a transplanted organ could carry out its normal function in the recipients. In 1935, a Russian surgeon attempted transplant of human kidney for the first time which could not carry out its normal function and the recipient died without restoring normal renal function. This event was termed as rejection of transplanted organ by the immune system of the recipient, occurring as a result of mismatched blood types of donors and recipients. Technically such an event in immunology is called hyperacute rejection. With the development of techniques to overcome such rejections, now-a-days, transplantations are being carried out successfully, dampening the intensity of the immune response against the foreign organ (transplanted organ).

However, there are certain barriers to transplantation. First, the number of donors for organs cells (blood transfusion) or tissues (skin graft) are less in number than the recipients, and secondly the foreign agents (transplanted organs, cells, tissues) are always subject to reaction by the immune system of the recipient resulting in acceptance (tolerance) or rejection, unless the donors and recipients are genetically identical. Introduction of mismatched blood into a recipient, who has antibodies (isoagglutinins) to the donor's blood-group antigen, causes an immediate often fatal response. Such an event occurs in blood transfusion carried out without matching the blood-groups. This matching of blood-groups between donors and recipients was first made known by Karl Landsteiner. The matching of blood groups holds vital in transplantation of cells (blood transfusion) and tissues (because the vascular endothelium expresses the blood-group antigens). This means that the blood vessels of the donor organ can be attacked by blood-group antibodies present in unmatched recipients. This phenomenon is however non-existing with only ABO blood group, because antigens are expressed only on red blood cells. The immune mechanism in accepting or rejecting a foreign organ, cells, and tissues is simply acting by recognizing or destroying altered self cells. The discovery of the major histocompatibility complex (MHC) was a direct result of studies on graft rejection in humans and tumor transplantation in animals.

Emerging out of the above studies the principal objective of transplantation in response to immune mechanism is to evolve techniques for preventing of rejection of grafts from genetically nonidentical donors, without exerting any immunosuppressive effect on the recipient.

This chapter as such deals with underlying immunologic basis of graft-rejection, reactions implicated in graft rejection, stages of graft rejection, types of graft rejection, general and specific immunosuppressive therapies for graft rejection, and some possible transplantation.

Characteristics of Grafts and Their Reasons for Acceptance or Rejection

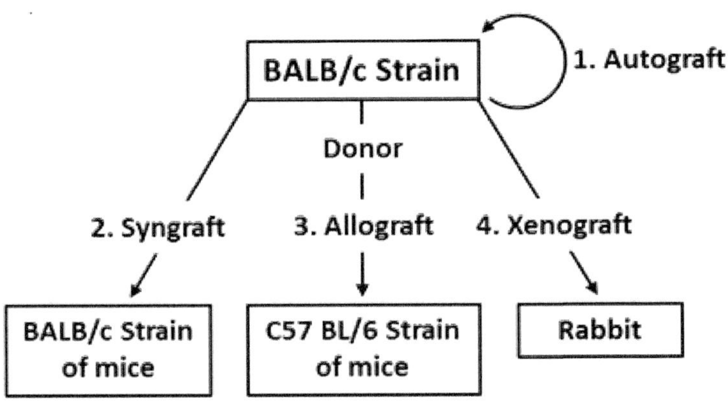

Fig.-17-1. Types of transplants. The four types of grafts (1,2,3 & 4) are presented.

On the basis of genetic similarity or dissimilarity there are four kinds of graft, and the degree of immune response varies to the particular kind of graft. The four kinds of graft are **(Fig. 17-1)**;

- **Autograft**– transfer of self tissue from one site to another in the body of the same individual, is called autograft and preferred in case of burns.

- **Isograft**– transfer of tissue between genetically identical individuals is called isograft and preferred in the case of twins.
- **Allograft**– transfer of tissue between genetically different members of the same species is called allograft and preferred in the case of organ grafts in humans (genetic similarity).
- **Xenograft**– transfer of tissue between different species is called xenograft (animal to human).

Autografts and isografts are assimilated in the body of the recipient, while the allografts and xenografts are rejected by the recipient recognizing them as foreign.

Immune Responses Involved in Rejection of Grafts

The rejection of graft and its relative time in rejection depends on the tissue grafted. The immune responses involved in the rejection of grafts include specificity, memory, antigens determining histocompatibility, HLA complex and blood group.

Skin grafts are rejected faster than other tissues such as kidneys or the heart. In the case of skin grafts, vascularisation of the grafted skin is decreased by infiltration with lymphocytes, monocytes, neutrophils and other inflammatory cells. This rapid decrease in vascularisation results in rejection of graft within 12-14 days. In these 12-14 days, an autograft is completely accepted whereas an allograft is completely rejected. Cellular infiltration, thrombosis and necrosis are the specific factors in graft rejection. The memory factor in skin graft is visible in terms of time taken by the immune system in rejection. When strain-A inbred mouse is grafted with skin from strain-B and simultaneously with skin of strain-C. The time taken in rejection of strain-C skin is surprisingly the same as that of strain-A. This happens because of attributes of memory cells. Figure 17-2 shows the graft reaction between donor and recipients.

Besides specific and memory cells, T cells are also involved in graft rejection. In studies with allograft rejection, on nude mice which lack a thymus and consequently lack functional T cells, it was found that they were unable to initiate an allograft rejection, and accepted even the xenograft. This supports the theory of cell-mediated response in graft rejection.

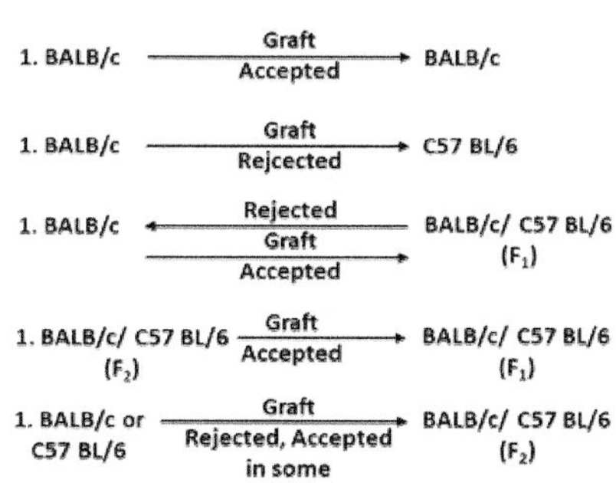

Fig.-17-2. The law of transplantation. There are five proposed laws of transplantation, governing donor-recipient reaction.

Collaboration of both $CD4^+$ and $CD8^+$ populations of T cells in rejection of allograft was observed in a study by injecting the recipient mice with monoclonal antibodies to deplete one or both types of T cells and then measuring the rate of graft rejection. It was found that the removal of both $CD4^+$ and $CD8^+$ T cells let the allograft survive from 15 to 30 days while that of $CD4^+$ only 15 days. This establishes the role of T cells in rejection of allografts.

The MHC inheritance is another immunologic response that leads to tissue rejection. Tissues that display significant antigenic similarity are called histocompatible. Tissues that display significant antigenic dissimilarity are called histoincompatible. These histoincompatible tissues induce an immune response, which leads to tissue rejection.

The MHC loci are inherited as a complete set called haplotype from each parent. In offspring having two different haplotypes developing out of mating between such two different haplotypes bearing parents, grafts will not be accepted, but in offspring inheriting haplotypes from a homogenous strain, grafts will be accepted.

Tissue compatibility is assessed to determine potential donor and recipient cells. The common ways of assessment are:

- ABO blood group compatibility,
- Microtoxicity test (HLA Test)
- Mixed-lymphocyte Reaction (MLR)

ABO blood-group compatibility: The blood-group antigens are expressed on RBCs, epithelial cells, and endothelial cells. Antibodies produced in the recipient to any of these antigens if present in the graft tissue will result in its rejection.

HLA Test: In this test, WBCs from the potential donors and recipients are distributed into a series of wells on a microtiter plate, and then antibodies specific for various class I and class II alleles are added to different wells. After incubation, complex is added to the wells, and cytotoxicity is assessed by the uptake or exclusion of various dyes (e.g. trypan blue or eosin g) by the cells. If the white blood cells (WBCs) express the MHC alleles for which a particular monoclonal antibody is specific, then the cells will be lysed upon addition of complement, and these dead cells will take up a dye such as trypan blue. The phenomenon is antibody based microcytotoxicity, implicated in allograft. It has been shown in figure 17-3 in serologic tissue typing, which needs before going to transplant any graft.

Mixed-lymphocyte Reaction (MHC compatibility test): Rejection of tissue can be lengthened in HLA compatible donor, as described in the HLA assessment test (by microcytotoxicity) above. Transplantation can be hopeful in the absence of a HLA compatible donor. The advantage of the MLR over HLA test is that the former gives a better indication of the degree of T^H cell activation generated in response to the class II MHC antigens of the potential graft, but the disadvantage of this test is the time taken in running the assay. In this test lymphocytes from a potential donor (stimulator cells) and lymphocytes from the recipient (responder cells) are taken.

Donor lymphocytes are X-irradiated or treated with mitomycin C. Proliferation of recipient T cells indicates their activation. T cell activation is measured by the uptake of [3H] thymidine into cell DNA. A greater uptake of thymidine indicates a greater difference between the donor and recipient cells. The assay is unsuitable for the reason of time taken in running assay, in case the donor is a cadaverand the organ needs to be used soon after removal from the cadaver. However, in case of kidney bone-marrow, liver and heart transplant MHC compatibility has been prudence and prominence, so far as survival rates of recipients are concerned. Genetically different individuals differ in various minor histocompatibility loci. Minor histocompatibility antigens are recognized only when they are presented in the context of self MHC molecules, so when tissue is transplanted between genetically different individuals, it is rejected.

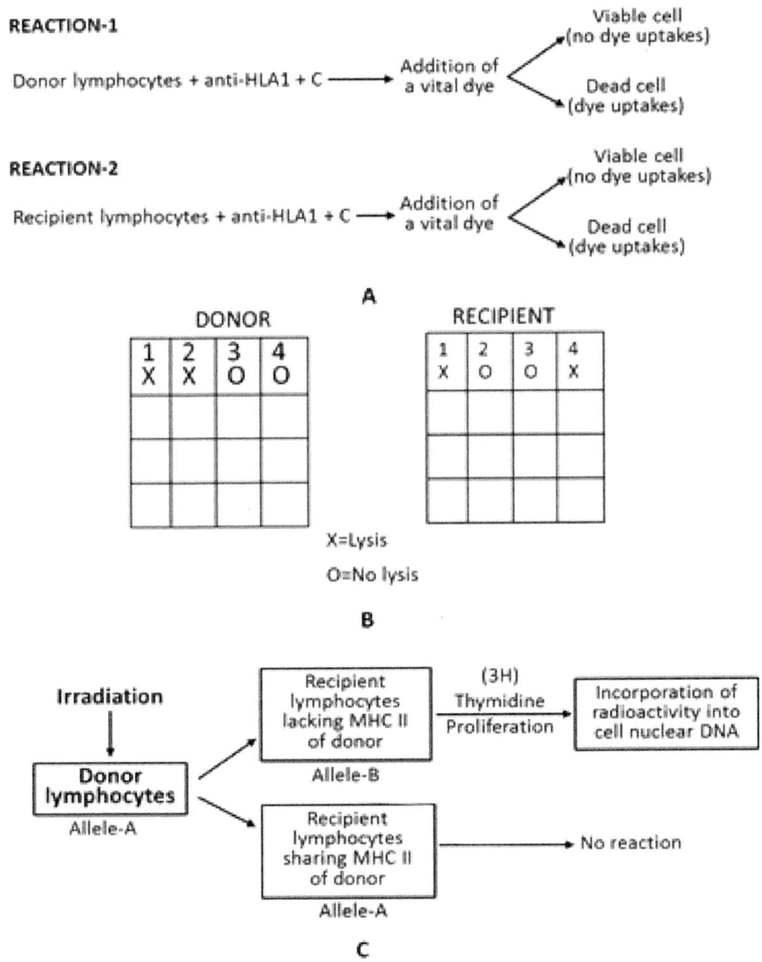

Fig.-17-3. Schematic representation of serological tissue typing for HLA-antigen.

Stages of Graft Rejection

Cell-mediated immune response to alloantigens (MHC molecules) is generated by the cells of the graft. If there are minor tissue differences (minor histocompatibility loci) between the cells of donor and recipient the graft is likely to be rejected because of the recognition of alloantigens presented in the context of self-MHC molecules. This is why immune suppression to some degree is required in cases of organ transplants.

There are two stages in the rejection of graft:

- A sensitization phase, and
- An effector phases

In the sensitization phase, $CD4^+$ and $CD8^+$ T cells recognize alloantigens expressed on cells of the foreign graft and proliferate. Alloantigens of both major and minor histocompatibility molecules are responded by self-MHC molecules in terms of recognition. A host T_H cell becomes activated when it interacts with an antigen-presenting cell (APC) that both expresses appropriate antigenic ligand-MHC molecule complex and provides the requisite costimulatory signal. Dendritic cells function as APCs in a graft tissue, as they express high levels of class II MHC molecules. APCs of host migrate in graft and endocytose the foreign alloantigens (both major and minor histocompatibility molecules) and present them as processed peptides together with MHC molecules.

Donor APCs of grafts of kidney, thymus and pancreatic islets, are called passenger leukocytes. These migrate from the graft to the regional lymph nodes. Passenger leukocytes are dendritic cells, which express the allogeneic MHC antigens of the donor graft and being foreign (of donor) activate T-lymphocytes in the lymph nodes. Donor's dendritic cells are non-existing in skin-grafts, but existing in other organ and tissue grafts. In the effector stage, a number of mechanisms come into play in the rejection of grafT cell-mediated reaction, CTL-mediated cytotoxicity, and antibody plus complement lysis and destruction by antibody dependent cell-mediated cytotoxicity (ADCC) - wherein cytokines secreted by T_H cells play a major role. Levels of cytokines increase, including vascular endothelial cells and myocytes to express class II MHC molecules, whereupon these cells are attacked by CTL.

Types of Graft Rejection

Depending upon the type of tissue or organ grafted and the immune response involved, rejection of graft is always related to time. That is to say that some grafts are rejected within the first 24 hours, some within the first few weeks (acute) and some are rejected within months or years after transplantation.

There are three main types of graft rejection;

- Hyperacute rejection occurs within the first 24h of graft, for want of vascularisation. This type of rejection is caused by serum antibodies present in the host. The complement system is activated by antigen-antibody complex in the host, as a result of which neutrophils infiltrate into the grafted tissue and concomitant blood clotting in the capillaries inhibit vascularisation of the graft.

Further, if the MHC antigens expressed on white blood cells of the transfused blood are the same as those on repeated transfusions, then some antibodies will develop in the host rejecting blood transfusion. This happens in most of the cases of repeated blood transfusions. With repeated pregnancies, women develop antibodies to the paternal allograft/alloantigens resulting in full term delivery. However, if the identical MHC antigens are confronted, there is likely to be rejection. Detection of antibodies in the host by tissue and blood group matching may result in avoidance of rejection of the graft.

- Acute rejection of allograft occurs within a week or so after the graft caused by massive infiltration of macrophages and lymphocytes at the site of tissue destruction of graft. The tissue destruction is largely done by cytokines, complement and ADCC (Fig. 17-4).
- Chronic rejection occurs by months and years after graft involving both humoral and cell mediated responses by the recipient. Long term survival of graft has been rare even when done in the conformation of HLA typing and administration of immunosuppressive drugs. Virtually chronic rejections require subsequent grafts, posing practical problems.

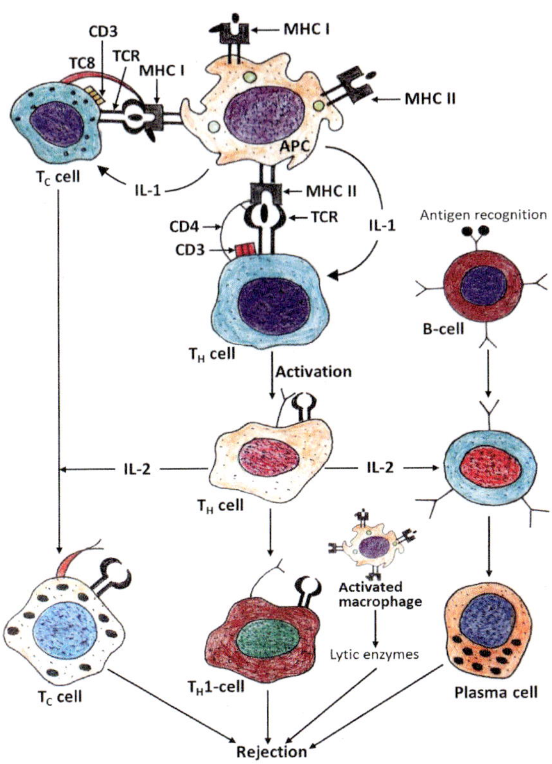

Fig.-17-4. Generation of effector mechanism for allograft rejection.

Immunosuppressive Therapies for Preventing Graft Rejection

Clinically graft rejection is suppressed by nonspecific immunosuppressive agents:

- Purine analogous (mitotic inhibitors),
- Dexamethasone, prednisone (corticosteroids),
- Cyclosporine A, FK506, rapamycin (fungal metabolites), and
- Total lymphoid X-irradiation.

Therapies conducted for the long-term survival of graft by above agents are nonspecific, as they aim at general suppression of immune response and not specific to the alloantigens of the graft.

The commonly used purine analogues are Azathioprine (Imuran), cyclophosphamide and methotrexate. Azathioprine, a potent mitotic inhibitor, acts on cells in the S phase of the cell cycle to block synthesis of inosinic acid (a precursor of the purine adenylic and guanylic acid). Immune assays have revealed a significant decrease in T cell proliferation. Cyclophosphamide (an alkylating agent) inserts into the DNA helix and disrupts its chain, besides blocking T cell proliferation. Similarly, purine biosynthesis is blocked by methotrexate (a folic acid antagonist).

The nonspecific nature of above mitotic inhibitors is established by the fact that they act on all rapidly dividing cells (proliferation), even which are not specific to the rejection of the graft. The nature of these mitotic inhibitors yields a deleterious negative effect on the immune system of the recipient rendering risk to increased infection.

Dexamethasone and prednisone (corticosteroids) are potent anti-inflammatory agents which exert their general effects on the entire immune response of the recipient, aggravated in combination with mitotic inhibitors, yielding similar effects.

Fungal metabolites (cyclosporine A, FK 506, Rapamycin) block activation of resting T cells by inhibiting the transcription of genes encoding IL-2 and the high affinity IL-2 receptor (IL-2R) essential for activation. FK506 and rapamycin bind to an immunophilin (cytoplasmic protein). These three drugs decrease the T_{DTH} cells, T_C cells, NK cells, macrophages, and B cells in specific relation to their activation. Failure of activation of these effector populations have rendered these drugs, prominence in uses against transplantation of kidney, liver, heart, heart-lung with 80% survived chances. However, drug induced nephrotoxicity and kidney failure (with other organs' transplantation) are some of the hazards usually run with these drugs as side effects.

Lymphocyte elimination by X-rays (200 rads per day for several weeks until a total of 3400 rads of administered) has been in use to render an immunosuppressed state for transplantation.

As the nonspecific immunosuppressive agents, as described above, increase the recipients' risk for infection in general, antigen-specific immunosuppressants have been developed recently on an experimental basis with animals; human trials await. In these experiments, antibodies or soluble ligands reactive with cell-surface molecules have been used successfully to suppress T cell activity (increase in subpopulations of T cells). Emerging out of experiments on animals two types of specific immunosuppressants have been developed.

Specific Immunosuppressant

The specific immunosuppressant, which have their promising implications on humans, initiated on animals, are

- Monoclonal Antibody Therapy, and
- Blocking Co-Stimulatory Signal Therapy

Monoclonal Antibody Therapy

It is used to deplete the recipient of a certain broad or specific cell population. That is to deplete immune cells involves use of a monoclonal antibody to the CD3 molecule of the TCR complex. The depletion factor involves a two way effect on immunity. First, when monoclonal antibodies are injected in the circulation, they bind to FC receptors on phagocytic cells, which then phagocytose and clear the T cells from the circulation. If a cytotoxic agent such as diphtheria toxin is coupled with the monoclonal antibody, the cell with which it reacts internalizes the toxin resulting in its death. Secondly, since the high-affinity IL-2 receptor is expressed only on activated T cells, exposure to anti-TAC blocks the proliferation of T cells activated in response to the alloantigens of the graft.

With these two ways effect on depletion of T cells or causing an immunosuppressed state of T cells, the targets are particular T cell subpopulations and cell-surface adhesion molecules, ICAM-1 and LFA-1. A monoclonal antibody therapy aiming simultaneously to both the above targets has been found most effective on mice for cardiac grafts.

As the monoclonal antibodies are generally of mouse origin, many human recipients develop an antibody response to the mouse monoclonal antibody rapidly cleaning it from the body. This problem has however been overcome by developing human monoclonal antibodies and mouse-human chimeric antibodies, involving monoclonal antibodies specific for the implicated cytokines (TNF-α, IFN-γ and IL-2). Bone-marrow transplants in mice and cardia transplants in rats have been successfully carried out and prolonged by this specific monoclonal antibody treatment, prior to transplantation.

Blocking Co-Stimulatory Signal Therapy

This therapy involves binding of inhibitors that interfere with the interaction of B7 and CD28 or CTL-4 or those involved in the reaction CD40 with its ligand CD40L. In the absence of this co-stimulatory signal, specific antigen-activated T_H cells become anergic but other responses are free.

By interaction of the B7 molecule on the membrane of antigen-presenting cells with the CD28 or CTLA-4 molecule on T cells, a co-stimulatory signal is the source of activation of T_H cell. For want of a co-stimulatory signal, antigen-activated T cells become anergic. Another source of activation of T cells is co-stimulatory signals mediated by CD40 molecule and CD40 ligand. In the in vivo studies on mice, it was observed that blocking the B7 molecule-mediated co-stimulatory signal with CTLA-4 after transplantation turned the host T cells anergic, letting the graft survive (Fig. 17-5). In the experiments, conducted by Allan Kirk, David Harlon and colleagues on kidney transplantation between monkeys mismatched at both class I and class II MHC antigens, treated with either CTLA-4-Ig or a monoclonal antibody directed against CD40L or both in combination, the grafts were retained for longer time than with a single treatment as above. A viable strategy has developed out of this experiment wherein without general immunosuppression and side effects of drugs, promising results have come out, paving the way for successful organ transplantation.

Treatment Possibility with Transplantation

As the research on transplantation of artificial vital organs has failed a transplant of natural organs, cells and tissues remains the only possible mode of treatment. There are certain problems in the transplantation of various cells, tissues and organs in human bodies, the most common problem being of shortage of organs. The problems of transplantation have been indicated under various sub-headings in the foregoing paragraphs, viz. side effects of immunosuppressive drugs, the careful treatment of antibodies directed against donor's cell mechanism etc. Despite the problems being faced in the field of transplantation, successful transplantation, and monitoring of the relative factors of the following cells, tissues, and organs have been reported worldwide.

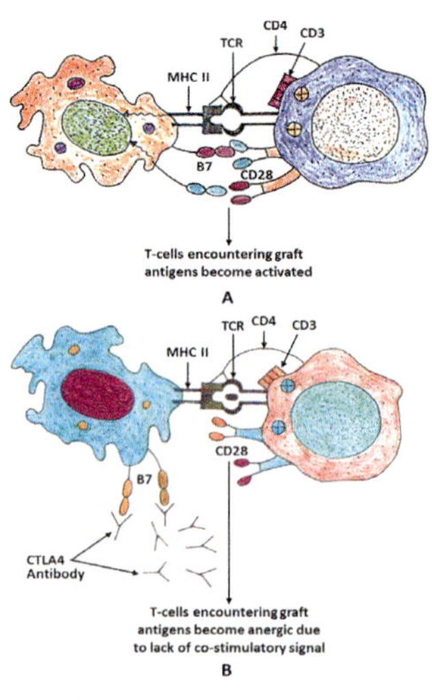

Fig.-17-5. Inactivation of T cells by blocking of co-stimulatory or secondary signal at the time of tissue transplantation.

Transplantation as carried out these days may be grouped under two groups; first being the transplantation of cells, tissues and organs requiring the most vigilant monitoring and maintenance of relative factors, and second of organs at privileged sites which do not require genetic matching because of the absence of lymphocytes.

Transplantation at unprivileged sites: The unprivileged sites of transplantation comprise such cells, tissues, and organs which require HLA balancing besides other factors that aid or hinder acceptance of the particular transplant. Some of such transplants are:

Kidney Transplantation: In case of failure of functions carried out by kidneys, caused by diabetes and various types of nephritis, kidney transplantation is required. A detailed HLA typing to ascertain that the patient has no antibodies or active cellular mechanisms directed against the potential donor's kidney is done before kidney transplantation. Moreover kidney transplantation has to be carefully made to survive with the expected span of life of the patient, so that no second implant is necessitated. The second implant has to be guarded against sensitization of the nutrient, which developed out of rejection of the first transplant. That any subsequent graft containing antigens in common with the first would be soon rejected. Balancing the survival of a transplanted kidney with the expected span of life of the recipient, does require the administration of some immunosuppressive drugs, which dreadfully have their side effects claiming on recipients' life, for example hypertension and metabolic bone disease.

Heart Transplant: The availability of the heart is the major problem in this type of transplantation. Accident victims however provide a minimal source of this vital organ. Another problem is the urgency of transplantation, to remove mechanical aids, which makes HLA matching almost practically unmanageable. A further problem is development of a new type of atherosclerosis in the coronary arteries of the implanted heart, guarded against antibodies- mediated injury to the vessels in the donated heart. The preservation of the heart (rhythmic beating) is again a barrier to its transplant beyond a limited period. However, heart transplants since 1964 (the first one carried out by Dr. Christian Bernard in South Africa and Dr. P. Venugopal of AIIMS in India on August 3, 1994) are being reported in greater number in the United States with 80% survival rate and varied longevity.

Lung Transplant: Cystic fibrosis caused by smoke inhalation requires lung transplantation, which being very little by itself or in conjunction with heart transplantation, is not widely known regarding the problems associated with it.

Liver Transplant: This large organ is damaged by viral diseases such as hepatitis or exposure to toxic chemicals such as alcohol. Once the causative factors are cleared, liver is capable of regenerating its functions, otherwise resulting in death of the patient. Mismatching of blood groups and lymphocyte-mediated reactions pose problems in liver transplantation. GVHD has also been reported in case of liver transplants, even with compatibility of blood groups. The United States has cases of liver transplants, though of a very limited rate of survival (one year) in about 65% of cases.

Bone-Marrow Transplants: Hematologic diseases such as leukemia, lymphoma, aplastic anemia, thalassemia major, and immunodeficiency disease specially SCID (severe combined immunodeficiency) require bone-marrow transplantation, obtained from a living donor by multiple needle aspirations (of erythroid, myeloid, monocytoid, megakaryocytic and lymphocytic lineages). The standard quantum of graft is 10^9 cells/kg of host body weight intravenously.

HLA typing is practical in case of bone-marrow transplantation, but finding a matched donor is a problem. GVHD has been commonly reported in bone-marrow transplantation, because of recognition of alloantigens on the host cells (by donor's T cells). GVHD in its severest form can cause generalized erythroderma of the skin, gastrointestinal hemorrhage and liver failure. Donor cells that survive the immunosuppressive treatment kill any host cells, so a low level donor T cell activity is desirable.

Pancreas Transplantation: Diabetes mellitus is caused by the malfunction of insulin-producing islet cells in the pancreas. Pancreas transplantation regulates the levels of insulin needed to make the diabetic individual normal. In pancreas transplantation, only islet cells are transplanted to restore function. The surgical and medical barriers to the procedure required in islet cells' transplantation are being reported from the states with increasing rates of survival of the patients.

Skin Transplantation: It is required in cases of severe burn as biologic dressings. This graft does not grow on hosts' body because cellular elements are no longer viable and require regular replacement. Allogeneic grafts have been reported in the presence of immunosuppressive drugs, which carry their own risks for the recipient.

Xenotransplantation: Xenograft is under trial for the reason of shortage of organs and imminent requirement of patients for graft. Although xenograft of kidney, heart, liver, and bone marrow has been attempted yet the requisite aim of restoring normal function of the related organ as graft, and recouping the weakened immune system (as in case of HIV infection, where baboon as donor was used) has desperately thwarted the attempts. In grafts of liver and bone-marrow, where baboons were used as donors, the survival rate of the patient and restoring normal immunity in the patient have been reported as failure.

With the rampant problem of rejection of graft and especially in the case of xenograft, however, because of rapid breeding, large litters, capacity of sharing a number of anatomical and physiological similarities with humans, pigs are under consideration as potential transplant donors. In xenograft, humoral antibody and complement mediated hyperacute rejection of graft has been the most challenging problem. The complement activity and hyperacute rejection are reduced in their severity by the presence of DAF (decay-accelerating factor) on the cells of the implanted organ. In transgenic donor pigs, the DAF can be produced, so there is a possibility of further graft trials on the line of this factor on humans. The fact of the survival rate corresponding to the natural make-up of the specific organ in relation to that of span of human lives is likely to pose further problems and complications. Xenograft poses the most threatening problem of the spread of pathogens in the host's body, which holds true in the case of species closely related to humans more than those, such as pigs which are not closely related to humans. It is this fact, which has provoked xenograft in humans, for some specific organs only.

Grafts at Immune Free Sites: There are some sites in humans and other species which are sequestered from the cells of the immune system. Grafts at these sites are therefore grafts at immune free sites. From this understanding there has developed a scientific concept to physically sequestering the grafted cells, so that the immune system cannot penetrate the sequestered phase. In a study, the pancreatic islets were encapsulated in a membrane fabricated from an acrylic copolymer and then grafted into the diabetic mice. The islet cells so encapsulated were not rejected, and the normal levels of insulin were observed as being produced. This has encouraged such human trials.

Further grafts into sites that are derived from lymphatic vessels and blood vessels, such as anterior chamber of the eye, the cornea, the uterus, the testes and the brain are suitable for allografts, because alloantigens of the graft are unable to sensitize the recipient's lymphocytes, increasing chances of establishment irrespective of HLA matching. Experiments with rats for pancreatic islet cells being allogeneic have supported this fact of grafts into immune free sites.

Since, further experiments on grafts in the frame of reference of various hypotheses are going on, insurmountable barriers of the immune system, coupled with immunosuppressants, appear to be handled in the near-future.

CHAPTER – 18

CANCER AND IMMUNE RESPONSE TO CANCER CELLS

Unregulated cell growth leads to altered self cells developing into cancer. In cancer, the production of new cells is not regulated nor responds to normal growth control mechanisms. These unregulated cells produce clones of cells, which increase in abnormal sizes called tumors. The tumors pass through various stages and finally reach the malignant stage where treatment is the most challenging phenomenon, carrying the disease into the top killer diseases. In view of varying records of killer diseases world-wide, cancer has assumed various rungs in various countries of the world amongst killer diseases.

This chapter discusses the nature and types of cancer, cancer related genes, multiple mutations in the development of cancer, immune mediated tumors, tumor immunology (tumor antigens and immune response), immune evasive mechanisms of tumors, and immunologic manipulation and monitoring of cancer.

Nature and Types of Cancer

The life cycle of cells under normal conditions is a regulated phenomenon, under which they are produced and decayed in a given life span. In cancer, the cells become unregulated and produce clones of cells that increase in abnormal sizes. This growth of cells into an abnormal mass is called a malignant tumor or cancer. From this perspective, cancer cells are altered self-cells. Evidently, these altered self-cells have escaped or evaded normal cell growth-regulating mechanisms.

In their life cycles, these malignant tumors are no longer malignant, since they do not assault the neighboring tissues. The malignant tumors pass through a stage called metastasis, in which small clusters of cancerous cells separate from the malignant tumors and proceed to the circulation through which they are carried to other tissues (Fig. 18-1). The proliferation of these separated cancerous cells continues in the body unabated unless partly inhibited by some immunologically active devices or drugs. The alarming factor about the characteristics of cancer is that for long in its primary stage (so long as they do not assault the neighboring tissues), it goes undiagnosed for the specific symptoms often confused for want of a set parameter with other common ailments. However, early diagnosis of cancer is being worked out extensively to check the disease at its primary stage, to which this chapter will refer in the coming headings.

Generally there are three types of cancer-

- Carcinomas- found in the skin or the epithelial lining of internal organs and glands (colon, breast, prostate, lung).
- Leukemia/Lymphomas- found in bone marrow. Leukemia develops in a single cell whereas lymphoma grows as a tumor mass. These both are malignant in nature and awfully life threatening.
- Sarcomas- found in bone, fat, and cartilage as the most frequenting disease.

As cancer shows a malignant form of cells, several chemical agents have been tested to induce cell transformation that leads to the establishment of cancer cells. Carcinogens are those chemical and physical agents which have the capacity of transforming a cell into a malignant one. The morphology and growth properties of cells can be relatively affected by chemical carcinogens, physically by irradiation and introduction of certain viruses (through any carrier or system of the body). In vitro, the properties of transformed cells and cancerous cells are similar. As the malignant state (transformation of a cell into malignant) is inducible, the chemical or physical agents initially cause changes in the genome and then stimulate cell division. This cell division is categorically malignant transformation of cells.

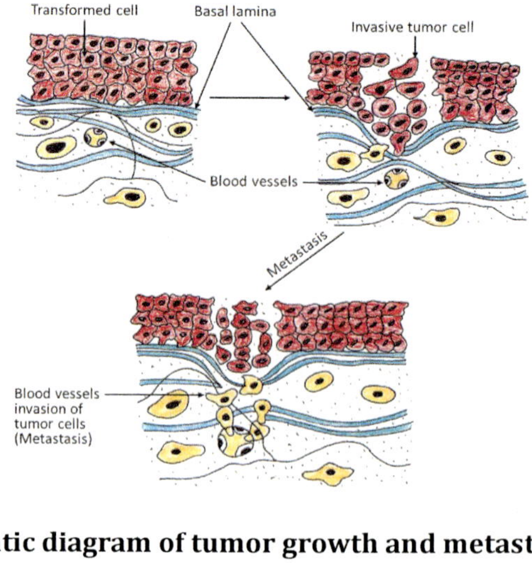

Fig.-18-1. Schematic diagram of tumor growth and metastasis through blood vessels.

DNA and RNA viruses have also been shown to induce malignant transformation of cells, as stated above, besides chemical and physical agents. Two DNA viruses SV40 and polyoma are known to cause malignant transformation by encoding early proteins called large T, middle T and little T, each of which causes malignant transformation of cells being infected by these viruses. Retroviruses carry oncogenes, which also lead to malignant transformation of cells. These retroviruses are called transforming retroviruses- HIV-1, HIV-2 and Rous Sarcoma Virus, as they transform cells into malignant ones. Rous Sarcoma virus carries an oncogene called v-src which when cloned and transfected into normal cells in culture, the cells more transformed into malignant cells.

Two types of oncogenes were later identified in studies by Howard Themin, J.M. Bishop and H.E. Varmus. They concluded that a virus might acquire oncogenes from the genome of an infected cell called cellular oncogenes (c-one) or proto-oncogenes. A DNA sequence in normal chicken cells similar to v-src of Rous Sarcoma virus was termed c-src. For vividness, these oncogenes were termed c-one and v-one, consisting of a series of exons and introns with high degree of homology. Most of the oncogenes, both viral and cellular are derived from cellular genes that encode various growth-controlling proteins and exhibit similar functions.

Cancer Related Genes

There are a number of genes, which contribute to the process of cellular proliferation balanced by cell death. At any stage, whether it be of cellular proliferation or cell death, if there occurs imbalance, a state of malignancy or cancer will develop. The two genes which maintain this balance are oncogenes and tumor suppressor genes.

By the function, the above genes perform, three categories of cancer related genes have been identified:

- Proto-oncogenes for cellular proliferation,
- Tumor suppressor genes or anti-oncogenes for inhibition of cellular proliferation, and
- Anti-apoptosis gene for regulation of programmed cell death.

Proto-oncogenes and their oncogenic counterparts encode proteins that induce cellular proliferation. Proteins, which function as growth factors or growth factor receptors are sis (encodes a form of platelet-derived growth factor), fins, erbB and neu (encode growth factor receptors). Normal cells express well regulated growth factors and their receptors, resulting in secretion from one population of cells acting on another population of cells. Thus, unless there is an appropriate expression from either growth factors or their receptor, the normal proliferation of cells goes on. Failure of either the growth factor or their receptors, in secretion, results in uncontrolled proliferation. The state of uncontrolled proliferation is likely to develop into a transformed cell. Thus, proto-oncogenes are cancer related genes. Table 18-1 shows the various antigens known to be associated with cancer.

In order to inhibit excessive cellular proliferation there are a second category of cancer related genes called tumor-suppressor genes or anti-oncogenes, which encode proteins that inhibit excessive cell proliferation. Failure or inactivation of these genes results in unregulated proliferation. This gene is named as the retinoblastoma gene (Rb), characterized by an immature retina. This type is purely hereditary, causing genetic abnormality.

The third category of cancer related genes encode proteins that either block or induce apoptosis, hence called anti-apoptosis genes. It is associated with B cell follicular lymphoma, and regulates cell survival during hematopoiesis and in the survival of selected B cells and T cells during maturation. Its prototype is bcl-2 and has sequence homology to Epstein-Barr virus.

In a study of cells from various types of human cancers, there have been found cellular oncogenes that cause increased prodution of oncogene products. This suggests that proto-oncogenes may be changed by cellular oncogenes. Further, some cancer cells exhibit movement of a proto-oncogene from one chromosomal site to another, causing synthesis of the c-Myc protein (a transcription factor). This movement of proto-oncogenes increases the synthesis of c-Myc protein. A proto-oncogene may be changed into a transforming oncogene by itself from viral integration into the host cell genome, such as avian leukosis virus (a retrovirus), which changes B cells into lymphomas. This particular retrovirus is capable of integrating into c-Myc gene and increases its synthesis. Proto-oncogenes are also transformed into a cancer-inducing oncogenes by chemical carcinogens and X-irradiation as found in carcinomas of bladder, colon, and lung.

Table 18-1: Functional classification of cancer-associated genes

Gene	*Nature of gene product*	*Function*
Growth factor		
sis	A type of platelet-derived growth factor (PDGF	Cell proliferation
Growth factor receptors		
fms	Receptor for colony-stimulating factor (CSF-1)	Cell proliferation
erbB	Receptor for epidermal growth factor (EGF)	Cell proliferation
neu	Protein (HER2) related to EGF receptor	Cell proliferation
erbA	Receptor for thyroid hormone	Cell proliferation
Signal transducers		
src	Tyrosine kinase	Cell proliferation
abl	Tyrosine kinase	Cell proliferation
Ha-ras	GTP-binding protein with GTPase activity	Cell proliferation
K-ras	GTP-binding protein with GTPase activity	Cell proliferation

Gene	Nature of gene product	Function
Transcription factors		
jun	Component of transcription factor AP1	Cell proliferation
fos	Component of transcription factor AP1	Cell proliferation
myc	DNA-binding protein	Cell proliferation
Apoptosis regulator		
bcl-2	Suppressor of apoptosis	Cell death
Tumor suppressors		
Rb	Suppressor of retinoblastoma	Inhibit proliferation
p53	Nuclear phosphoprotein that inhibits formation of small-cell lung cancer and colon cancers	Inhibit proliferation
DCC	Suppressor of colon carcinoma	Inhibit proliferation
APC	Suppressor of adenomatous polyposis	Inhibit proliferation
NF11	Suppressor of neurofibromatosis	Inhibit proliferation
WT1	Suppressor of Wilm's tumor	Inhibit proliferation

Multiple Mutations Cause Cancer

Physical mutations have a dominant role in converting a normal cell into a cancerous cell. This has been observed in human colon cancer, which exhibits a series of morphologic stages in development from a benign tumor in the colorectal epithelium through a disorganized cellular mechanism until growth into a malignant tumor. This gradual development of a precancerous cell into a cancerous cell is again related to gene changes or myriad chromosomal abnormalities, led by inactivation or loss of tumor suppressor genes (APC, DCC and p53), and activation of one cellular proliferation oncogene. Actually, cancer, as stated herein, is a multi step phenomenon of cell proliferation based on chromosomal abnormalities, as observed in studies with transgenic mice. It was observed that transgenic mice that have a c-myc translocation to the immunoglobulin H-chain locus and high levels of Bcl-2 develop lymphomas very rapidly. The same was the case with double-transgenic mice, as leukemia was developed by mating bcl-2^+ with myc$^+$ transgenic mice.

Cancer of the Immune System

Some types of cancer are generated in the immune system itself and cause bloom of immunopotent cells. These are very deleterious and completely paralyze the immune function. These are as follows:

Leukemia: In it, single cells proliferate and increased cell number in the blood or lymph is observed progressing from mild symptomatic disease to aggressive disease involving mature cells. Leukemia in its initial stage proliferates and forms as liquid tumors. These may be acute or chronic according to the clinical sign and the progression of tumors. Generally, chronic leukemia is originated from mature lymphoid cells unlikely to the acute leukemia. Acute leukemia may be acute lymphocytic leukemia (ALL) and acute myelogenous leukemia (AML) according to the origin of cells. These are very aggressive and at any age, aggravate cell numbers in the blood or lymph.

Lymphoma: In it also, single cell proliferates and cell number in the blood or lymph is observed progressing. Lymphoma grows and forms as a solid tumors within the lymphoid tissues such as bone marrow lymph nodes or thymus. However, several lymphoma types such as Dalton's Lymphoma have been well reported to grow as ascites in the peritoneum in a single cell suspension and as solid forms in the lymph nodes. The lymphoma may be Hodgkin's lymphoma or non-Hodgkin's lymphoma. Hodgkin's Lymphoma is characterized by the presence of peculiar cell type, large in size called Reed-Sternberg cells, while non-Hodgkin's lymphoma does not have this particular cell type. Both of these may be of B and T cells. When the B cells become malignant, then it is called as B cell lymphoma and when the T cells become malignant, then it is called as T cell lymphoma. These may be Hodgkin's lymphoma or non-Hodgkin's lymphoma, designated according to the presence of Reed-Sternberg Cells.

In both leukemia and lymphoma, translocations of c-myc to the Ig heavy chain gene cluster on chromosome 14 have been found the causative factor. Morphologically, the translocation removes the c-myc coding exons from the regulatory mechanisms operating in chromosome 8 and places them in the Ig region. Thus, myc expression mediates the development of cancer of immunological origin.

Immune Response to Tumor

Immune response to antigens on tumor cells is the subject matter of tumor immunology. There are two types of antigens on tumor cells, which are responded by the immune system- tumor-specific transplantation antigens (TSTA) and tumor associated transplantation antigens (TATA). The tumor-specific transplantation antigens are specific to tumor cells and do not affect the neighboring normal cells. A cell mediated response by tumor-specific CTLs is directed to these tumor specific transplantation antigens. The tumor-specific transplantation antigens may result from mutations in tumor cells that generate altered cellular proteins. Cytosolic processing of these proteins gives rise to novel peptides that are presented with class I MHC molecules.

The tumor associated transplantation antigens are proteins that are expressed on normal cells during fetal development, encoded by embryonic genes, which are later expressed on the fully differentiated tumor cells.

There may be any of the following types of tumor antigens recognized by human T cells:

- antigens encoded by genes with tumor specific expression,
- antigens encoded by variant forms of normal genes altered by mutation,

- antigens normally expressed only at certain stages of differentiation or differentiation lineages, and
- antigens that are over expressed in particular tumors.

Origin of Tumor Specific Antigens

Tumor specific transplantation antigens (TSTA) originated from chemical or physical carcinogens and some viruses. By use of methycholanthrene (MCA), a chemical carcinogen, lines of tumor cells may be generated. Two ways to develop tumor cell lineages have been under study with the use of chemical methycholanthrene. One by injecting killed cells derived from a chemically induced tumor cell lineage of an animal, and the other by tumor-specific transplantation antigens of chemically induced tumors, a immune response can be generated.

In chemically induced tumors, the specific cell lineage can protect against the live cells of tumor on the same lineage but other tumor cell lines do not express any such preventive output. This means that the immune response to one cell lineage is not capable of protecting against the other cell lineage, if any.

Tumor specific transplantation antigens of chemically induced tumors are not easily characterized but the study of their gene encoding (involving T cell mediated rejection) (Figure 18-2). Two methods have been evolved to characterize tumor-specific transplantation antigens (TSTAs). In the first method, peptides bound to class I MHC molecules on the membranes of the tumor cells are eluted with acid and purified by high pressure liquid chromatography (HPLC). In the second method, high level expression of the tumor-cell DNA is established by preparing cDNA library of tumor cells. Usually, TSTAs differ from normal cellular genes by a single point mutation and characteristically they are short peptides.

X-irradiation is a potent carcinogen to induce tumors in the cell lineages, which are responsive to radiation, though there always run the risk of damage to the neighboring cells in radiation.

Tumor specific antigens are also induced by viruses. Tumors induced by viruses express tumor antigens shared by all tumors induced by the same virus. It was observed that injection of killed cells from a particular polyoma induced tumor protected the recipients from live cells from any polyoma induced tumors. Burkitt's lymphoma cells express a nuclear antigen of the Epstein-Barr virus which may be a tumor specific antigen for such a tumor. Cervical cancers are induced by human papilloma virus (APV) encoding E6 and E7 proteins. That tumor specific antigens trigger rejection of all tumors induced by the virus.

Origin of Tumor Associated Antigens

Tumor antigens are also present on normal cells of the fetus in a dormant state, or at low levels, which express at higher levels in normal adult cells. When the tumor antigens are express at higher levels in normal adult cells they are called growth factor-related antigens, since they serve as tumor associated antigens. The higher levels of expression of tumour antigens in adults' cells are a transferrin growth factor, designated p97. Normal cells express less than 8000 molecules of p97 per cell melanoma cells express 50,000-500,000 molecules of p97 per cell. Experimentally, it has been observed that a recombinant vaccinia virus prepared from gene that encodes p97 on injection into mice induced both humoral and cell mediated immune response protecting the mice against live melanoma cells expressing the p97 antigen.

Tumor antigens found on normal fetal cells are called oncofoetal tumor antigens, and appear early in the fetal embryonic stage. If these antigens appear later on cancer cells, they are likely to induce and immunologic response. Alpha-fetoprotein (AFP) and carcinoembryonic antigen (CEA) are the two oncofoetal antigens. The concentration of elevated AFP is present in patients of liver cancer, and CEA is present in patients of gastrointestinal tract cancer besides liver cells expressing increased serum levels. As the tumor grows, the serum level of CEA is monitored to inhibit the growth. Thus, the CEA may be present in normal adults, who may have colorectal cancer. If by surgery this colorectal carcinoma is removed, the CEA will have to be monitored to vigil the resumption of tumor growth. Table 18-2 gives a comprehensive picture of AFP and CEA concentration in the serum of patients suffering from various diseases.

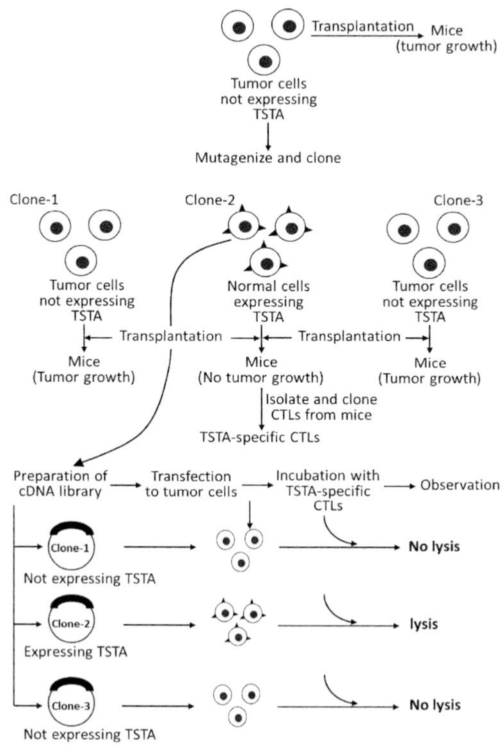

Fig.-18-2. Procedure for identifying a gene encoding tumor-specific transplantation antigen (TSTA) by the preparation of cDNA library.

Some tumor associated antigens are also present in normal cells encoded by the proto oncogene, which express no qualitative difference from the oncogene. The immune system recognizes the oncogene products. Human breast cancer cells exhibit elevated expression of the oncogene encoded Neo protein (a growth factor receptor). Normal adult cells express only trace amounts of neo protein. In breast cancer cells and normal adult cells, the amounts of neo level are recognized by anti neo-monoclonal antibodies whereby cancer cells are eliminated and the neighboring cells remain unaffected.

Table 18-2: Increase in Alpha-fetoprotein (AFP) and Carcinoembryonic antigen (CEA) level in the serum of patient with various diseases

Disease	*No. of patients tested*	*% of patient high AFP (>400µg/ml)*	*% of patient high CEA (>10mg/ml)*
Alcoholic cirrhosis	120	0	2
Hepatitis	100	1	0
Hepatocellular carcinoma	100	69	0
Breast carcinoma	125	0	14
Colorectal carcinoma	544	0	35
Gastric carcinoma	79	0	19
Non-carcinoma malignancy	228	0	2
Pancreatic carcinoma	55	0	35
Pulmonary carcinoma	181	0	26
Pulmonary emphysema	49	0	4
Ulcerative colitis	146	0	5

Besides the above, several tumor associated antigens are found on human melanomas. These antigens are oncofoetal type antigen and are expressed in human melanoma in significant proportion. Common tumor antigens offer hope for the design of therapeutics and for also anti-tumor immunization.

Response of Immune System to Tumor Cells

Response to tumor cells by the immune system is, however, interesting, as affected in early stages of cancer claiming lives in later stages of development. Both the humoral and cell-mediated immune responses cause the elimination of tumor cells, but the role of CTLs is major in the elimination process as they recognize tumor antigens presented by class I MHC on the tumor cells. Sometimes, the role of CTLs is limited when the expression of class I MHC molecules is decreased in a number of tumors.

The immune response to tumors includes;

- CTL-mediated lysis

- NK-cell activity
- macrophage-mediated tumor elimination
- ADCC mediated elimination of tumors
- Cytotoxic factors (TNF-a and TNF-b) mediated tumor elimination.

NK cells and macrophages are not restricted by MHC expression exhibited by some tumor cells. Both express Fc receptors, enabling them to bind to antibody on tumor cells and mediate ADCC. Tumor Necrosis Factor (TNF-a) is a potent antitumour cytokine secreted by macrophages. Injection of TNF-a in animals having tumor resulted in necrosis.

The cell-mediated branch of the immune system incessantly keeps a patrolling of the body and eliminating cancer cells, which often arise in the body. Under this type of continual patrolling (a type of surveillance), tumors arise in case of escape from the surveillance or in case of impairment of the immune system by any reason. As the immunosuppressive drugs exert on impairing effect on the immune system, transplantation patients are likely to show an increase in cancer, but this is not always the case. Patients of common cancers of the lung, breast, and colon, do not show an increase in the cancers of the immune system. There are other concepts to counter this postulate of immune surveillance, which render this postulate unprove.

Immune Evasive Mechanisms of Tumor

The ever-increasing toll of death by cancer world-wide suggests that tumor cells must be evading the immune system at the very early stage of their development, seemingly so powerful and effective. The tumor cells must be adopting some mechanisms to evade the immune response. The tumor cells may evade the immune response by any or all of the following mechanisms:

- by modulating their tumor antigens,
- by reducing their expression of class I MHC molecules, and
- by antibody-mediated or immune complex-mediated inhibition of CTL activity.

Since, it was observed that antibodies could be produced to tumor-specific antigens, tumor growth in animals were subjected to passive immunization or active immunization, but the tumor growth could not be inhibited, rather tumor growth following immunization increased. This observation led to the supposition that tumor antigens must be adopting any mechanism to evade immunity induced passively or actively. For reaching a conclusion, serum from animals with progressive tumor growth was studied *in vitro* to confirm, if cell-mediated lympholysis was the mediating factor. It was found that serum taken from animals with regressing tumors did not block the cell mediated lympholysis, whereas serum taken from animals with progressive tumor blocked the cell-mediated lympholysis (CML). Later in the sera of children with progressive neuroblastoma, some kind of blocking factor was noticed by K. and I. Hellstrom. This observation led to the conclusion that some blocking factors let the tumor grow in human. Supposedly, antitumour antibody masks the tumor antigens from cytotoxic T cells, and so do antibody complexes, inhibiting ADCC by binding to Fc receptors. Thus, this is one mechanism (iii above) by which tumor cells evade the immune response.

Another mechanism by which certain tumor specific antigens evade the immune response is known as antigenic modulation (i above). In adopting it, certain tumor-specific antigens no longer appear on the surface of tumor cells in the presence of serum antibodies. Mice immunized with a leukemic T cell antigen, when injected with leukemic T cells develop high titers of anti-TL antibody which binds to the T cell antigen on the leukemic cells inducing capping, endocytosis and shedding of the antigen-antibody complex. As a result of this binding, as long as antibody is present, these leukemic T cells fail to display the T cell antigen. By this modulation T cell antigens evade killing.

By reducing the expression of class I MHC molecules, tumor cells evade the immune response (as in the above). Malignant transformation of cells is often associated with a reduction (or absence) of class I MHC molecules. Besides above, absence of sufficient numbers of antigen presenting cells in the nearness of a tumor, the T cells may not receive a co-stimulatory signal, thwarting the T cells activation and thus resulting in clonal energy.

Cancer Immunotherapy or Immunologic manipulation of Cancer

Immunologic manipulation and monitoring of cancer aim at increasing or supplementing the natural defense mechanisms against cancer. The manipulative and monitoring devices are exclusively experimental envisaging clinical trials. These include:

- Increasing the co-stimulatory signals.
- Using recombinant cytokines.
- Using monoclonal antibodies.
- Combining antibodies against tumor antigens with toxins, chemotherapeutic agents or radioactive elements.
- Vaccination against clearly identified tumor viruses.

The T cell is activated by an activating signal. When a peptide MHC molecule complex is recognized by the T cell receptor and B7 ligand on antigen presenting cells interacts with CD28 on T cells, a costimulatory signal is triggered. Both these signals are required to activate T cells. Co-stimulatory signals activate CTL precursors (CTL-Ps). Incubation of mouse with melanoma cells in vitro for want of costimulatory signals do not cause proliferation of CTL-Ps and effector cells do not differentiate. But melanoma cells transfected with the gene that encodes the B-7 ligand causes differentiation of CTL-Ps into effector CTLs. This observation has suggested the possibility of CTL response in vivo by using B-7 transfected tumor cells. Figure 18-3 is illustrative of this phenomenon i.e., use of transfected tumor cells for cancer immunotherapy.

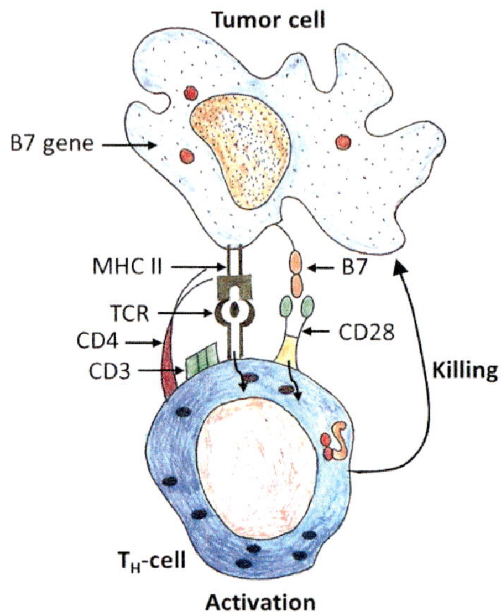

Fig.-18-3. Application of transfected tumor cells with B-7 co-stimulatory molecule for cancer immunotherapy.

Using recombinant cytokines for increasing the immune response against cancer has been tried in respect of IFN-a, b and g, IL-1, IL-2, IL-4, IL-5, IL-12, GM-CSF and TNF, since each of these enhance a particular branch of the immune response. But some cytokines act antagonistically and this posed an obstacle in using this type of device for manipulating immune response. However, partial or complete tumor regression has been observed with usage/ administration of recombinant cytokines. Hematologic malignancies (leukemia, lymphomas, and myeloma) and solid tumors (melanoma, Kaposi's sarcoma, renal cancer and breast cancer) have regressed with daily injections of recombinant IFN-α. Similarly IFN-γ has been reported to increase the activity of TC cells, macrophages and NK cells, all of which have destructive effect on tumor cells. Likewise, TNF-α or TNF-β has exerted hemorrhagic necrosis and regression effect on tumor antigens.

The flow of blood and oxygen to the vascular endothelium cells in the vicinity of a tumor causes progressive growth of tumor. TNF-α has partially or completely inhibited the flow of blood and oxygen, thereby inhibiting vascularisation. Interferon may act by restoring MHC expression, thereby increasing CTL activity against tumors, and inhibiting cell division of both normal and malignantly transformed cells in vivo. Thus, interferons inhibit tumor cell proliferation. In vitro activation of lymphocytes with antitumour specificity has been shown acting against tumor antigens. In animal studies, in vitro culturing of lymphocytes with X-irradiated tumor cells in the presence of IL-2 and added tumor antigens have shown destructive effects on tumor in tumor-bearing animals. In this method of sensitizing lymphocytes to tumor antigens if tumor antigens were not added, the presence of high concentrations of cloned IL-2 was enough to exert an elimination effect on tumor cells sparing the normal cells. High concentrations of cloned IL-2 activate large numbers of lymphoid cells called lymphokine-activated killer cells (LAK cells, after S. Rosenberg). Rosenberg found that LAK cells plus recombinant IL-2 mediate the destruction of tumor cells in tumor bearing animals, because of TCR-bearing cells in LAK populations.

The tumoricidal activity of LAK cells can be generated in vitro and these cells are active against a wide variety of tumors, their impact has been confirmed in several trials on human tumors. High concentrations of cloned IL-2 needed for LAK cell activity have ignited side effects, though sparing normal cells, such as vascular leak syndrome in which lymphoid cells and plasma emigrate from the peripheral blood into the tissues leading to shock. Figure 18-4 shows the mechanism of tumor growth regulation by the cytokines.

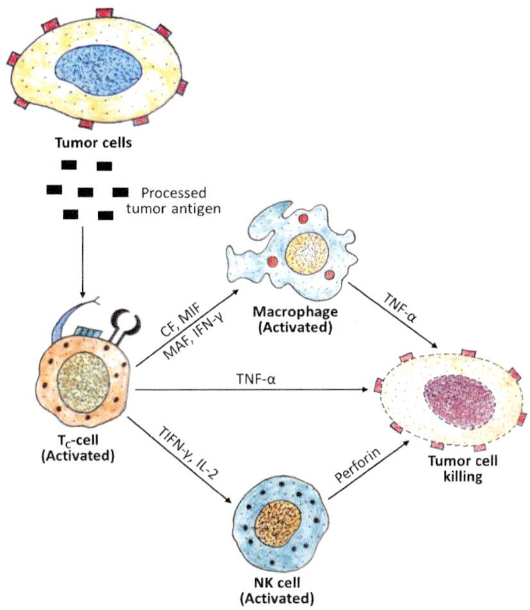

Fig.-18-4. Role of cytokines in tumor cell killing.

Biopsy samples of tumors have shown lymphocytes that have infiltrated the tumor and acting as an antitumour agent. If these lymphocytes are expanded *in vitro* with IL-2, a similarity in action to that of LAK cells is observed. Lymphocytes of tumors activated *in vitro* with IL-2 are called tumor-infiltrating lymphocytes (TILs). TILs bear the cytolytic activity against their autologous tumor i.e., they are tumor-specific and require much lower levels of IL-2 for their activation than LAK cells. Renal cell carcinomas and malignant melanomas have shown partial regression with injection of TILs plus recombinant IL-2. TILs populations for the purpose of administration in tumor patients are expanded *in vitro* from biopsy samples taken from patients of solid cancer.

Using monoclonal antibodies as immunotherapeutic agents have been in use in treating human B cell lymphomas and T cell leukemia. R. Levy and his colleagues in treating a case of terminal B cell lymphoma, which then had metastasized to the liver, spleen, bone marrow and peripheral blood, produced mouse monoclonal antibody specific for the B-lymphoma idiotype (cancerous cells express their particular idiotypes). As the case was that of B cell lymphoma, a monoclonal antibody was produced for the B-lymphoma idiotype which bound specifically to the B-lymphoma cells. Since B-lymphoma cells are susceptible to complement mediated lysis, the monoclonal antibody activated the complement system and lymphoma cells were lysed sparing the other cells. The patient enjoyed remission of B cell cancer for a long period of time, as the tumors had shrunk with successive administration of anti-idiotype monoclonal antibodies. As the raising of monoclonal antibody for each case is not practical, in view of large numbers of patients diagnosed of B-lymphoma, patients are immunized with their own tumor-specific antigens, the idiotypically unique immunoglobulins produced by their own lymphomas. This resulted in the development of anti-idiotype antibodies against their own tumors. However, complete remission in a large number of patients was not seen. This anti-idiotype approach is patient-specific and used to prepare tumor-specific anti-tumor agents. The mechanism of antibody dependent tumor cell lysis is shown in **Figure 18-5**.

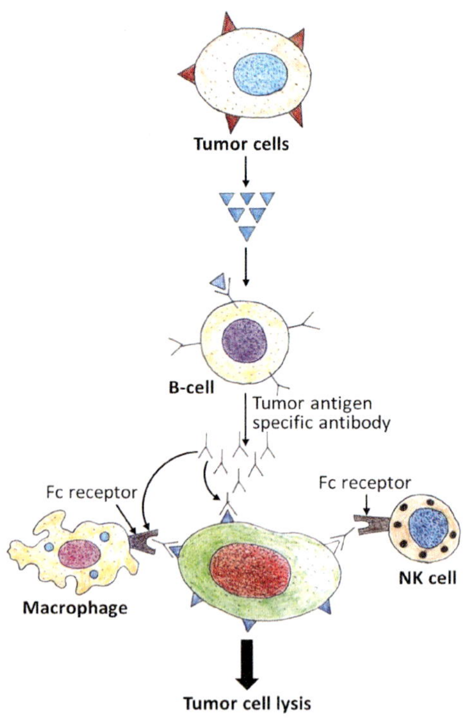

Fig.-18-5. Schematic representation of tumor immunity by antibody-dependent cell mediated cytotoxicity (ADCC).

In preparing tumor specific antibodies or tumor associated antibodies, toxins, chemotherapeutic agents (drugs) and radioactive isotopes have recently been tried with leukemia and lymphoma patients. The toxins are delivered specifically to the tumor cells, while normal cells remain unharmed. Large tumor cells though have not been successfully targeted by this joint approach, a variety of hematologic malignancies and solid cancers have been treated with varied period of remission.

Vaccination against clearly identified tumor viruses- papillomavirus (in cervical cancer), hepatitis virus (in liver cancer), since from a global perspective, viruses are a significant cause of cancer, is under active consideration in clinical trials. Vaccination against the cancer producing viruses is likely to be the most effective strategy in combating cancer. The identification of significant tumor antigens by genetic or biochemical approaches, the development of strategies for the effective presentation of tumor antigens, and the generation of activated populations of helper or cytotoxic T cells will be added strategies against cancer.

CHAPTER – 19

EXPERIMENTAL SYSTEMS AND PRACTICAL APPROACH

Generally two types of systems are used to study the complex cellular interactions of the immune system. One is called the in vivo system in which the whole animal provides the most natural experimental conditions. The results obtained by the in vivo systems, owing to their complexity at times render ambiguity to the interpretation of data. This is the limitation of this type of experimental study carried out on animals. The other system is known as in vitro, in which individual cellular interactions are studied under controlled and repeatable conditions. The artificiality is the remarkable limitation of this system, such as, if antigens are to be provided to purify B cells, unless T cells are present, maximal antibody production will not be stimulated. This implies that in an artificial *in vitro* study, a lack of T cells may thwart correct conclusions.

The experimental systems used in the study of the complex immune system, mostly animal models are used, for the reasons of non-availability of human species, involving risks of life and many other indirect socio-legal factors. The various techniques employed in experimental study of the immune system are the subject matter of this chapter.

Experimental Model Animal

The choice of animals for experimental study corresponds to the type of study to be carried out. For example, in a study of antiserum, a goat, rabbit, sheep or horse prove as suitable ones; in preparing a vaccine the animal chosen must be susceptible to the particular infectious agents, so that the efficacy and route of the vaccination may be determined; but in case the efficacy is to be studied in reference to humans, and primates, the animal chosen must be monkeys, chimpanzees or baboons. However, it is carefully conducted to ensure that the endangered species (monkeys, chimpanzees, and baboons) used in experimental study for the benefit of humans, are protected from extinction. However, mouse strains have been the most suitable models for various types of experimental studies.

Among the animals as an experimental system used in research and teaching, and testing of drugs, mice comprise a majority of all experimental mammals. The remarkable genetic similarity of mice to humans, combined with great convenience, perhaps accounts for mice so often being the experimental model of choice in research. When the use of mice in scientific research was started, is not certain. But it has been documented that mice have continued to be an experimental model of research since 1894, when it was first used for cancer research. Initially, mice were used for same-species tumor transplantations and drug treatment studies. In 1921, at first, inbred strains that were predisposed to getting tumors, were started and disseminated among cancer researchers by Leonell Strong. Many more strains of mice were originated beginning in 1929 with the founding of Jackson Laboratory in Bar Harbor, Maine, USA, now the largest supplier of mice. In 1962, the discovery of a mutant mouse with low immunity led to human tumor transplantation, a valuable breakthrough for cancer research that enable cancer researcher to study the human cancer in the experimental animal models. A further breakthrough in the late 1980s led to transgenic mice, those whose genes have been altered to produce a desired characteristic; oncogenes could then be studied in greater detail. Thus, now-adays, the importance of mice as laboratory animals in experimental cancer research, experimental immunology, and other biomedical research is undoubtedly indisputable. However, the choice of animal models and techniques adopted in studies is corresponding to the specific type of immune response to be studied.

Why Mice are the Primary Selection? Mice are the mammals (Taxonomic position is given in Box 1) having great genetic similarities with humans. Besides being genetically similar to humans, mice are small and inexpensive to maintain. Significant numbers can be housed in relatively little space because of their small body size. Mice adapt well to laboratory housing and can be housed socially and individually. Their short life span and rapid reproductive rate make it possible to study disease processes in many individuals throughout their life cycle and make then the mammal of choice for fine-tuned genetic manipulation. Further, mice with many different special features have been created and bred.

Box 1. Taxonomic Position of Mice.
Class: Mammalia
Order: Rodentia
Family: Murinidae
Sub-family: Murininae
Also termed as Murine animal

How Mice be Handled? Quality of scientific research is very much dependent on the care and handling of mice or laboratory animals. This is an obligatory factor when animals are used as a source of biological materials such as cells, antibodies and serum for laboratory use or as experimental subject. Research on animal is, therefore, best served by a coordinated, multidisciplinary approach that addresses program management, ethics of animal experimentation, animal environment maintenance, and sound animal research methodology.

Box 2. Mouse by Design: Some Strain Types.

Inbred: Genetically identical population of mice due to inbreeding.

Transgenic: Mice that have been genetically engineered and altered by injection of one or more genes.

Immunodeficient: Mice that have minimal immune functions, including nude mice and mice with severe combined immune deficiency (SCID).

Knockout: Mice that are engineered to lack a specific gene.

Germfree: Mice that is free from all detectable viruses, bacteria and parasites.

Therefore, investigators, caregivers and those responsible for resource management must work together to establish a program of animal care and use that reduces unwanted variables in animal research, and assures humane treatment of laboratory animals under appropriate ethical consideration. Frameworks for the humane and ethical use of experimental animals have been proposed by Russel and Burch in 1959 and still it has been continued to be applicable. Experimentation with animals should be replaced with non-animal methodology, or experimental methodology and techniques should be refined to minimize the experience of pain or distress to laboratory animals. For experimental model animal handling following major points should be noted;

- **Caging:** Animals must be kept in minimum space recommendation to ensure freedom of movement and normal posture adjustments. Crowd may lead to undue stress and therefore variations in results.
- **Feed:** Standardized and balanced pelleted diets made up of natural, purified, or chemically defined ingredients, containing the nutrient requirements should be used.
- **Bedding:** Animals can be kept either in solid-bottom cages with contact bedding, or in suspended-wire cages with non-contact bedding beneath the cages. For mice, solid bottom cages with contact bedding made up of usually rice husk is maintained.
- **Water:** Continuous access to fresh uncontaminated water must be provided to avoid suffering from any water borne diseases.
- **Sanitation:** Regular and frequent sanitation of the animal's living space is necessary to minimize the impact of environmental contamination. The

bedding must be changed 2 to 3 times per week, and cage should be thoroughly washed with detergent at least ones in a week. The floors of the animal's room should be mopped daily with a disinfectant solution.

- **Environmental Monitoring and Maintenance:** The monitoring and maintenance of controlled temperature, humidity, ventilation, and illumination are critical to the stabilization of animal metabolism and behavior that must be taken into care.
- **Health monitoring:** Regular health monitoring is required to minimize the variations in results obtained from laboratory animals.

Blood

Blood is viscous liquid connective tissue playing an important role in survival. In the average human adult male, there are approximately 4.75 liters of blood or 5 quarts (1 quart is equal to 0.95 litre). It is composed of 60% (2.85 liters or 3 quarts) of plasma and 40% (1.9 liters or 2 quarts) of blood cells. The major blood cells are red blood corpuscles (erythrocytes), white blood cells (leukocytes) and platelets (thrombocytes). These blood cells are suspended in water medium called as plasma, which is made up of water (~98%) and rest is dissolved solid materials (~2%) including hormones, antibodies, enzymes, waste materials, minerals and gases. The red cells are delicate, anucleated, round, concave bodies containing hemoglobin that transports oxygen and carbon dioxide. The white blood cells or leukocytes are nucleated and have primary function is to fight infection. Platelets are small fragments of special cells that play an important role in blood clotting.

Blood collection from mice: Blood is most frequently collected for experimental analysis, evaluation of serum antibodies, and analysis of surface markers on peripheral blood cells. Thus, blood collection an constitutes essential requirement for most of the studies. The amount of blood sampled is strictly dependent on the species from which blood is collected. Mostly, blood is collected from the auricular vein or artery such as from axillary plexus/choroids plexus or by cardiac puncture, which are unstressful or has the least stress to the animals. Here, we will discuss about three procedures; blood collection by retro-orbital bleeding, venipuncture of the tail, and cardiac puncture.

Blood Collection from Orbital Sinus or Retro-orbital Bleeding: It is a simple blood sampling technique mostly applied to the laboratory animal such as to rodents (rats and mice). In this procedure, about 0.5 ml of blood can be collected from a mouse without killing the animal. For the sampling, firstly, a mouse is anesthetized either by injectable anesthesia such as pentobarbital or ketamine drug or simply by inhalation (e.g. ether or methoxyflurane). The microhematocrit tube is introduced slowly at the medial canthus of the orbit of the anesthetized animals (mice or rats). Slowly, with axial rotation, the top of the microhematocrit tube is gently advanced towards the rear of the eye-ball socket until blood flows into the tube as shown in Figure 19-1. After the blood is collected, the microhematocrit tube is gently removed from the orbit and dabbed the excess blood from the site with a swab moistened in saline or PBS.

Fig.-19-1. Retro-orbital bleeding of a mouse by microhematocrit tube.

Blood Collection by Venipuncture of Tail: It is a procedure for sampling blood, where maximum 0.5 ml of blood is safely withdrawn from a 25 g mouse; and can be used for multiple sampling of blood. However, the preferred technique for the collection of blood is retro-orbital bleeding. In this procedure, firstly, an animal is manually restrained and anesthetized by common anesthesia. The Tail is warmed with a heat lamp or by immersing in warm water to dilate the tail vein. The vein to be punctured is visualized approximately at the midpoint on the length of the tail. Tail is extended with one hand, and with other hand needle is inserted 3 to 4 mm with slight inclination into the lateral dilated vein. The blood starts to flow. Then blood is collected from the hub of the needle with the microhematocrit tube. After blood is collected, the needle is removed and a gauze sponge is applied with gentle pressure on the bleeding site to ensure hemostasis.

Blood Collection by Cardiac puncture of Animal: However, this method is not preferred over retro-orbital bleeding. This technique allows the collecton of maximum blood (~1.5 ml) from the experimental animal, and may refer to as exsanguination. By this technique, one can not collect blood for multiple periods due to death of animals, which may occur after collection of blood. The blood is usually collected with 3-ml syringe having 22-G needle. Firstly, the animals are anesthetized and placed in dorsal recumbancy. A needle is inserted just below and slightly to the left of the xiphoid cartilage at the base of the sternum at a 15° to 20° angles in the case of mice and 200 to 300 angles in the case of rats (Fig. 19-2). The needle is slowly advanced together with applying very slight negative pressure on the barrel of the syringe. Blood starts to flow into the hub of the needle. Then, blood is gently aspirated until blood flow ceases. After the collection of blood, the animal is killed by painless cervical dislocation.

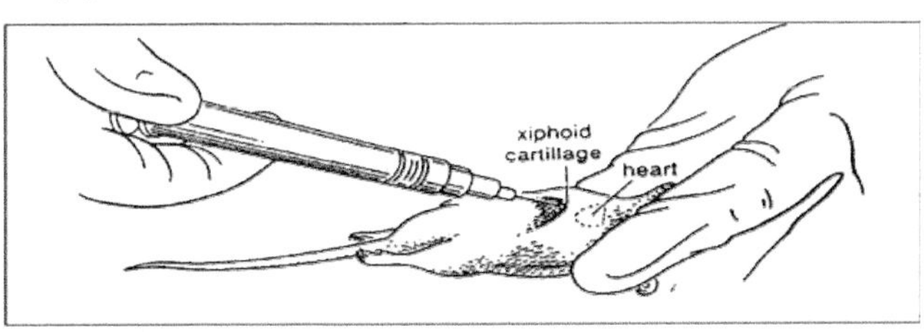

Fig.-19-2. Blood collection by cardiac puncture of a mouse using 3 ml syringe with 22 Gauge needle.

Constraints of Blood Collection from Rodents: The amount of blood that can be safely withdrawn depends on the size of the animals. Most animals go into shock if 25 - 30% of their blood volume is rapidly removed; over 50% die if 30 - 40% is removed; and nearly all die if more than 40% is removed. For most species, the blood volume in milliliter (ml) is approximately 6 - 8% of their body weight in gram. The average total blood volume of a mouse is about 77 – 80 ml/kg (1.9 to 2.0 ml of about 25 g mice). Without fluid replacement, about 0.18 to 0.2 ml of blood from a 25 g mouse can be safely withdrawn. But, to withdraw more than this limit (about 0.35 to 0.4 ml blood from 25 g mouse) fluid replacement is required, which usually is given subcutaneously. For multiple sampling of blood, generally 7.5% of the total blood volume (0.15 ml of 25 g mouse) per week should be withdrawn. The low amount of blood withdrawn in multiple sampling is due to the fact that after the withdrawal of blood, blood volume is rapidly restored, but other blood components may not be restored for several weeks. Therefore, for multiple sampling, hematocrit and/or hemoglobin level should be monitored regularly.

Serum and Plasma

Serum refers to blood plasma in which clotting factors such as fibrin have been removed naturally by allowing the blood to clot. In this manner, the essential difference between plasma and serum is that plasma retains fibrinogen (a clotting factor), on the other hand, fibrinogen is removed from serum. Serum is obtained from clotted blood that has not been mixed with an anticoagulant (a chemical that prevents the clotting of blood). Plasma is obtained from blood that has been mixed with an anticoagulant in the collection tube or vacutainer. Plasma is the pale yellow or grey yellow protein containing fluid, while normal serum is straw in color. Sometimes, plasma or serum may appear as cloudy or milky, called lipemic serum or plasma. The lipemic serum or plasma may result from bacterial contamination or transient high lipid levels in the patient's blood.

Plasma Preparation from Whole Blood: Plasma is a liquid part of blood in which blood cells are suspended. Therefore, plasma is separated easily by centrifugation before clotting. Therefore, it becomes necessary that sufficient volume of blood must be withdrawn to get the required amount of plasma because 1 ml of blood only gives less than 0.6 ml of plasma. For plasma preparation, blood is collected in a vacutainer coated with anticoagulant or before collecting the blood, the tube and syringe should be carefully rinsed with anti-coagulant such as Sodium Citrate or Heparin. Blood is gently mixed by inverting the tube for six to ten times immediately after collection and centrifuged at 2,000 rpm for 5 to 10 minutes. After centrifugation, plasma is carefully removed without disturbing the red blood cells at the bottom.

Serum Preparation from Whole Blood: For analysis of serum immunoglobulin level or for estimation of other dissolved substances (e.g., hormones, cytokines etc.), the serum is commonly used. For serum preparation, mixing of anti-coagulant is avoided. Firstly, blood is collected in vacutainer tubes without anti-coagulant and placed at 370 for 30 – 60 minutes to clot. After clot formation, the clot is "ringed" by a long Pasteur pipette to separate it from the sides of the glass tubes. As an alternative, the tube is inverted. Tubes are then placed at 40C overnight to allow the clot to contract or shrink. The serum is decanted away from the clot into a transfer tube, which is centrifuged at 2000 rpm for 5 to 10 minutes to remove the remaining clots/red blood cells fraction and other insoluble materials, and stored at -200C. Serum is straw in color (Fig. 19-3), but the hemolysed serum is pink or red due to the pressure of hemoglobin.

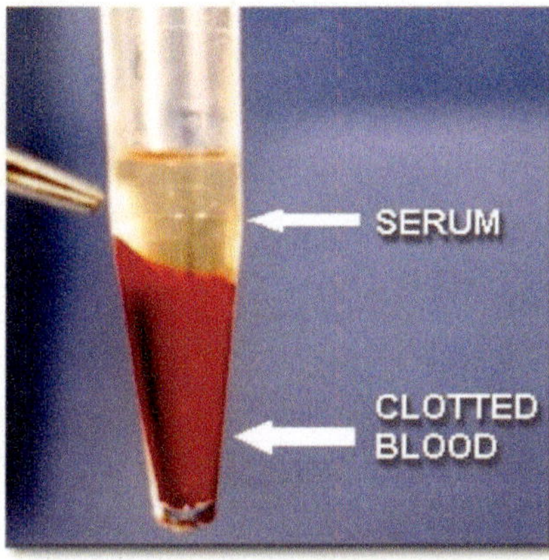

Fig.-19-3. The eppendorf tube showing serum at upper and blood cells below after centrifugation.

Types of Tubes Used in Blood Collection:

Red-top Tube: Contains no anti-coagulant or preservative, and is used for serum preparation.

Mottled Red/Grey or Cherry Red-top (gel barrier): contains clot activator but not anticoagulant and gel for separating serum from whole blood.

Lavender-top Tube: contains K2EDTA, and is used for plasma preparation or to store whole blood samples.

Grey-top Tube: contains Sodium fluoride as a preservative and Potassium oxalate as an anticoagulant. Used for plasma preparation or to store whole blood sample.

Blue-top Tube: contains Sodium citrate (anti-coagulant), and have yellow diagonal stripes on the label and used for plasma preparation or to store whole blood sample.

Green-top Tube: contains Sodium heparin or Lithium heparin. Used for storing whole blood sample.s

Yellow-top Tube: contains acid citrate dextrose (ACD) solution, and used for storing whole blood sample.

Royal Blue-top Tube: contains Sodium EDTA for trace metal studies. Used to store whole blood sample.

Plasma Preparation Tube (PPTTM): contains EDTA. Used to store whole blood sample for molecular diagnostic tests (e.g., PCR, bDNA techniques etc.).

Lymphoid Organs

Lymphoid organs are the places in which different lymphocyte populations differentiate and develop. There are two types of lymphoid organs; primary lymphoid organs and secondary lymphoid organs. Details of these lymphoid organs are given previously in Chapter 2. Here, we discuss differential practical approaches dealing with the handling of lymphoid organs and the manipulation of different lymphocytes population.

Removal of Mouse Lymphoid Organs

Lymphoid organs are removed mainly for in vitro study of B- and T-lymphocytes. Typical uses include providing cells for cytotoxicity or proliferation assays, for analysis of cytokine production or effect of cytokines, and for the isolation of B, T, and adherent cell sub-populations for culture or reconstitution of other animals. The procedure started with sacrificing the animal in a humane manner i.e. by painless cervical dislocation. The animal is placed on its back on clean and dry absorbent paper. The fur is washed with 70% ethanol to sterilize the area and to reduce the possibility of contamination. With sharp iris scissors, a middle incision is made and the skin above the head and below the thighs is retracted by pulling it with gloved fingers. The animal is turned to expose its left side for spleen and lymph node removal, while animal is kept on its back for thymus removal. After that following different steps are taken;

Removal of Spleen

To remove the spleen, firstly the spleen is located as stated above and 1 inch incision is made at the left of the peritoneal wall with surgical scissors. The spleen is found to be attached to the greater curvature of the stomach by connective tissues. The spleen as much as possible is grasped with curved iris forceps, and gently pulled free of the peritoneum, tearing the connective tissue behind the spleen. The removed spleen is then kept in chilled PBS or HBSS or other culture medium dependent upon the protocol of the cell types.

Removal of Thymus

By placing the animal on its back as stated above, an incision is made in the chest, beginning at the xiphoid and extending to the neck with surgical scissors. With curved forceps, the ribs are retracted. Sometimes, ribs are broken for effective retraction. The thymus is seen as a yellowish-white, bilobed organ just under the ribs, attached above the heart in the midline. Each lobe of the thymus is grasped with curved forceps and is gently pulled to avoid tearing. The removed thymus is then kept in chilled PBS or HBSS or an other culture medium dependent upon the protocol of the cells.

Removal of Lymph Nodes

The major lymph nodes are found in the axillary, cervical, inguinal, and mesenteric regions. In normal, non-immunized animals, the lymph nodes are usually small and can easily be overlooked, while in infected or immunized mice, the lymph nodes are large and swollen mass. The axillary nodes can be readily identified upon dissection in this region. Cervical nodes are flanked over the trachea and are beneath the large salivary gland. Inguinal nodes are found in the groin region, while mesenteric lymph nodes lie deep in the intestinal mesentery. After, visualization, the lymph node of interest is grasped with curved forceps and pulled them free of attached tissue. After the lymph node is removed, it is placed in chilled PBS or HBSS or other tissue culture medium.

Lymphocytes

Lymphocytes are the effector cell populations namely B- and T-lymphocytes (derived from yolk sac, fetal liver and bone marrow going through different stages of life), which migrate to the central lymphoid organs i.e., the thymus and bone marrow (bursa fabricius in birds). When their differentiation is completed, lymphocytes are migrated to and populate the secondary lymphoid organs such as blood, spleen, lymph nodes, Peyer's patches and lymphoid appendix in intestine, etc. At this stage, the animal is immunologically mature and able to respond to antigenic challenge. Details of structure and function of lymphocytes are dealt in the previous chapter. Here, we will discuss some practical approaches;

Morphology of Blood Leukocytes and Thymus

Cells from the mouse thymus (a central lymphoid organ) and blood (a peripheral lymphoid organ) are stained for morphological examination.

Preparation of Blood Film

For the preparation of blood film, a mouse is grasped firmly by the nape of neck using thumb and forefinger, holding the tail between little and third fingers. Then, by cutting the tip from its tail with a scissor, 1 drop of blood is squeezed and placed it at one end of a clean glass slide. A second 'spreader' slide is touched with the drop of blood with an angle 45°, and pushed away from the drop to obtain a film of cells. The slide is then air-dried by waving and fixed in 95% methanol for 30 to 60 seconds.

Single Cell Suspension of Thymocytes or Splenocytes

Mouse is killed by painless cervical dislocation and thymus or spleen is removed. Removed thymus or spleen is kept in chilled PBS or culture medium in a Petri dish, and either teased with forceps or needle, or kept in between two frosted slides, minced gently, and washed by micropipette, so that the connective tissue capsule is burst. The connective tissue capsule is removed from the Petri dish.

The cells, thus, obtained are centrifuged at 1000 rpm for 5 minutes to remove dead cells and other tissue fragments. Now, cell suspension has been prepared. One drop of this cell suspension is then placed on a clean slide and with a second 'spreader' slide, the smear is formed as before. The Slide is air-dried, fixed in 95% methanol for 30 to 60 seconds, and visualized under a phase contrast microscope.

Giemsa Staining

After the preparation of single cell suspension or smear on slide, the smear of cells is fixed in 95% methanol as stated above, and immersed in Giemsa buffer (0.1 M citric acid and 0.2 M Na_2HPO_4). The slides are then transferred to May-Grünwald or Giemsa stain freshly diluted as 1:2 with Giemsa buffer for 5 to 15 minutes. The stained slides are rinsed properly to washout excess stain. If the cells are overstained, cells are allowed to stand in the buffer. The slides are then dried in air and examined under the microscope. On the basis of the morphology of the nucleus and granules, lymphocytes are differentially counted and percent of lymphocytes are calculated.

Total Lymphocyte Count

Single cell suspension is prepared as mentioned above and counted on an improved Neubauer ruling haemocytometer. When counting, it is most convenient to use a 40^{x} objective lens, and the lymphocytes should be counted in the central triple ruled area of the haemocytometer (also used for RBC counting) **(Fig. 19-4).** Usually, cells are counted in the large triple-ruled squares until; a minimum of 100 cells have been counted, and calculated as mentioned below;

Calculation of total lymphocytes:

$$\text{Number of Total Lymphocytes per ml} = \frac{\text{Number of Lymphocytes Counted}}{\text{Number of Triple Ruled Square Counted}} \times 25 \times 10^4 \times \text{original dilution (if any)}$$

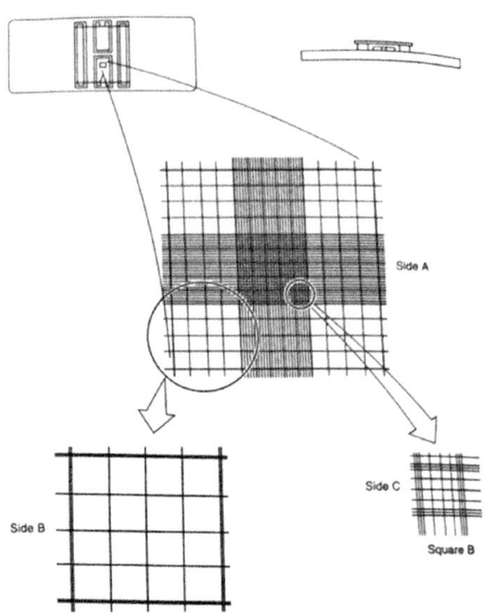

Fig.-19-4. Diagrammatic representation of haemocytometer and its counting area.

Note: 25 is the total number of triple ruled square, and 104 come for total volume in ml-1. The length and width of 25 triple ruled areas is 1 mm and depth is 0.1 mm, therefore volume in centimeter is (0.1 x 0.1 x 0.01) cm3.

Percent of Viable Lymphocytes

This test is performed to know the percent of viable cells in cell suspension. For this, vital dye such as trypan blue or nigrosin is used to stain the dead cells as these are becoming permeable to vital dyes. 0.1 ml of cell suspension is mixed with 0.1 ml of 0.2% trypan blue or nigrosin solution and incubated at room temperature for 5 minutes. After staining, cells are counted by haemocytometer under a phase contrast microscope, and percent of viable cells are calculated as mentioned below;

Calculation of Viable Lymphocytes:

$$\text{Percent of Viable Lymphocytes} = \frac{\text{Number of viable Lymphocytes}}{\text{Total number of Lymphocytes}} \times 100.$$

Enumeration of T cell by E-rosetting methord

Cell suspension from the thymus or spleen is obtained as described above. By adherence purification, phagocytic cells are removed, and washed three times by centrifugation at 1000 rpm for 5-10 minutes. 0.1ml of sheep red blood cells (SRBCs) is added and mixed. After mixing, the cells are centrifuged at 1000 rpm for 5-10 minutes at 4 °C. Again 0.1ml SRBCs is added, centrifuged and counted the number of rosettes formed by a haemocytometer under a phase contrast microscope. It is found that B cells do not form rosette, only T cells form a rosette and therefore only T cells are counted.

Preparation of lymphocytes from blood

Usually, differential centrifugation on a density gradient is performed to get high-purity lymphocyte preparation. It is a relatively rapid method. First, 9.2% w/v Ficoll solution is made in distilled water and autoclaved. 43.4ml of Ficoll solution and 6.6ml of Triosil 75 are mixed under sterile conditions. Blood is diluted 1:1 with sterile tissue culture medium with serum, and 5ml of diluted blood is layered onto 2ml of the gradient mixture and centrifuged at 1500 rpm for 10-15 minutes at 4°C. The lymphocytes are found at the plasma-density gradient interface as a white band. Above the white lymphocyte band the medium is discarded, and the lymphocyte band is removed. The lymphocytes are then washed three times by centrifugation at 1000 rpm for 5-10 minutes at 4°C.

Separation of T and B cells from PBMC by Nylon Wool method

Due to the adherence nature of peripheral blood mononuclear cells, T and B-lymphocytes are purified from the blood by the column tightly packed with sterile nylon wool. Firstly, 60 mg of sterile nylon wool (approximately 6 ml) is packed into a 20 ml syringe barrel and washed with tissue culture medium supplemented with 5% fetal bovine serum (FBS). The column is sealed and incubated for 1hour at 37°C. Cell suspension prepared as described above, is adjusted to 5 x 10^7 lymphocytes/ml. To equilibrate the pH of column, 5ml of the warm tissue culture medium is flushed, and then 2 ml of cell suspension is added dropwise to the top of the column. Finally, 1 ml of warm tissue culture medium is added, and column is incubated for 45 to 60 minutes at 3°C. The unbound cells are collected by washing the column with 25 ml warm tissue culture medium, and cells are then concentrated by centrifugation at 1000 rpm for 5 to 10 minutes at 4°C. Number of viable cells is counted by haemocytometer under phase contrast microscope.

Macrophages

Macrophages are myeloid phagocytic cell populations. It is a large terminally differentiated cell population developed from mononuclear cells. Mononuclear cells after entering into the tissue from blood vessels readily differentiated into macrophages. The details of macrophages have been given in the previous chapters, here, we will discuss about some practical approaches.

Isolation of Peritoneal Macrophages

Peritoneal macrophages are prepared by washing the peritoneal cavity of mice or rats. However, by eliciting local inflammatory response with Starch, Sodium triolet, paraffin oil, thioglycollate etc., it is possible to get very large number of peritoneal macrophages as peritoneal exudates cells (PECs). The PECs are not normal macrophages and contain macrophages a with engulfed particles. For isolation of normal peritoneal macrophages, mouse or rat is killed by painless cervical dislocation, placed on its back and pinned the legs. The abdominal skin is opened to expose the peritoneum, and 6 ml of chilled incomplete tissue culture medium (8 ml in rats) is injected. The abdomen is lavaged and using a syringe with 22 gauze needle, gently aspirates the peritoneal fluid containing peritoneal macrophages carefully avoiding the possibility of blockage of the needle by fat bodies. The cells are centrifuged at 1000 rpm for 5 to 10 minutes at room temperature. The number of phagocytes is estimated by the uptake of 1% neutral red solution and counting with a haemocytometer under a phase-contrast microscope. However, it has been estimated that only 50% of cells obtained in this way are macrophages, and rest is lymphocytes, polymorphonuclear cells etc. Therefore, the cells are purified by adherence purification method. The cell suspension is incubated for 2 hours in culture medium at 37°C. Non-adhered cell population suspended in culture medium is removed, and adhered cell population is collected by rubber policeman or cell scraper to get highly purified cell suspension of macrophages (more than 98%).

Non-specific Esterase Staining for Identification of Macrophages

Non-specific esterase staining is performed to confirm the particular cells are macrophages. Macrophages are stained strongly with non-specific esterase staining. However, normal peritoneal macrophages stain more strongly than macrophages and monocytes isolated from elsewhere from the body (elicited peritoneal macrophages stain more strongly than normal peritoneal macrophages). Firstly, stock solution of the stain is prepared by mixing 1 g pararosaniline with 5 ml concentrated HCl (10 M) and 20 ml distilled water. The solution is heated to 70oC, allowed to cool to room temperature, filtered and stored at 4°C in the dark colored bottle. A smear of cell suspension is prepared on a microscopic slide. Cells are fixed for 30 second at 4°C in 30 ml of 0.1 M phosphate buffer, pH 6.6, mixed with 45 ml of acetone and 25 ml of formaldehyde solution. Slides are washed three times with distilled water and air-dried. Immediately before use, active diazonium salt, hexazotised pararosaniline is prepared by mixing freshly prepared 6 ml of 4% w/v sodium nitrite with an equal volume of the stock solution of stain and diluted to 200 ml with 0.067 M phosphate buffer, pH 5.0. The pH of activated stain solution is adjusted to 5.8 with 0.1 M sodium hydroxide, because optimum pH for esterase activity is 5.8. 50 mg of naphthyl acetate is dissolved in 2 ml acetone and added to the staining solution. The smear on slide is then incubated for 45 to 60 minutes at 37°C; excess stain is washed and counterstained in 0.4% aqueous solution of methyl green dye for 1 to 2 minutes at room temperature. The Slide is then washed and observed under the microscope.

Isolation of mononuclear cells from peripheral blood

Blood mononuclear cells are isolated by the Ficoll density gradient method. Blood is collected in anti-coagulant rinsed tubes and diluted with PBS at making total of 50 ml. Diluted blood is then carefully layered on 10 ml of Ficoll and centrifuged at 1600 rpm for 30 minutes. Resultant layers are approximately from top to bottom: Plasma – platelets -- PBMC – Ficoll – red blood cells (with granulocytes). Buffy coat with PBMCs (peripheral blood mononuclear cells) are carefully aspirated avoiding the aspiration of plasma and Ficoll and PBS added to make 50 ml, while the other upper layer is discarded. Resuspended PBMCs are centrifuged at 1200 rpm for 10 minutes to make a cell pellet of PBMCs. The supernatant is decanted and loosened cell pellet is washed again in PBS by centrifugation. The Supernatant is again decanted, loosen cell pellet is resuspended in ~30 mL of PBS and mixed well. Resuspended cell population is mixed with trypan blue or nigrosin to stain as described above for viability test for lymphocytes (dyes exclusion methods) and counted on haemocytometer chamber under the microscope.

Immunologic Assays

By taking advantage of antigen-antibody interaction, many of the biochemical as well as immunological assays have been developed, whereby diseases can be diagnosed, the level of humoral immune response can be monitored and molecules of biological or medical importance can be identified. The assays are of qualitative and quantitative nature.

Precipitation Reaction

In an aqueous solution, a lattice is formed by the interaction between an antibody and a soluble antigen (Fig.-19-5). This lattice is visible as a precipitate, and usually develops in a day or two. The valency of both the antibody and antigen determines the formation of an Ag-Ab complex/Ag-Ab lattice. For the valency the antibody must be bivalent and the antigen also must be bivalent. There are some principles that account for precipitation in various media viz. fluids, gels brought out by major types of assays because both antibody and antigens are for more sensitive. Many Ag-Ab reactions, antigen excess and antibody excess are highlighted by the process of quantitative precipitation, reaction in fluids. In it a constant amount of antibody is placed in a series of tubes and antigens are added to the tubes. After the formation of precipitate each tube is centrifuged to pellet the precipitate, the supernatant is poured off and the amount of precipitate is measured. **Fig. 19-6** illustrates a precipitation curve for a system of one antigen and its antibodies. Excess of either antibody or antigen interfered with maximal precipitation occurring in the equivalence zone. In this zone the ratio of antibody and antigen II is optimal. The two other zones of precipitation are zone of antibody excess and zone of antigen excess. In the region of antibody excess unreacted antibody is found in the supernatant along with small, soluble complexes consisting of multiple molecules of antibody bound to a single molecule of antigen.

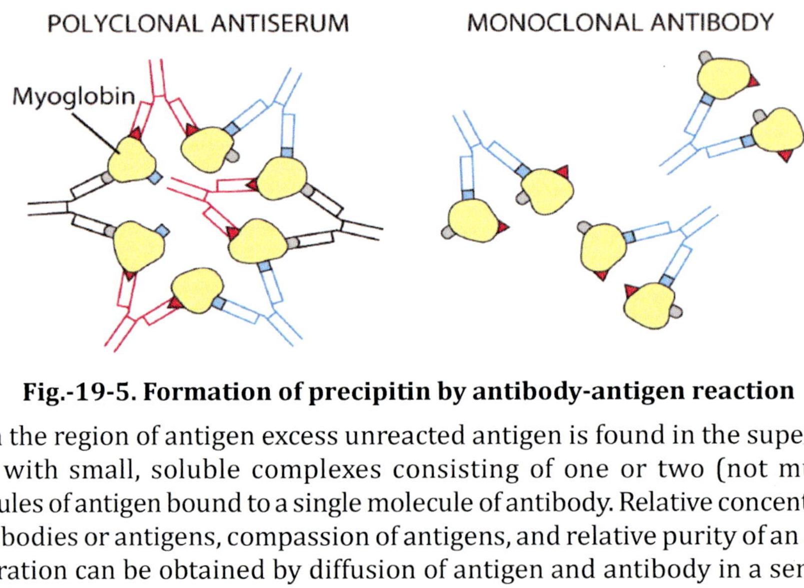

Fig.-19-5. Formation of precipitin by antibody-antigen reaction

In the region of antigen excess unreacted antigen is found in the supernatant along with small, soluble complexes consisting of one or two (not multiple) molecules of antigen bound to a single molecule of antibody. Relative concentrations of antibodies or antigens, compassion of antigens, and relative purity of an antigen preparation can be obtained by diffusion of antigen and antibody in a semi-solid medium such as agar. The methods are referred to as radial immunodiffusion (Mancini method) and double immunoduffusion (Ouchterlony method). By employing these methods, visible precipitation occurs in the zone of equivalent whereas no visible precipitate forms in the zones of antibody or antigen excess. The methods in detail are:

- **Mancini method**: An antigen sample is placed in a well and allowed to diffuse into agar containing a suitable dilution of an antiserum. Consequent upon diffusion of antigen into agar the zone of an equivalence is formed (precipitin ring) around the well (Fig. 19-7, upper panel). The concentration of antigen sample is established by comparing the area of the precipitin ring with the precipitin areas of known concentration of the antigen. This method is also known as radial immunodiffusion.

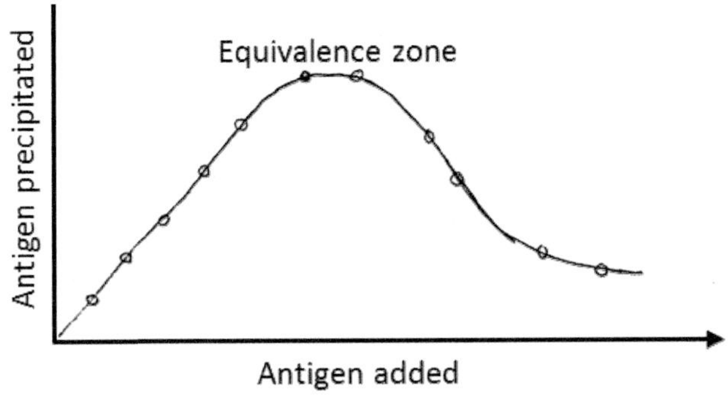

Fig.-19-6. Precipitation reaction between antigen and antibodies in different conditions; antibody excess,antigen and antibody equal and antigen excess.

- ♦ **Ouchterclony method**: In this method, both antigen and antibody diffuse radially from wells toward each other. A visible line of precipitation (precipitin line) shows the equivalence. The lower panel of Fig.-19-7 is illustrative of this method. The pattern of lines which form on the diffusion of antigen and antibody indicates whether the two antigens (placed in adjacent wells) have identical epitopes, partly identical epitopes, or no epitopes.

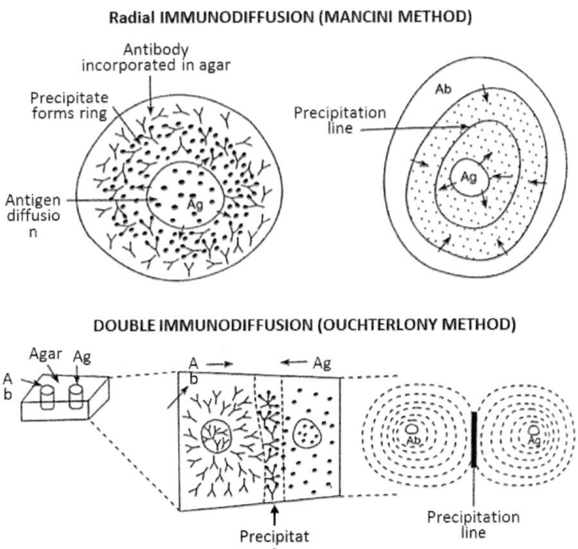

Fig.-19-7. Immunodiffusion (a) depict the radial immunodiffusion (Mancini Method), and shows the distincts precipitation line between the antigen and antibody zone; (b) depict the double immunodiffusion (Ouchterlony Method) and shows the distinct precipitation line in the agar matrix.

Immunoelectrophoresis

It is a strictly qualitative technique that can detect relatively high antibody concentration, and is useful in detecting quantitative anomalies in immune deficiency and immuno-proliferative disorders. In this method, an antigen mixture is first electrophoresed to separate its components by charge. Troughs are then cut into the agar gel parallel to the direction of the electric field, and antiserum is added to the troughs. Antibodies and antigens then diffuse toward each other and produce lines of precipitation where they meet in appropriate proportions. Fig. 19-8 shows the separation and identification of immune deficiencies. A related technique called rocket electrophoresis (the precipitate formed between the antigen and antibody resembles the shape of a rocket) provides for the measurement of antigen levels. In this method, immunoglobulins cannot be analysed since they cannot be negatively charged.

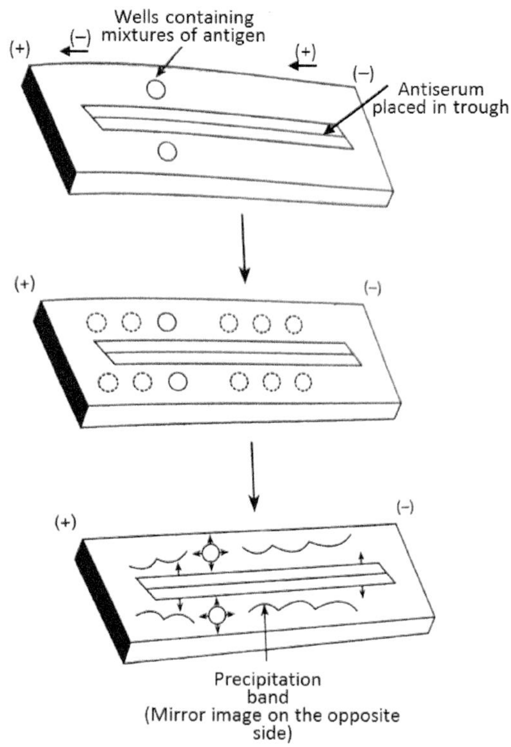

Fig.-19-8. Immunoelectrophoresis of an antigen mixture.

Agglutination Reaction

The visible clumping of interaction between antibodies and antigens is called agglutination. In principle, this method is like precipitation reactions being dependent on cross linking of polyvalent antigens. Like precipitation reaction, in this method also an excess of antibody inhibits precipitation reaction (known as prozone effect). There are certain limitations of this method of agglutination viz. a) the antiserum of antibodies with high concentration, b) agglutinating monoclonal antibodies c) the density of epitope distribution or the location of some epitopes in deep pockets of a particular antigen fails to produce the desired effect of concentration in Ag-Ab. Agglutination reactions provide ways to type red blood cells (RBCs), bacteria and detect small amount of antibody (very low concentration as nanogram/ml.). Thus the matching of blood types (RBCs) known as hemagglutination, obtained by mixing RBCs on a slide with antisera to the A or B blood group antigens (Fig. 19-9). Different species of the bacterium *Salmonella* can be distinguished by agglutination reaction with a panel of typing antisera. Soluble antigens may be identified by a method complementary to agglutination reaction called passive hemagglutination wherein serum containing antibody is serially diluted into a microtiter plate (the agglutinin titer is defined as the reciprocal of the greatest serum dilution that elicits a positive agglutination reaction). Now a day, antigens are coupled to the latex beads as matrices for agglutination reactions. The synthetic beads can be read rapidly, often within 3 to 5 minutes of mixing the beads with the test sample. The advantage with latex beads is their immediate use or storage for later use.

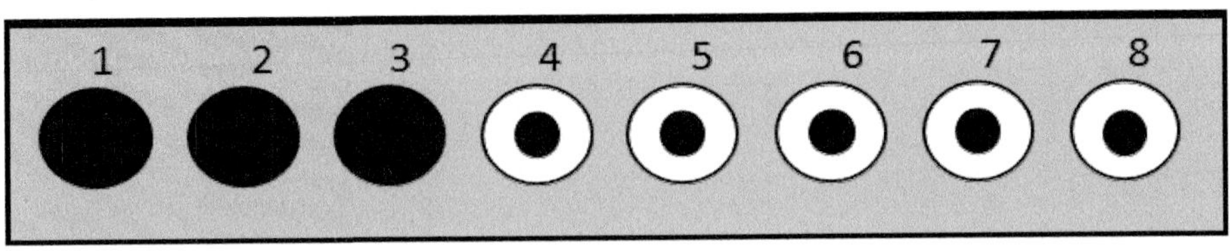

Fig.-19-9. Demonstration of hemagglutination using antibodies against sheep RBCs. The spread pattern of hemagglutination is visible from the experimental table-4 shows positive hemagglutination.

Drug addition, usage for certain facts, exposure to certain types of viruses, and immunity to rubella may be determined by a technique called agglutination inhibition assay. The agglutination inhibition technique provides a highly sensitive assay for small quantities of antigens. Addition of RBCs (or other particles) coated with the suspected dong to the sample (urine or blood) incubated with antibody specific for the suspected drug inhibits agglutination, thereby confirming use of illicit drug. The immune status of women with respect to rubella virus can be determined by the titer of the serum (1:10 dilution).

Agglutination reaction is simple, inexpensive and sensitive. It may be of the following types;

Hemagglutination of RBC's

The tests done in the blood bank are serological, i.e., they involve the interaction of antibody (in serum) with antigens (on red cells) and result in hemagglutination. Hemagglutination will result if serum contains an antibody specific for a red cell antigen, and the two are brought together under optimal test conditions.

Hemagglutination tests done in test tubes are graded (from 4+ to 1+). They are read macroscopically, often with the aid of a lighted magnifying mirror. Some laboratories also read negative tests (especially antiglobulin tests) microscopically.

In blood banking, antibodies to red cell antigens may be IgM, IgG, or IgA. IgA antibodies are rare, and if present, usually occur with either IgM or IgG. Sometimes red cell antibodies with a single specificity are just IgM, sometimes they are just IgG, and sometimes they are both IgM and IgG.

All tube tests are graded (e.g., 4+, 3+, 2+, 1+, -) where 4+ is one solid clump of red cells and 1+ is the presence of many small clumps (agglutinates).

Reasons for Grading Reactions:

- Helps detect the presence of more than one antibody. For example, if some cells react 2+ and others react 4+, the 2+ reaction may be one antibody reacting with one antigen, whereas the 4+ reaction may be two antibodies reacting (both antigens are on the same cell).
- Helps detect dosage effect of antigens. Dosage is a phenomenon in which having a gene present in the homozygous state results in an increased number of antigens per red cell than if the gene was only present in the heterozygous state. For example:
    - anti-M + rbc MM (homozygous): 4+
    - anti-M + rbc MN (heterozygous): 2+
- Helps detect abnormal reactions in the ABO BGS. When ABO grouping, we expect strong agglutination (3+ or 4+). If weak reactions are present (2+ or less), this may indicate a weak subgroup (e.g., A3) or a weak antigen due to a disease state (e.g., weakened A phenomenon due to leukemia or acquired B phenomenon due to infection).
- Helps detect spurious (false positive) reactions. Antigen typings of patient cells are read in comparison to positive/negative controls. If a test on patient cells is noticeably weaker than the positive antiserum control, the patient test may be a false positive. For example, if a positive control (antiserum plus heterozygous cell) is 3+, but the test using the patient's cells is 1+, we should be suspicious and ask, "Why is the test so much weaker?"

Bacterial Agglutination

It is used to identify specific bacterial antigens. Because the clumping reaction occurs quickly and is easy to produce, agglutination is an important technique in diagnosis. Bacterial agglutination is performed using either *P. aeruginosa* cells autoclaved at 120°C for 1 hour, then centrifuged and resuspended in saline or live *P. aeruginosa* cells grown to mid log phase, then centrifuged and resuspended in saline. (For either method resuspend the bacteria so the concentration is 5×10^8 bacteria/ml.) In both method the cell preparations are incubated with serial dilutions of antiserum for 1 hour at 37°C. The agglutination of live cells is characterized by coarse granular bacterial clumping, as opposed to the finer granular clumping of heat-killed *P. aeruginosa* with homologous LPS O-antigen-specific serum. Bacterial agglutination is scored on a scale of +1 (weak agglutination) to +4 (strong reaction). The agglutination can also be performed on glass slides using a drop each of antiserum and bacterial preparation, and the results are observed after 3 to 5 minutes.

Passive Agglutination

The agglutination of inert particles by antibody directed against antigen bound to their surface is called passive agglutination. Passive agglutination is performed using tanned sheep red blood cells and P. *aeruginosa* LPS (heated for 1 hour at 100oC prior to use). Sheep erythrocytes are washed three times in saline (0.9% NaCl), adjusted to 4% erythrocytes (v/v) in PBS, pH 7.5 and 2.5mg of tannic acid in 50ml PBS is added to 50ml of the cell suspension. After a 15 minutes incubation at 37oC with occasional mixing, the cells are centrifuged (100 x g for 20 minutes) and washed with 100ml PBS. One half of the cells are kept as a control in PBS containing 1mM NaN3. To the other half of the cells, 20μg/ml LPS or protein is added, and the mixture is incubated for 1 hour at 37oC with very gentle agitation at regular intervals. The antigen-coated cells are washed three times in saline and 1mM NaN3 is added for storage at 4oC. The cells are made up to 1% (v/v) in saline before use.

Passive agglutination is performed in 96 well conical bottom plates using 50μl of antiserum serially diluted in saline and 50μl of 1% (v/v) antigen-coated sheep erythrocytes. Control wells contain the antiserum and tanned, non-antigen-coated sheep erythrocytes. The plates are incubated at 37oC for 1 hour. The titer is the inverse of the highest serum dilution showing agglutination. Non-agglutinated cells give a tight button of cells at the bottom of the well, while the agglutinated cells form a mat at the well bottom.

Agglutination Inhibition

Agglutination-inhibition is a test used to identify unknown antigens. It is used in some pregnancy tests, tissue matching, blood grouping and diagnosing infections. Antibodies against the viral protein responsible for haemagglutination can prevent haemagglutination; this is the basis behind the haemagglutination-inhibition test (HAI). The specificity of the HAI test varies with different viruses. With some viruses such as influenza A, the haemagglutination antigen is the same as the antigen responsible for virus adsorption and thus virus neutralization, and therefore the HAI test is highly specific for the different strains of the virus. With other viruses, the HAI test is less specific eg. flaviviruses, where HAI antibodies against one flavivirus may cross-react with other related flaviviruses. HAI tests are more sensitive than complement-fixation tests but are less sensitive than EIAs and RIAs.

The HAI test is simple to perform and requires inexpensive equipment and reagents. Serial dilutions of the patient's sera are allowed to react with a fixed dose of viral haemagglutinin, followed by the addition of agglutinable erythrocytes. In the presence of antibody, the ability of the virus to agglutinate the erythrocytes is inhibited. The HAI test may be complicated by the presence of non-specific inhibitors of viral haemagglutination. and naturally occurring agglutinins of the erthrocytes. Therefore, the sera should be treated before use or false positive or negative results may arise. HAI tests are widely used for the diagnosis of rubella and influenza virus infections.

The advantages of HAI tests are that they are relatively easy and inexpensive to perform. The disadvantages are that HAI tests are not as sensitive as EIAs or RIAs, the actual reading of results is subjective and the reagents should be fresh or else abnormal agglutination patterns may arise which makes the reading and interpretation of the test very difficult. As a result the HAI test for rubella had been replaced by more sensitive and reliable EIA and RIA tests for rubella IgG in many virus diagnostic laboratory.

Radioimmunoassay (RIA)

This technique is used for measuring hormones, serum proteins, drugs and vitamins at concentrations of 0.001 micrograms/ml. or less. In this technique, there is competitive binding of radio labeled antigen (the generally used labels are gamma-emitting isotope such as 125I, though beta-emitting isotopes such as tritium (3H) are also used) and unlabeled antigen to a high-affinity antibody, Fig. 19-10 is amply illustrative of the technique. The labeled antigen is mixed with antibody at a concentration that saturates the antigen binding sites of the antibody molecule and then increasing amounts of the test samples containing unlabeled antigen of unknown concentration are added. The antibody does not distinguish labeled from unlabeled antigens, and thus the two kinds of antigen compete for available binding sites on the antibody. The radio labeled antigen is the part of the assay mixture; the test sample may be a complex mixture (serum or other body fluids), which contains the unlabeled antigen. A small amount of unlabeled antigen added to the assay mixture of labeled antigen and antibody will cause a decrease in the amount of radioactive antigen bound, proportionate to the amount of unlabeled antigen added. The amount of antigen in the test mixture may be precisely determined by using unlabeled antigen samples of known concentration, instead of test samples.

In RIA bound antigens are separated from the free antigen by any of the following methods:

- by precipitating the Ag-Ab complex with a secondary anti-isotype antiserum, or
- by precipitating the Ag-Ab complex of an IgG antibody in a mixture of formalin-killed Staphylococcus aureus.

With either of the above methods the amount of free labeled antigen (remaining in the supernatant) can be measured in a radiation counter, by subtracting this value from the total amount of labeled antigen by cross linking the antibody covalently to sepharose beads. The procedure is termed as solid-phase RIA. After centrifuging and washing the beads the amount of radio labeled antigen bound to the beads can be measured.

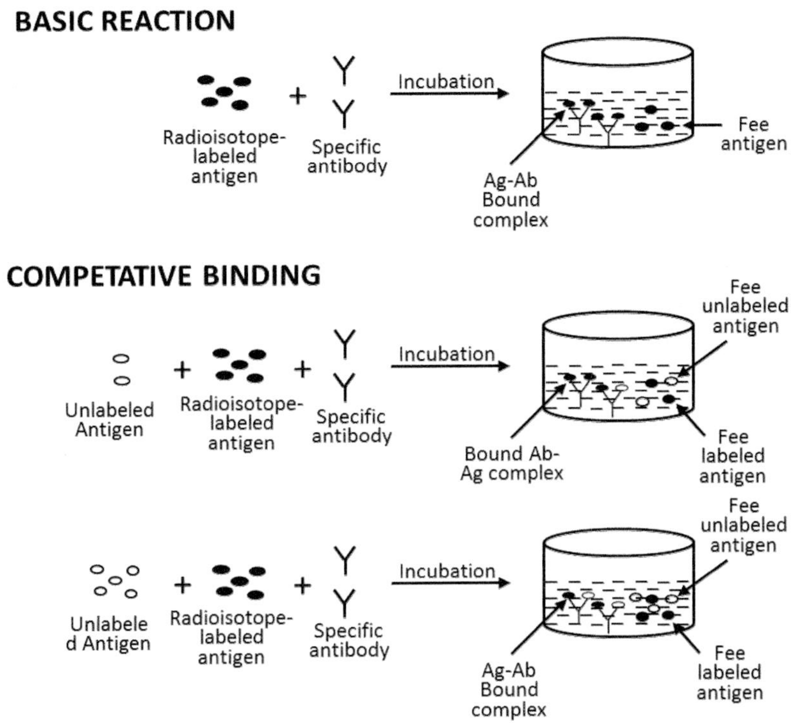

Fig.-19-10. A solid-phase radio-immunoassay (RIA). It shows the basic reaction between radio-isotope labeled antigen with specific antibody and competitive binding with an excess of labeled antigen and equal amount labeled and unlabelled.

Enzyme-linked Immunosorbent Assay (ELISA)

ELISA or EIA uses an enzyme unlike RIA and is comparatively safe and less costly than RIA. There are many versions of ELISA for qualitative detection of the presence of antibodies or antigens, as also their quantitative measurement. Alkaline phosphatase, horseradish peroxidase, b- galactosidase are some of the enzymes used in this technique. Fig. 19-11 shows the various options of ELISA viz. indirect ELISA, Sandwich ELISA and Competitive ELISA. Usually, in ELISA, an enzyme conjugated with an antibody reacts with a colorless substrate to generate colored reaction products. Substantially, the producers of ELISA under different options, as above, are directed to detect antibodies or antigens, where the case may be. A highly sensitive alternative procedure, known as chemiluminescence (measurement of light produced during certain chemical reactions) is widely employed these days. The options under ELISA as above, in brief are:

Indirect ELISA

This procedure is directed towards the detection of HIV, the causative agent of AIDS. For a quantitative determination of antibody serum or some other sample containing primary antibody (Ab1) is added to an antigen coated microtiter well and allowed to react with the antigen attached to the well. Subsequent to washing any free Ab, the presence of antibody bound to the antigen is detected by adding an enzyme-conjugated secondary antiisotype antibody (Ab2), which binds to the primary antibody. Then after washing any free Ab2, a substrate for the enzyme is added and the colored reaction is measured by spectrophotometric plate readers. This yields measurement of the absorbance of the entire well (96 well plates) rapidly in a single platform.

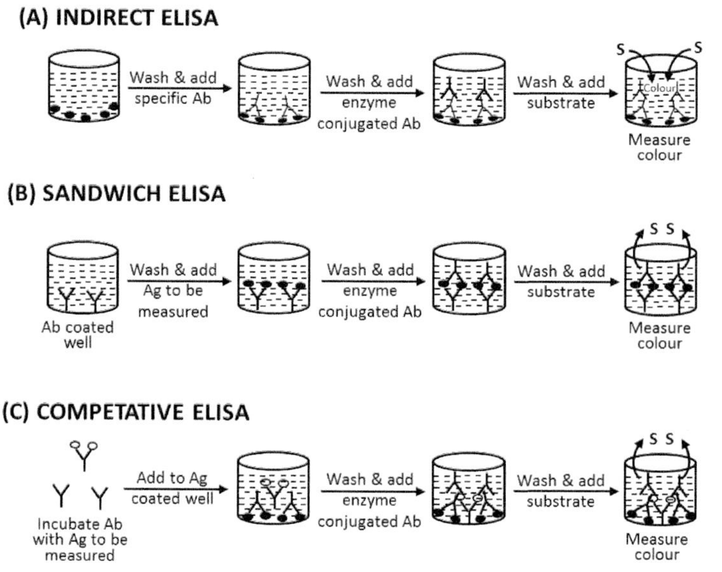

Fig.-19-11. Variation in the enzyme-linked immunosorbent immuno-assay (ELISA) technique allows the determination of antigen or antibodies. (a) Indirect ELISA, (b) Sandwich ELISA and (c)Competitive ELISA.

Sandwich ELISA

This procedure is directed toward the detection of antigens. In this method, the antibody is immobilized on a microtiter well. The immobilized antibody reacts to an antigen containing sample. Subsequent to washing the well a second enzyme-linked antibody specific for a different epitope is added to the antigen and allowed to react with the bound antigen. The colored reaction is measured following the washing of any free second antibody.

Competitive ELISA

This technique is a variable to the earlier techniques. In it, first the antibody is incubated in solution with a sample containing antigen. This mixture is then added to an antigen coated microtiter well. Enzyme conjugated secondary antibody (Ab2) specific for the isotype of the primary antibody is added to determine the amount of primary antibody bound to the well. In procedure and sensitivity, this technique is like indirect ELISA, but for the concentration of antigen in the original sample, which if higher indicates lower absorbance.

The above options of ELISA are, however, surpassed in sensitivity with the chemiluminescence assay, whereby the detection limit can be increased. By employing this technique as little as 5 × 10-18 moles of target antigen have been detected. This technique is commonly known to have improved over chromogenic substrate being luxogenic (light generating) substrate.

ELISPOT Assay

Ag-Ab specificity for each can be quantitatively determined by the modified form of ELISA assay. In this method, the plates are coated with the antigen recognized by the antibody specific for the antigen under assay. A suspension of the cell population under assay is then added to the coated plates. On incubation, the cells settle onto the surface of the plate and secreted molecules reactive with the sample molecules are bound to the antibody. This is visible as a ring of antigen-antibody complexes around each cell producing the molecule of interest. The plate is then washed and an enzyme-linked antibody specific for the secreted antigen or specific for the antibody is added and allowed to bind. A point of colour or light (by chromogenic or chemiluminescence treatment) reveals the position of each antibody or antigen producing cell (Fig. 19-12).

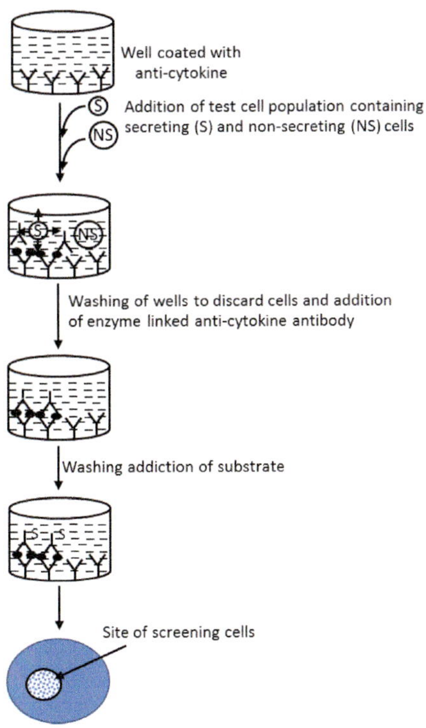

Fig.-19-12. Demonstration of ELISPOT-assay.

Western Blotting

By this technique a specific protein or a specific antibody in a mixture can be identified. In this technique, a protein mixture is electrophoretically separated on a polyacrylamide slab gel infused with sodium dodecyl-sulfate (SDS). The protein bands are transferred to a nitrocellulose membrane by electrophoresis and the individual protein bands are identified by flooding the nitrocellulose membrane with radio labeled or enzyme-linked polyclonal or monoclonal antibody specific to the protein under investigation. An appearance of colour (by chromogenic procedure) or light (by chemiluminescence procedure) at the site of the target antigen leads to identification of a specific protein or antibody in a mixture.

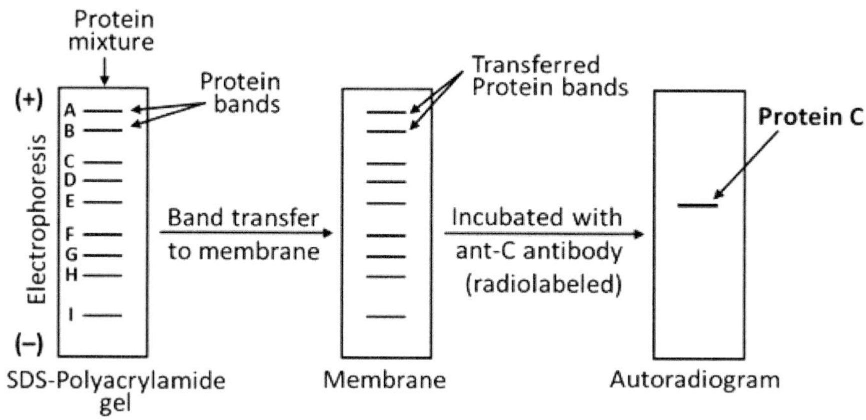

Fig.-19-13. Immunoblotting for the expression and separation of proteins of the specific interest.

In identifying a specific antibody in a mixture of known antigens is separated by sodium dodecyl sulfate (SDS) on a polyacrylamide slab gel and blotted onto nitrocellulose (SDS-PAGE). The Antibodies is identified by its reaction with a band (of known antigens) either by using radio labeled or enzyme-linked secondary antibody that is specific for the species of the antibodies in the test sample.

Immunoprecipitation

Despite limitations, this technique is used for isolation of the antigen of interest and there, it gives a sensitive assay for the presence of a particular antigen in a given cell or tissue type. An antigen antibody complex is formed by mixing of disrupted cells or tissues. This mixture is then allowed to precipitate. This precipitation is very uncertain as can take hour, or even days, and the point of limitation is thus enhanced to isolate the small amount of formed immunoprecipitate. Some measures have been devised to minimize the time of precipitation, as follows:

- Attaching the antibody to a synthetic bead, as it allows the antigen-antibody complex to be collected by centrifugation.

- Attaching the secondary antibody to a bead, as it is collected by centrifugation.
- Coupling the secondary antibody to magnetic beads and collecting them against the side of the tube (Fig. 19-14)
- By growing the cells in the radioactive medium and washing the Ag-Ab complex, free of unincorporated radio labeled amino acids and other impurities. The amount of synthesized protein is counted in a scintillation counter.

The immunoprecipitation procedure allows the concentration of proteins that are not very abundant or difficult to detect using western blot. IP can concentrate proteins up to 10,000-fold as it can take large volumes of cell lysates and concentrate protein of interest into a small precipitate which can then be used for western blot analysis or other methods.

Immunoprecipitation can be applied to;

- Determination of the molecular weight and quantity of immunoprecipitated protein by SDS-PAGE.
- Immunoprecipitation is used to assess for protein-protein interactions. This is done by immunoprecipitation for one protein, and then blotting for another protein.
- Quantification of the rate of synthesis of a protein in cells by determining the quantity radiolabeled protein made during a specific amount of time.
- Immunoprecipitation enables to the concentration of proteins that are otherwise difficult to detect.

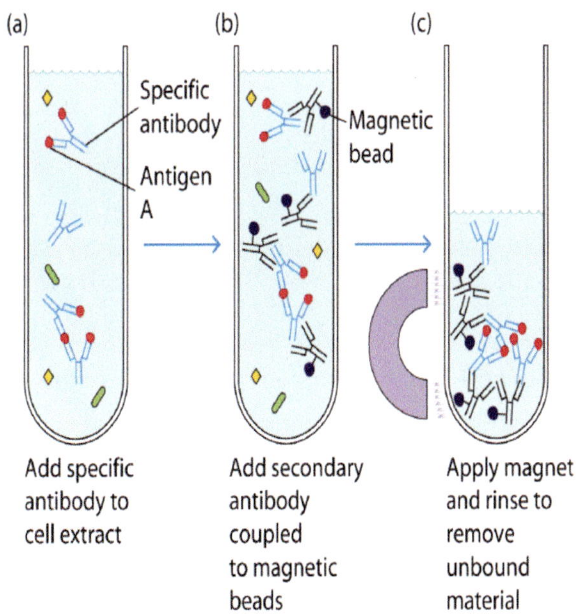

Fig.-19-14. Separation of antigen using magnetic beads conjugated with secondary antibody.

Immunofluorescence

Fluorescent molecules absorb light of one wavelength and emit light of another wavelength. Thus, antibody molecules if tagged with a fluorescent dye or fluorochrome, immune complexes can be detected by colored light emission following excitation by the light of the appropriate wavelength. The emitted light is viewed with a fluorescence microscope equipped with a UV light source. This technique is known as immunofluorescence. The widely used dyes are phycoerythrin and phycobiliproteins, besides fluorescein and rhodamine. The characteristic features of these organic dyes are as follows:

- Phycoerythrin and Phycobiliproteins are efficient absorbers of light and brilliant emitters of light (red fluorescence).
- Fluorescein and Rhodamine. Fluorescein is an absorbers of blue light and emitters of green fluorescence. Rhodamine is an absorber of yellow green light and an emitter of deep red fluorescence.

Cell membrane molecules or tissue sections are stained directly or indirectly. When the specific antibody is directly conjugated with an organic dye the process is known as direct staining, and if the primary antibody is unlabeled and stained with reagents (fluorochrome labeled anti-isotype antibody viz. fluorescein labeled goat and anti-mouse immunoglobulin) the process is known as indirect staining (Fig.-19-15). The gross anatomy of molecular arrangement of tissues and organs, since can be spotted by fluorescence microscopy, this technique (immunofluorescence) is greatly useful in identifying bacterial species, detecting Ag-Abcomplexes in autoimmune disease, detecting complement compounds in tissues and localizing hormones and other cellular products stained in situ.

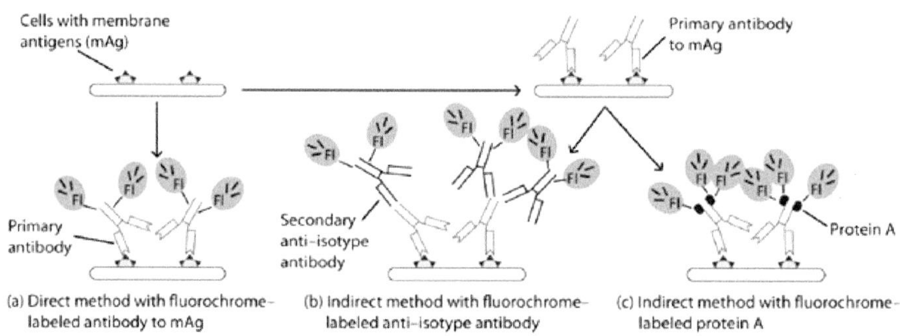

Fig.-19-15. Direct and indirect immunofluorescence.

Immunofluorescence is often used to detect the presence and amount of antinuclear antibody (ANA). These are antibodies that the body makes against itself in some autoimmune diseases like *systemic lupus erythematosus* (SLE). The presence of these antibodies aids in the diagnosis and management of these diseases.

The test may also be used to detect and measure the quantity of antibodies against microorganisms such as *Coxiella burnetii*, rickettsia or Colorado tick virus.

A positive result in this case indicates that the body's immune system has seen the pathogen at some point in the past. Variations on this test can be used to better define the timing of the past exposure.

This test may also be used to identify various microorganisms in fresh samples or after culturing them in the laboratory. In this case, antibodies that are known to be specific for particular pathogens are used on infected cells. This test is also known as a direct fluorescent antibody (DFA) test.

Flow Cytometry

The foregoing techniques of fluorescent antibody-antigen complexes determination are major qualitative in character. Fluorescence-activated cell sorter (FACS) is a device wherein lasers beam and light detection is used to count (sort) single intact cells in suspension. The technique is named flow cytometry and its version is FACS (Fig. 19-16). As a cell passes the laser beam, the instrument records the level of fluorescence the cell emits, and an attached computer generates plots of the number of cells as the ordinate and their fluorescence intensity as the abscissa. Flow cytometry and monoclonal antibodies are widely used for immunophenotyping, to design the most appropriate therapy for Leukemia (unlimited and unchecked proliferation of an abnormal clone or hematopoietic cells) since the techniques give an indication to what is the lineage of the abnormal cells and what is their maturational stage. FACS analysis is also done in identifying T cell sub-populations which is an indicator of AIDS's prognosis.

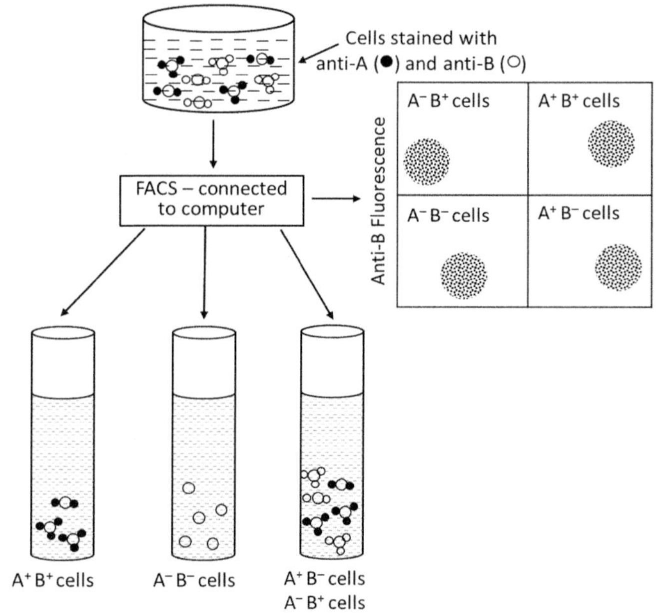

Fig.-19-16. Separation of fluorochrome-labeled cells with fluorescence-activated cell sorter (FACS).

Immunohistochemistry

Specific intracellular tissue components may be visualized by incubating enzyme conjugated polyclonal antibodies and thereafter reacting with chromogenic substrate to produce colored products. The colored products are visualized under electron microscopy as the confirmation of products. However, alternatively, as the electron dense label absorbs electrons, in this technique, the electron dense label (ferritin and colloidal gold) is either conjugated to the Fc portion of a specific antibody (direct staining) or conjugated to an immunoglobulin reagent (indirect staining). In a later procedure, an electron microscope shows the antigens as small black dots.

GLOSSARY

Abzyme (catalytic antibody). Abzymes is a monoclonal antibody with catalytic activity. Abzymes are usually artificial constructs, but are also found in normal humans (anti-vasoactive intestinal peptide autoantibodies) and in patients with autoimmune diseases are antibodies with variable regions possessing enzymatic activity.

Activated Macrophage. Under the influence of cytokines, particularly IFN-γ and IL-4, released by antigen-stimulated Th1 lymphocytes, resting macrophages are activated whereby they become larger, more motile, stickier, express more MHCII proteins on their surface, contain more lysosomes and lysosomal enzymes, and secrete a variety of substances including IL-1, TNF; they have increased phagocytic activity and increased killing via reactive oxygen intermediates, collagenases and lysosomal enzymes.

Actively acquired immunity. Immunity is acquired by exposure to an immunogen and subsequent lymphoid response.

Acute inflammation. Any inflammatory process of relatively short duration and clear termination, or characterized by infiltration of tissues by neutrophils.

Acute phase proteins. Acute-phase proteins are a class of proteins whose plasma concentrations increase (positive acute-phase proteins) or decrease (negative acute-phase proteins) in response to inflammation. In response to injury, local inflammatory cells (neutrophils granulocytes and macrophages) secrete a number of proteins into the bloodstream, most notable of which are the interleukins IL-1, IL-6 and IL-8, and TNF-α.

Acute rejection. A form of allograft rejection based upon both T cell and B cell activities. This form of allograft rejection shows a predominant T cell infiltration within the parenchyma of the graft and these usually develop within 1-2 weeks and probably represent the T cell response against an entirely 'new' set of antigens presented by the donor organ against which the T cells were not tolerized.

Acquired autoimmune hemolytic anemia (AIHA). AIHA is a disease that is a result of antibody reactivity against the Rh antigens (CDE/cde) on the red blood cell. Usually this occurs after transfusion with Rh incompatible blood or in conditions such as systemic lupus erythematosus. Approximately 0.3% of patients developing anti-e antibodies progress to AIHA.

Acquired immune deficiency syndrome (AIDS). A clinical disorder resulting from infection by the human immunodeficiency virus (HIV-1 or HIV-2) and representing the late stage of that infection. Once defined by the Center for Diseases Control (CDC) as HIV-infection plus an opportunistic infection or certain neoplasms, but now defined as any HIV+ individual with a decrease of CD4+ T cells below 200/ mm^3.

Adjuvant. A pharmacological or immunological agent that non-specifically enhances the immune response against an antigen.

Adoptive immunity. The transfer of immune cells from pre-immunized animal to host or unimmunized animal.

Affinity. The Binding strength of a univalent interaction between an epitope and a paratope. Assumes equilibrium binding. Usually determined by the dissociation constant.

Affinity selection. Selection of high-affinity clones which results in the maturation of an immune response (such as in antibody maturation).

Agglutinin. Antibody (or other molecule) that clumps bacteria or other particles.

Agranulocytosis. When the neutrophil blood count falls below 500/mm3. Contrast with "neutropenia." Agretope. That part of an antigen binds an MHC class I or class II molecule via a desetope.

Alloreactivity. The recognition and reaction of immune system cells of one individual against allotypic differences expressed by cells of a second individual is called alloreactive.

Allotype. The allelic differences among individuals of a given species which may be detected immunologically. Usually, two alleles of any given gene, one from Mom and one from Dad are found. Sometimes, more than two alleles are also found.

Alpha granules. Granules found in platelets which contain beta-lysin (a low molecular weight icrobiciD substance), fibronectin, fibrinogen, platelet-derived growth factor and others which play important role in hemostatis by clot formation. Alpha granule proteins (PDGF) are chemotactic for leukocytes whereas, Beta -lysin may be important as a microbicide.

Alternative pathway (of complement activation). This pathway of complement activation involves the formation of a C3 convertase by the spontaneous hydrolysis of the internal thioester bond of a C3, leading to its association with factor B, resulting in the generation of a fluid phase C3 convertase. This fluid phase C3 convertase enzymatically cleaves C3 into C3a and C3b. The C3b can then be hydrolyzed by a cell membrane OH or NH2 group, which initiates the binding of Factor B. This results in the formation of the membrane bound C3 convertase.

Anaphylactic shock. An observation made by Richet and Portier, 1902, that shock could be induced in hyperimmunized animals upon injection of the antigen to which the animal was immunized.

Anaphylaxis. Vasodilation and smooth muscle constriction (especially important are those of the bronchus) mediated by mast cell degranulation. Bronchial smooth muscle constriction can leap to asphyxiation. Various substances can cause anaphylaxis, including anaphylatoxins and antigens reactive with IgE.

Anaphylatoxin. A substance that results in an anaphylaxis-like response (ie., produces inflammatory changes via mast cell activation). Complement split products C3a, C4a, and C5a (especially C5a) are anaphylatoxins in man.

Angiogenin. A non-glycosylated peptide consisiting of 123 amino acids induces blood vessel growth. It is produced by lymphocytes, fibroblasts, and epithelium.

Antibody. An immunoglobulin which recognizes (binds via its Fab portion) antigen.

Antibody-dependent cell-mediated cytotoxicity (ADCC). Process of killing target cells specifically by cells with receptors for the Fc portion of antibodies. Antibodies bind to target cell via antigen-combining site.

Antibody-enhanced infection. Antibody-enhanced infection is the increase in infectivity of a microorganism as a result of antibody. It may be complement-dependent indicate the participation of complement receptors (such as CR1 or CR3), whereas or complement independent suggests a role for Fc receptors.

Antibody maturation. Antibody maturation probably results from both the selection of high affinity clones (affinity selection) and more importantly, "hypermutation." Single point mutations have been observed which may increase the antibody affinity by one log order of magnitude.

Antigen. An antigen is a molecule or molecular fragment that can be bound by a major histocompatibility complex (MHC) and presented to a T cell receptor or bound to the antibody.

Antigen blockade. The ability of an antibody to bind to antigen and thereby prevent the interaction of that antigen with other antigen receptors.

Antigen-induced apoptosis. The apoptotic death of a lymphocyte after recognition of an antigen. There are two areas where antigen-induced apoptosis is important: "clonal deletion," in which unwanted lymphocytes are eliminated in primary lymphoid organs, and "proprіociD apoptosis," in which T cells are eliminated in peripheral tissues as a means of limiting an immune response.

Antigen-presenting cell (APC). Any cell possessing MHC class I, MHC class II, or CD1 molecules can be an antigen- presenting cell. However, usually this term is reserved for cells which present antigen to T cells via MHC class I and II molecules. This would include Dendritic cells, monocytes/macrophage, B cells, Langerhans cells and certain other myeloid cells.

Antigen presentation. Antigen presentation is a process by which macrophages, dendritic cells and other APCs capture antigens and then with the help of MHC class I and II or CD1 molecules enable their recognition by T cells or NK cells.

Antigen processing. The processing of antigen is the partial digestion of an antigen such that it may associate with an MHC molecule and can be presented to T cells.

Antigen recognition activation motif (ARAM). A 20-25 amino acid motif found in all subunits of CD3, the α-chain associated with the TCR, and the Ig-α and Ig-β chains associated with the BCR. This tyrosine based motif possesses a pair of YXXL/I sequences separated by about 10 amino acid residues probably important for phosphorylation by protein tyrosine kinases.

Antigenic determinant. An epitope or that portion of a larger molecule (antigen) which is recognized by antibody.

Anti-idiotype. An antibody or TCR which specifically recognizes an epitope within the idiotype of another antibody or TCR.

Antiserum. Serum containing a large mixture of antibodies (many B cell clones involved) against an antigen. Apoptosis. Apoptosis is the process of programmed cell death (PCD) that may occur in multicellular organisms. Biochemical events lead to characteristic cell changes (morphology) and death. These changes include blebbing, loss of cell membrane asymmetry and attachment, cell shrinkage, nuclear fragmentation, chromatin condensation, and chromosomal DNA fragmentation.

Arthus Reaction. It is a type of local type III hypersensitivity reaction which is mediated by immune complex, and involve the deposition of antigen/antibody complexes mainly in the vascular walls, serosa (pleura, pericardium, synovium), and glomeruli..

Associative cognition. Recognition of both self and non-self-antigens by the same receptor (such as the T cell receptor).

Attenuation. Reducing the virulence, toxicity, or activity of an organism, toxin, or other factor.

Autocrine. When a substance is secreted and effects the cell which secreted it, the effect is referred to as "autocrine." Contrast with "paracrine," see also "hormone," and "cytokine."

Autoimmunity. Autoimmunity is the recognition of self as non-self by the immune system.

Avidity. Strength of binding between two molecules determined by both the affinity and valence of the reaction.

B cell. They are the lymphocyte which produces and express membrane immunoglobulins (such as surface IgD and surface IgM). The membrane immunoglobulin acts as the B cell antigen receptor (BCR).

B cell antigen receptor (BCR). Membrane immunoglobulin complexed with Ig-α and Ig-β to form a functional receptor which binds antigen with great specificity. The BCR is used to bind specific antigens for endocytosis and processing by the B cell and associated with CD45.

B cell enhancement. Increased activation of antigen-selected B cell as a result of concomitant binding of complement metabolite C3dg by CR2 (CD21). B cell enhancement is also called "costimulation."

Bcl-2. A 26 kD integral membrane protein encoded in a proto-oncogene termed bcl-2 on chromosome 18q21 and expressed on the membranes of intracellular organelles including mitochondria, nucleus, and smooth endoplasmic reticulum.

Cadherin. A family of "cell adhesion molecules (CAM)" which possesses five homologous extracellular domains, two calcium-binding domains and short cytoplasmic tails which interact with the cell skeleton via linkage proteins called catenins.

Calcineurin. A calcium-dependent, calmodulin-dependent cytosolic phosphatase in T cells which is important in the dephosphorylation of the transcription factor, NF-AT (nuclear factor of activated T cells). Dephosphorylated NF-AT crosses the nuclear membrane and activates the transcription of IL-2, IL-3, IL-4, GM-CSF, TNFα, and possibly, IFNγ.

Calnexin. An 88-90 kD transmembrane, calcium-binding, phosphorylated protein which localizes exclusively to the nuclear membrane and the rough endoplasmic reticulum.

Calprotectin. A cytosolic S-100 family protein complex found in the cytosol of neutrophils and oral/vaginal epithelium which exhibits zinc-reversible microbiostatic effects.

Calreticulin. A 60 kD Ca++ binding protein associated with tubulovesicular structures believed to be important in the sequestration of calcium within a cell. Such structures accumulate near the neutrophil phagosome, and are believed to be important in the calcium-dependent phagolysosomal fusion process, chemotaxis, and phagocytosis.

Carrier. That part of an immunogen which can be bound by the MHC class II molecule and is recognized by the TCR of the T-helper cells. The carrier is necessary for the elicitation of T cell dependent immune responses against hapten.

Carrier effect. Require for a carrier molecule which is identical to that which was used to immunize against hapten in order to elicit secondary responses.

Carrier priming. It is priming of the animal against a carrier molecule 'without hapten' in order to elicit a secondary response against a hapten-conjugated to that carrier.

Cell Adhesion Molecule (CAM). Molecules which mediate the adhesion between two cells. There are five main families of cell adhesion molecules: IgSF and the related cadherin superfamily, integrin superfamily, the selectins, and poorly defined "addressins."

Cell cooperation. The collaboration between two or more cell types in for the generation of specific immune responses.

Cell-mediated immunity. Immunologic reaction mediated by sensitized lymphocytes and mononuclear phagocytes (monocytes and macrophage). Usually involves recognition of target cell by a variety of interactions, most notably, the TCR interaction with presented antigen. Target cells are then killed by apoptosis or cytolysis.

Chaperones. A heterogenous group of intracellular proteins which guide the development and proper folding of nascent polypeptide and prevent aggregation of unfolded or denatured proteins in the ells. The two main families of proteins involved in chaperone function are the "heat shock protein 60 (HSP 60) and Hsp 70 families.

Chaperonins. Chaperonins are protein complexes that assist the folding of nascent, non-native polypeptides into their native and functional state. These proteins belong to a large class of molecules that assist protein folding, called molecular chaperones. Proteins such as GroEL and GroES are the major chaperonins.

Chediak-Higashi syndrome (CHS). CHS is a rare, autosomal recessive disease, mapping to a 5-cM interval in the CHS1 gene on chromosome 1q42.1-q42.2. CHS is characterized by abnormal intracellular protein transport to and from lysosomes resulting in defects in immune processes requiring degranulation: including neutrophil killing and chemotaxis and NK-cell or CD8+ CTL killing. Other systems are also impacted by such packaging disorders, thus CHS is associated with defective pigmentation, blood clotting, and neurologic function.

Chemokines. A family of small of pro-inflammatory proteins which play a fundamental role in inflammation by attracting and stimulating specific subsets of leukocytes. Chemokines participate in three phases of leukocyte recruitment. First, they foster rolling arrest of circulating leukocytes to the vascular endothelium by activating leukocytic integrins. Second, they are chemoattractans which guide leukocytes through tissues to inflammatory foci. Finally, they activate effector functions of leukocytes, including production of reactive oxygen intermediates and exocytosis of degradative enzymes.

Chemokinesis. Chemokinesis is the increased motility of a cell in response to a chemical that does not involve the sensing of a concentration gradient or any directed movement.

Chemotaxin. A molecule that can form a gradient which can be sensed by a leukocyte and attracts that leukocyte along the gradient. Also referred to as a chemoattractant. Certain chemotaxins can also stimulate phagocyte secretion and respiratory burst activity.

Chemotaxis. Chemotaxis is the directed movement of a cell along a chemical concentration gradient.

Chronic inflammation. Inflammation which is persistent and usually characterized by a cellular infiltrate of monocytes and lymphocytes.

Chronic rejection. This is the most common manner in which allografts are rejected nowadays since the advent of tissue matching with respect to the major histocompatibility complex. Allograft matches are based on MHC class I molecules, HLA-A and HLA-B. HLA-C apparently is not that important in determining graft survival (why do you think that may be?). Chronic rejection usually occurs over a time-scale of years and includes parenchymal fibrosis and vascular thickening.

Classical pathway (of complement activation). The classical pathway is the formation of a C3 convertase via C1, C4, and C2. The classical pathway is traditionally measured by the "total hemolytic complement assay."

Clonal anergy. A decrease in the reactivity (specific immunologic responsiveness) of an antigen-specific lymphocyte clone due to intrinsic (intracellular) changes. This is a functional inactivation of a cell or cell clone expressing reactivity to a specific antigen (not a deletion of the clone). Clonal anergy occurs when a T cell is presented with antigen in the absence of a costimulatory factor, such as membrane-IL-1 or B7.1- B7.3. This process occurs outside of the primary lymphoid organs and is an important mechanism of Tolerance.

Clonal deletion. The destruction of lymphocyte clones expressing TCR or BCR reactive to inappropriate antigens. This apoptotic process occurs in the primary lymphoid organs, the bone marrow, thymus, and fetal liver, where over 95% of immature lymphocytes are destroyed. An important mechanism of tolerance.

Clonal expansion. Immunojargon for the proliferation T cells or B cells of which have been selected on thebasis of the specificities of their antigen receptors (ie., TCR or BCR, respectively). Expansion is usuallybetween 100-5000 fold over a period of about 7 days, after which there is an ensuing period of cell deathby apoptosis (7-30 days) in which ~95% of the cells die.

Clone. A group of cells arising from one common progenitor and all cells of the clone possess an identical genetic make-up.

Cluster determinant/differentiation. An antigen on a cell surface which reacts with a number of different monoclonal antibodies. In order to simplify terminology, such antigens (originally named after the clones of the monoclonal antibody which recognized them) were given a single, unifying name accepted by convention. Some examples include "CD2, CD3, CD4, and CD8."

Collectin. A term referring to a group of calcium-dependent lectins which form part of the innate immune system. Collectins include mannose-binding protein (man-BP), surfactant proteins A and D (SP-A and SPD), bovine conglutinin, and the human conglutinin-like lectin, CL-43. In man, man-BP, SP-A and SP-D have been mapped to chromosome 10q21-23. All collectins bind to carbohydrates with equatorial 3 and 4-OH groups, including glucose (Glc), N-acetyl glucosamine (GlcNAc), mannose (Man), maltose (Mal), fucose (Fuc), and N-acetyl mannosamine (ManNAc).

Colostrum. Colostrum is a secretion of mammary glands which is made for about two days after birth. Colostrum contains high levels of sIgA, 10-20 mg/ml, as well as other elements of the immune system, including neutrophils, monocytes, and lymphocytes.

Complement. 20 or so plasma glycoproteins produced by the liver(mainly) but also by monocytes which perform diverse biological functions when activated. The complement system serves five mainfunctions: (1) killing microorganisms, (2) opsonizing microorganisms, (3) recruiting phagocytes, (4)mediating inflammation, (5) assisting in immune homeostasis of self antigens.

Complement activation. Changes in conformation and cleavage of complement glycoproteins resulting inthe formation of biologically active hormones, opsonins, and bacteriolytic membrane attack complexes. The two modes of complement activation are the alternative and classical pathways.

Congenic mice. Using an outbreeding/ inbreeding technique called back-crossing, one can obtain congenic mice; that is, mice that have different alleles in only a few selected genes. For example, mice are bred such that they are genetically identical except for differences in the MHC molecules (congenic). Alloreactivity based upon the selected allelic differences can then be analyzed.

Conglutinin. A C1q-like serum protein with lectin activities which causes the precipitation of red blood cells which have been coated with antibody and complement.

Constant domain. Domains which exhibit little variation in primary structure within a given isotype. These domains impart biologic function to the antibody molecule (including complement activation, placental transfer, external secretion, binding to Fc receptor-bearing cells. There is one constant domain in immunoglobulin light chains, and three to four constant domains in immunoglobulin heavy chains.

Costimulation. Costimulation refers to the signal which is essentially required by lymphocytes to proliferate in response to an antigen. That is, the presentation of antigen alone is not enough to lead to proliferation, and an additional cue, a "costimulatory signal," is required by lymphocytes to proliferate in response to the antigen.

Costimulatory signals. Costimulatory factors are usually membrane-bound glycoproteins found on antigen presenting cells (see table,below) which prepared $CD4^+$ T cells to differentiate further.

Coreceptor. Can refer to molecules which accompany a receptor and are important in bringing the receptor close to its ligand. For example, CD4 and CD8 are coreceptors for the TCR.

Cross-reaction. (1) The binding, by a single antibody or cell clone, of an epitope which is structurally related to (but not identical to) the original immunogen. (2) The binding, by antiserum or a polyclonal group of cells, of epitopes on a complex antigen possessing a number of epitopes which it has in common with the original immunogen.

Cytokine. A protein or glycoprotein produced by activated cells which mediates local inflammatory processes. The term is broad, and encompasses lymphokines, monokines, chemokines, or growth factors. Different cytokines are produced by different cells, and virtually every cell in the body has the ability to produce a one cytokine or another.

Cytotoxic T lymphocyte (CTL). A fully differentiated $CD8^+$ T cell which can kill target cells that it recognizes via its TCR and the targeT cell MHC class I molecule. Killing appears to involve two distinct categories or mechanisms: (1) the granule-release mechanism involving the release of perforin and granzymes and (2) the Fas-based mechanism, involving the a "cell-death-transductory element" present on target cells referred to as "Fas" (APO-1) and the presence of a Fas ligand on the CTL. Some cytotoxic Tcells are $CD4^+$, but these generally appear later and may be involved in immune regulation.

Death receptors (DR1 through DR5). Refers to certain members of the TNFR superfamily (TNFRSF). DR1 is TNFRI, DR2 is Fas, DR3 is TRAMP, DR4 (TRAIL R1, TNFSFR10A) and DR5 (TRAIL R2, TNFSFR10B) are TRAIL receptors. They all appear to function by binding ligand and subsequently trimerizing, recruiting an adaptor protein to their death domain through homophilic interaction. It is the adaptor protein which binds a caspase and activating apoptosis.

Decoy Receptors (DcR). A term which has come to mean receptors for members of the TNF superfamily which do not appear to transduce a signal across the cell membrane, either due to GPI linkage or truncation of the cytosolic domain. Decoy receptors for TRAIL include DcR1 (TNFRSF10C), and DcR2 (TNFRSF10D).

Delayed-Type Hypersensitivity. The classical type IV cell-mediated immunopathologic reaction, involving the infiltration of immunized (sensitized) T cells into a subdermal area in response to the presence of antigen. The primary T cell phenotype is $CD4^+$ and Th1. This mechanism usually takes 48 hours (time required for the processing and presentation of antigen and subsequent lymphocyte infiltration), and for this reason, was contrasted against IgE-mast cell-mediated specific responses, which were immediate (within minutes).

Dendritic cells (DC). Professional antigen-presenting cells (APC), $CD45^+$, that have a bone marrow origin and are specialized in antigen presentation to T cells.

Diapedesis. Interendothelial cell-junction diapedesis is the migration of a leukocyte between the endothelialcells and egress of that leukocyte through the capillary wall.

Diversity. Diversity means repertoire of immune specificies and responses. Immunology is concerned with three main types of diversity. Allotypic diversity refers to pleomorphic differences, and means the repertoire or variability of responses within a species (most notable example of allotypic diversity is the MHC encoded products). Isotypic diversity means the repertoire of molecules that all members of a given species possess.

Double negative T cell. Any pre-T cell (bearing surface b chains) lacking either CD4 or CD8.

Double positive T cell. An immature (naive) TCR-CD3 bearing T cell expressing both CD4 and CD8.

Endogenous pyrogen. An old term referring to IL-1 and TNFα and their ability to induce fever. The fever induced by these two pyrogens can be blocked by ibuprofen (a nonsteroiD antiiflammatory drug which blocks cyclooxygenase activity), suggesting that the effect of these pyrogens is indirect and dependent upon the endoperoxides (prostaglandins and thromboxanes) formed by the cyclooxygenase pathway.

Enzyme-Linked Immunosorbent Assay (ELISA). This is a primary binding assay similar to the radioimmunoassay.Briefly, a first antibody may be immobilized in the wells of a polystyrene microtiter plate. Excess sites are blocked with a protein such as BSA. The sample (containing antigen) is added for a standardized period to permit equilibrium binding. Bound antigen is quantified using a second antibody, followed by a reagent which is composed of an antibody (that binds the second antibody) conjugated to an enzyme such as horseradish peroxidase or alkaline phosphatase. A chromogenic substrate is then added, and the absorbance is measured to provide an estimation of the concentration of antigen.

Eosinophil. A granulocyte with granules which take-up eosin due to its arginine-rich major basic protein and other cationic granule proteins. This cell possesses low affinity IgE receptors (FceRII) and appears to be geared for anti-parasite activity.

Eosinophil peroxidase (EPO). Eosinophil peroxidase is a 67 kDl cationic glycoprotein most closely related to myeloperoxidase (68% amino acid homology). It localizes to the matrix of eosinophil granules. It can oxidize a number of halides in the presence of hydrogen peroxide, forming substances which are toxic to parasites.

Epidermal growth factor (EGF). A 6.1 kD peptide which stimulates keratinocyte proliferation, once called "urogastrone" for its ability to inhibit gastric secretions. EGF is a member of a large family of growth factors which includes TNFα, amphiregulin, neu-differentiation factor (NDF), heparin-binding EGF-like factor, and heregulins. EGF is produced by platelets, and monocytes/macrophage.

Epidermal growth factor/tumor necrosis factor receptor (EGF/TNF receptor). A single-chain 180 kD transmembranous glycoprotein expressed primarily on the basal keratinocytes. The EGF/TNFα receptor has intracellular tyrosine kinase activity.

Epitope. That structure which binds specifically to the paratope of an antibody, B cell antigen receptor, or Tcell antigen receptor.

Erythrocyte. A myeloid cell specialized in the transport of oxygen. It contains over 280 million molecules of hemoglobin, which give it its red appearance. The erythrocyte differentiates terminally in the bone marrow from the normboblast lineage. In circulation, the lifespan of the erythrocyte is about 120 days.

Erythrocyte chemokine receptor. A receptor which binds for both Cys-X-Cys a subfamily chemokines, including IL-8 and PF-4, and Cys-Cys b subfamily chemokines, including MCP-1 and RANTES, but not MIP-1 α or MIP-1β.

Erythropoietin (EPO). A 165 amino acid glycoprotein (MW 30 kD) encoded in a 5 exon/4 intron gene on chromosome 7. EPO works in concert with other growth factors to stimulate the proliferation and differentiation of erythroid precursor cells from hematopoietic stem cells. EPO also probably stimulates the growth and differentiation of megakaryocytes (including cytoplasmic process formation), and as such, probably stimulates thrombocytopoiesis.

Fab. That part of an antibody which contains the paratope, the antigen-binding site, consisting of light chain and a part of heavy chain. It is produced by limited digestion of immunoglobulin with papain (an enzyme from papayas).

Factor B. Factor B is a 93 kD zymogen encoded within the MHC, found in serum at 200-225 μg/ml, which can form an active enzyme when bound to C3b. The enzyme converts C3 to C3b and C3a and was once referred to as "C3 proactivator."

Fas (CD95, APO-1, TNFRSF6). A widely-distributed 42-48 kD transmembrane (325 amino acids) originally designated as "Fibroblast-ASsociated" glycoprotein, is a member of the tumor necrosis factor receptor/nerve growth factor family. Fas is expressed on the surface of activated target cells which is important in transducing a death signal into a target cell upon trimerization after exposure to "Fas ligand" present on cytotoxic T-lymphocytes.

Fc. That part of the immunoglobulin molecule which, after papain digestion, is bound by Fc receptor on cells. It includes crystallizable fragment of the carboxyl terminus of the heavy chain.

Fc receptor. A cell membrane-bound member of the IgSF which recognizes and binds the Fc part of an immunoglobulin molecule.

Follicular Dendritic Cells (FDC). Dendritic cells which are long-lived, often multinucleated, and reside within the germinal centers of regional lymph nodes and the spleen. These cells possess complement receptors (CR3) and Fc receptors and therefore, bind native antigen conplexed with complement or antibody. Their long protrusions 'embrace' B cells and are believed to present unprocessed, native antigen to B cells. In turn, B cells are activated and increase the expression of MHC class II molecules and the costimulatory signal, B7.

Gall Body. A subcellular organelle complex found only in resting T cells. It is characterized by a large fat droplet and lysosomes. Only one Gall body is found per lymphocyte, and when stained by the esterase stain, shows up as a big red dot on one side of the resting T cells.

Germinal center. An area within a lymph node of intense B cell activation; contains plasma cells, macrophages, B cells, and follicular dendritic cells (FDC).

Granulocytes. Cells of the myelocyte lineage possessing prominent cytoplasmic granules and multilobed nucleii. Includes basophils, neutrophils, and eosinophils.

Granulocyte colony stimulating factor (G-CSF). A 19,600 D glycoprotein encoded on chromosome 17 (17q11-21) which stimulates erythroid and granulocyte progenitors is a hematopoietic growth factor which is inducible (by LPS) in bone marrow stromal cells. Peripherally, this cytokine is produced by fibroblasts, endothelium, epithelium, osteoblasts, and LPS-activated macrophages.

Granulocyte-macrophage colony stimulating factor (GM-CSF). A 22 kD glycoprotein encoded on chromosome 5 (5q23-31) which stimulates production of cells which give rise to neutrophils, monocytes, erythroid cells, and eosinophils. Peripherally, GM-CSF is produced by fibroblasts, endothelium, epithelium, mast cells, B cells, and T cells.

Granzyme. It is a heterogenous group of neutral serine protease isolated from both cytotoxic T cells and NK cells. Granzymes appear to play an important role in cell mediated cytotoxicity.

GroEL chaperonin. It belongs to the chaperonin family of molecular chaperones, and is found in a large number of bacteria. To function properly, GroEL requires the lid-like cochaperonin protein complex GroES. In eukaryotes the proteins Hsp60 and Hsp10 are structurally and functionally nearly identical to GroEL and GroES, respectively.

Gut-associated lymphoid tissue (GALT). Gut lymphoid tissue consisting of Peyer's patches, appendix, and solitary lymphoid nodules.

Hapten. A molecule which is large enough to contain an epitope but or lacks the suitable properties to evoke an immune response unless coupled to a "carrier." The property it lacks is the ability to be bound within the antigen binding cleft of MHC class II molecules.

Heat shock proteins (HSP). Highly conserved, peptide-binding proteins found in the prokaryotic and eukaryotic worlds which were originally observed to increase in expression in cells which have been stressed.

Heavy chain. In immunology, this term is used to refer to the larger subunit of the immunoglobulin molecule.

Hemocytoblast. Pluripotent hematopoietic stem cell. It first appears after the third week of gestation, arising from mesenchymal cells within the yolk sac.

Hinge region. Portion of the immunoglobulin between the Fab and Fc regions which permits flexibility of the molecule.

Histiocytes. Quiescent cells found in connective tissues derived from blood monocytes.

HLA Complex. HLA is the abbreviation for "human leukocyte antigen." The major histocompatibility gene complex in man. Class I molecules are encoded within regions designated A, B, and C. Class II molecules are encoded for in regions referred to as DP, DQ, DR, DM and others.

Host. An organism which can be infected by a parasite. On a broader sense, any organism in which there is an immune system functioning to control antigen levels.

Hyperacute rejection. A rejection that usually develops in less than one hour from the implantation of a vascular graft, a form of antibody-mediated, usually irreversible damage to a transplanted organ, particularly the kidney, manifested predominantly by diffuse thrombotic lesions, usually confined to the organ itself and only rarely disseminated, for skin allograft rejection of this type, see white graft.

Hyperimmunoglobulinemia E (Job's syndrome, HIE). HIE is a rare, complex disorder characterized by the marked elevation of IgE, chronic dermatitis (eczematoid rash), "coarse facies," and serious, life-long bouts of recurrent infections by opportunistic organisms (Staphylococcus aureus and Candida albicans) which result in skin abscesses.

Hypervariable regions. Those parts of the variable domain of antibodies, BCR, and TCR which exhibit variations in primary structure. There are usually three to four of these hypervariable regions per variable domain, and these regions dictate the specificity of the antibody molecule.

Idiotope. That part of the variable domain of a given antibody, BCR, or TCR which is unique to that antibody and can be recognized as an epitope itself.

Idiotype. The collection of idiotopes expressed by an antibody, BCR, or TCR, immunologically detectable, which confer the ability of a class of receptors to bind many ligands.

Immediate-type hypersensitivity (Type I). An antigen-specific mmunopathologic reaction mediated by IgE antibodies, high affinity Fc receptors for IgE (FcεRI) on mast cells, and mast cell degranulation rsulting from the binding of antigen by IgE bound to the mast cell surface by the FCεRI.

Immune complex. Antibody and antigen bind together to form "immune complex". Complexes can become quite large if the antigen is multivalent. The largest complexes are formed near the zone of equivalence.

Immune regulation. Immune regulation as the process of limiting immunologic responses (or lternatively, limiting the antigens of the immune responses); including antibody, cell-mediated (cytotoxic), inductive (proliferative/ differentiative), and even suppressive responses.

Immune system. It is a system of biological structures and processes within an organism that protects against disease by identifying and killing pathogens and tumor cells. It detects a wide variety of agents, from viruses to parasitic worms, and needs to distinguish them from the organism's own healthy cells and tissues in order to function properly.

Immunity. Resistance to infection or changes in antigen levels.

Immunopathology. Immunopathology is the (study of the) destruction of host tissues by the immune system.

Immunogen. An antigen capable of eliciting a specific immune response.

Immunoglobulin. A glycoprotein, produced by lymphocytes of the B lineage, with a general structure of two light chains and two heavy chains. Immunoglobulins which bind to antigens are called "antibodies."

Immunological synapse. The immunological synapse is an area formed between the antigen presenting cell (APC) and the T cell. On the T cell side of the immunological synapse are cholesterol-sphingolipid "rafts," which have coalesced to form the synapse.

Immunotoxin. Immunotoxins are antibodies coupled to a poison (toxin). Toxins may be derived from a number of sources, including plants and microbes (Eg., ricin A lectin, Pseudomonas exotoxin, and diphtheria toxin [DT]).

Innate immunity. Nonspecific immunity resulting from physiologic and anatomic features. Includes the activity of complement and phagocytes.

Integrin. Integrins are transmembrane, heterodimeric (α, β) glycoproteins many of which recognize the RGD tripeptide motif (Arg-Gly-Asp) and serve to bind cells to various substrates (substrates such as laminin and fibronectin).

Intercrines See "Chemokines."

Interferon (IFN). Human interferons are a group of small proteins of which there are three main groups, IFN-α, IFN-β1, and IFN-γ. The interferons were originally defined by their anti-viral activity.

Interleukin. Interleukins are a group of cytokines that were first seen to be expressed by white blood cells (leukocytes). The term interleukin derives from (inter-) "as a means of communication", and (-leukin) "deriving from the fact that many of these proteins are produced by leukocytes and act on leukocytes".

Isotype. Usually synonymous with immunoglobulin subclasses. More accurately, all immunologically distinct members of a class of molecules found in every individual of a species encoded by a distinct genetic element. For immunoglobulins, the isotype is dictated (arbitrarily) by the products of the heavy chain genes (μ, δ, γ 3, γ 1, a1, γ 2, γ 4, ε, and a2).

J-chain. The immunoglobulin fragments found in the light chain fractions of secretory IgA and pentameric IgM joining heavy chains to each other to form dimers of IgA and pentamers of IgM.

Junctional diversity. Alterations in primary amino acid sequences into the V region of the TCR and antibody which results from reading frame shifts in the codon and anticodon brought about by the imprecise union of the V and J or VD and J gene segments.

Killer cell inhibitory receptors (KIR). A functional term referring to two very different types of receptors which stop NK cells (and certain T cells) from killing target cells by recognizing MHC class I moleculepeptide antigen complexes.

Killer lymphocyte. A lymphocyte with Fc receptors which exerts antigen-specific killing by antibody and its FcR.

Kinin. A protein which causes smooth muscle contraction and increases vascular permeability.

LAK cell. See "Lymphokine-activated killer cell."

Langerhans cell (LC). LC are dendritic cells found in the suprabasilar portions of squamous epithelium which have a dendritic appearance (ie., possessing cytoplasmic projections ranging from pseudopodia to very long dendrites and cytoplasmic veils) possessing complement and Fc receptors.

Large granular lymphocyte. A relatively large lymphocyte with a granular cytoplasm. This is a morphologic term and doesn't truly indicate a subset of lymphocytes.

Lectins. It is proteins, glycoproteins, or metalloproteins isolated from plants, viruses, fungi, animals, and bacteria bind to carbohydratemoity present in the cell.

Leukocyte. General term for white blood cell.

Levamisole. Levamisole is an antihelminthic drug that possesses nonspecific immunostimulatory properties. In deficient animals and humans, it restores many different immunologic functions, suggesting that it acts on multiple populations of cells, including neutrophils, macrophages, and T cells (but not B cells).

Light chain. In immunology, this term has traditionally referred to the smaller, 25 kD subunit of the immunoglobulin molecule.

Lipopolysaccharide (LPS). Lipopolysaccharides are the only lipid-containing moieties found in the outer leaflet of the outer membrane of Gram-negative bacteria. It acts as a stimulant in the cells when administered exogenously.

Live vaccine. Live vaccines are those in which an infectious agent is used as the vaccine.

LPS-binding protein (LBP). A 60 kD acute phase serum glycoprotein, synthesized by hepatocytes, required for the transport of LPS and cholesterol esters. With LPS, it binds to CD14 on monocytes (and neutrophils), stimulating the monocyte to release IL-1β, TNF-α, and PGE2.

Lymphoblast. An activated lymphocyte which has entered a proliferative cell cycle. The lymphoblast exhibits an increase in cytoplasmic volume, cell size, synthesis of nucleic acids, and an increase in protein biosynthesis relative to the normal lymphocyte.

Lymphocyte. A leukocyte derived from lymphoid precursors. The majority of these (B cells and T cells) have undergone "somatic" recombination and express antigen receptors (TCR or sIg) with idiotypic differences.

Lymphocyte activation gene-3 (LAG-3). The product of this gene, encoded in the CD4 locus on chromosome 12p13 near the NK gene complex, is transmembrane receptor of the IgSF found on activated NK cells and activated T cells. LAG-3 is most closely related to CD4 which can lead to downregulation of Tcell activation.

Lymphocyte recirculation. A term used to describe several pathways whereby lymphocytes leave the blood into various organs and then return to circulation, an estimated 50 times per day.

Lymphokine. A protein or glycoprotein produced by lymphocytes which mediates local inflammatory processes.

Lymphokine-activated killer (LAK) cell. A TCR-CD3-negative cell, usually identical to an NK cell, which has been activated, ex vivo, usually by exposure to interleukin-2. LAK cells kill more rapidly and kill a broader spectrum of target cells than NK cells.

Lymphopenia. Lymphopenia is regarded as a peripheral blood lymphocyte count less than 500/mm3. Normal levels are about 1500/mm^3.

Lymphopoietin-1. IL-7.

Lysosome. Membrane delimited subcellular organelle containing pro-enzymes, enzymes, cidal substances, hormones, and receptors, as well as structural components such as acidic proteoglycans.

Lysozyme. A cationic protein secreted by phagocytes encoded on human chromosome 12 with the ability to cleave the b1,4 linkage between N-acetylglucosamine and N-acetylmuramic acid of bacterial cell walls. However, lysozyme also has non-enzymatic bactericidal and fungicidal effects.

Macrophage. Macrophages are white blood cells within tissues, produced by the differentiation of monocytes. Monocytes and macrophages are phagocytes, acting in both non-specific defense (innate immunity) as well as to help initiate specific defense mechanisms (adaptive immunity) of vertebrate animals.

Macrophage colony-stimulating factor (M-CSF; CSF-1). A cytokine belonging to a group of proteins collectively referred to as "hematopoietic growth factors," produced by many cells including bone marrow stromal cells, keratinocytes, activated macrophages, activated T cells, activated B cells, and many others. It supports their survival anddifferentiation and enhances the ability of mature mononuclear phagocytes to kill microbes and tumor cells.

Major histocompatibility complex (MHC). The major histocompatibility complex (MHC) is a large genomic region or gene family found in most vertebrates. In man, it is located on chromosome 6p21.3 and encodes the MHC class Ia, Ib, II, and III molecules. It plays an important role in the immune system and autoimmunity.

Mast cell (mastocyte). It is a secretory cell rich in histamine and heparin, derived from hematopoietic stem cells in the bone marrow, bears receptors for C3a and C5a, and a high affinity Fc receptor for IgE and an Fc receptor for IgG. Apart from their role in allergy and anaphylaxis, mast cells play an important protective role as well, being intimately involved in wound healing and defense against pathogens.

Membrane immunoglobulin. Immunoglobulin portion of the B cell antigen receptor (mIg).

Membrane immunoglobulin. The antigen-specific portion of the B cell antigen receptor (BCR).

Memory. Immunologic memory refers to the heightened response to an antigen upon second exposure to the antigen. The cellular basis for immunologic memory is clonal expansion.

MHC-restriction. The requirement of MHC class I and class II molecules for recognizing self-antigens in order to stimulate an immune reaction is called MHC restriction. Although, the molecular basis of MHC restriction is not clear, but it is accepted that the TCR is restricted to recognize antigen in association with a pleomorphic part of the MHC molecule or the pleomorphic part of the MHC molecule.

Minor histocompatibility antigen (mHAg). A functionally defined term meaning those antigens which are of secondary significance in dictating allograft survival. Minor histocompatibility antigens appear to be peptides derived from intracellular self-proteins which are presented by only certain alleles of MHC class I molecules (hence, they are of secondary significance to the MHC class I molecules).

Mitogen. A chemical or physical stimulant which causes cells to divide.

Mitogen-activated protein kinase (MAPK). A 42 kD protein kinase which is phosphorylated on tyrosine residues in response to stimulation of CD30 and leads to the phosphorylation of myelin basic protein.

Monoclonal antibodies. Antibodies produced by a single B cell clone or hybridoma cell. A monoclonal antibody recognizes a single epitope or class of closely-related epitopes in contrast with "polyclonal antibodies."

Monocyte. White blood cell of the monomyelocyte lineage with an oval or kidney-bean shaped nucleus, capable of further differentiation and division in tissues into a cell such as a macrophage, histiocyte, osteoclast, alveolar macrophage, brain microglial cell, liver Kupffer cells.

Monokine. A protein or glycoprotein produced by monocytes/macrophage which mediates local inflammatory processes.

Monospecific antiserum. Antiserum containing a mixture of antibodies produced by many different B cell clones which reacts against epitopes on a single antigen.

Morphogenesis. The differentiation and growth of tissues and organs during development. Two cell adhesion molecule superfamilies are known to be important in morphogenesis -- the IgSF (including the distantly related cadherin superfamily) and the integrin superfamily.

Myeloma. It is a B cell lymphoma often characterized by the over production of a single antibody (ie., a monoclonal antibody) due to the excess proliferation of a single clone of B cells.

Naive. It means a lymphocyte that has not been exposed to antigen.

Naturalyimmunity. Non-specific immunity which exists in the absence of any known immunization. For example, we all have natural antibodies against blood cell antigens (Eg., the ABO system; or the M/N system).

Natural killer (NK) cell. Natural killer cells are lymphocytes which lack genetically rearranged antigen receptors -- ie., they are TCR-CD3-negative and BCR-negative and their TCR and BCR gene have not undergone productive rearrangement. NK cell is best known for its capacity to kill target cells and production of IFN-γ.

Natural killer T cell. A CD4 - CD8 T cell which recognizes CD1, shares a receptor (NK1.1) with NK cells, and produces IL-12. The NKT cell population possesses no clonal variation.

Necrosis. It is an accidental death of a cell induced by physical or chemical injury.

Neutrophil. A terminally-differentiated, polymorphonuclear granulocyte containing azurophil, specific, and tertiary granules, play important role in the acute phases of local inflammation.

Nitric oxide (NO). A reactive, gaseous, lipophilic molecule which has both cardiovascular effects and immunologic effects. Cardiovascular effects result from the generation of of cyclic GMP by endothelium in the presence of NO• , which leads to a decrease in blood pressure. Immunological effect results in the NO induced apoptosis, and inhibits adherence of neutrophils to endothelium and decreases neutrophil chemotaxis.

Nitric oxide synthase (NOS). A cytosolic heme-enzyme catalyzes the formation of nitric oxide from dioxygen and the guanido nitrogen of L-arginine. Enzyme requires NADPH, L-arginine, and tetrahydrobiopterin. It is enhanced by Mg++, FAD, and glutathione.

Opsonin. General term for molecules such as iC3b or IgM/IgG which enable a phagocyte to bind and engulf (phagocytose) a particle.

Opsonization. The process by which a pathogen is marked for ingestion and destruction by a phagocyte. It involves the binding of an opsonin, i.e., antibody, to a receptor on the pathogen's cell membrane, after which phagocytes are attracted to the pathogen. The Fa and Fb portion of the antibody binds to the antigen, whereas the Fc portion of the antibody binds to an Fc receptor on the phagocyte, facilitating phagocytosis.

Paracrine. When a substance is secreted and it affects nearby cells within the same organ or tissue, the effect is said to be paracrine.

Paratope. The part of an antibody, BCR, or TCR which specifically binds with the epitope of an antigen is called paratope..

Passively acquired immunity. It is the transfer of immunity from one host to another using antigen-specific lymphoid products, including antibody as well as whole cells.

Perforin. Perforin is a cytolytic protein found in the granules of CD8 T cells and NK cells. Upon degranulation, perforin inserts itself into the target cell's plasma membrane, forming a pore and as a result target cell die.

Peyer's Patches. Loose aggregates of lymphoid tissue in the gut which form a secondary lymphoid tissues. Phagocyte. A cell which ingests foreign particles are called phagocytes. Polymorphonuclear leukocytes (microphages) and monocyte/macrophages are the major phagocytes.

Phagocytosis. The process by which phagocytes engulf foreign particles, such as bacteria and enclosed in a vacuole (phagosome) in their cytoplasm.

Phagosome. A membrane-delimited structure formed by a cell after the engulfment of a particle.

Phagolysosome. A membrane-delimited structure formed by the fusion of lysosomal (granule) membranes with the membranes of the phagosome and the degraulation of the lysosomal contents into the phagosome.

Pleomorphism. Genetic, allelic differences among the molecules of individuals of a given species. Allotypic differences seen in histocompatibility antigens, immunoglobulins, complement proteins are well-known.

Polyclonal antibodies. Antibodies generated from more than one clone of B cells.

Polyfunctional antibodies. Antibodies which exhibit the ability to bind a number of small, unrelated epitopes.

Polymorphonuclear leukocyte. A polymorphonuclear leukocyte (PMN), or granulocyte, is a type of white blood cell whose nucleus has several different rounded parts referred to as lobes. The connections between these lobes are often so slender that the cell appears to have multiple nuclei.

Precipitin. Antibody that reacts with soluble, multivalent antigen and forms lattice-like aggregates which become so large that they come out of solution (precipitate). The precipitin reaction occurs optimally in the zone of antibody and antigen equivalence.

Prostaglandin. A group of inflammatory mediators derived from arachidonate metabolism via the cyclooxygenase pathway. The prostaglandins are important in chronic inflammation (contrast with "leukotrienes"), and are one of the main products of monocytes.

Proteasome. The proteasome core is a 700 kD, are found in all cells and are important in degrading proteins which are synthesized in the cytosol.

Protegrins. Cysteine-rich, arginine-rich antimicrobial peptides isolated from pig leukocytes. Protegrins exhibit sequence homology with the amino terminus of the human defensins.

Radioimmunoassay (RIA). This is primary binding assay is usually based upon the immobilization of antibody on polystyrene. It is a quantification method based on the size of the antigen.

RANTES. A chemokine of the β-subfamily produced by thrombin-stimulated platelets, macrophages, fibroblasts, endothelial cells, mesangial cells, and renal tubule cells. It is chemotactic for eosinophils, monocytes, and memory T cells.

Receptor. A cell surface molecule which binds to extracellular molecules and transmits signals intracellularly.

Regulatory T cells. A subpopulation of $CD4^+$ T cells (5-10% of all CD4+ T cells) which are $CD25^+$ and which appear to minimize inflammation and therefore, are important in controlling immunopathologies such as inflammatory bowel disease and autoimmune immunopathology.

Resident leukocytes. Leukocytes which egress from blood into the tissues in the absence of inflammation. Because they are within the tissues, resident leukocytes serve as sentinels of the immune system and help to attract the inflammatory leukocytes.

Respiratory burst. Consumption of dioxygen by stimulated professional phagocytes for the purpose of generating toxic, reduced oxygen metabolites such as superoxide anion and hydrogen peroxide.

Rh antigens. The Rh antigen system consists of three major determinants located on a single complex antigen: CDE/cde. Rh + refers to the "D" epitope, whereas Rh- refers to the "d" epitope. The term is derived from its presence in Rhesus monkey (Rhesus macaca).

Selectins. Selectins are a family of cell adhesion molecules (or CAMs). All selectins are single-chain transmembrane glycoproteins that share similar properties to C-type lectins due to a related amino terminus and calcium-dependent binding.

Self-tolerance. Self-tolerance is acquired actively (not preprogrammed), and appears to occur as a result of both thymic T cell clonal deletion and peripheral T cell clonal anergy. The peripheral form of self-tolerance is believed to be due to the lack of costimulation.

Septic Shock. An often fatal condition, mediated by an overwhelmingly high release of cytokines and enzymes by macrophages in response to exposure to lipopolysaccharide (LPS).

Severe combined immunodeficiency (SCID). A disease resulting from the inability to produce functional T and B cells. In some instances (documented in mice), it may be related to a single genetic lesion in the VDJ recombinase enzyme system (RAG-1, RAG-2).

Single positive T cell. A mature T cell or thymocyte expressing CD4 or CD8, but not both.

STAT. Signal transducer and activator of transcription. It is important in cytokine signal transduction and early induction of gene expression.

Stromal cell. A term which refers to the location and function but not the derivation of cells which induce the proliferation and differentiation of immune cells.

Superantigen (SAg). A protein that can stimulate T cells by binding to both the TCR and to MHC class II molecules, forming a trimolecular complex. Example of superantigen is Staphylococcus aureus enterotoxin C2.

Suppression. Suppression is the active inhibition of a cell's reactivity by some extrinsic regulatory component of the immune system.

Switch recombination. The process of genetic rearrangement resulting in isotype switching. Switch recombination is preceded by cytokine induced DNA-protein binding; increased susceptibility of the heavy chain locus to DNAse I digestion (i.e, the chromosome is opening up in the area); specific demethylation, and the synthesis of a "switch transcript."

T cell antigen receptor (TCR). A heterodimeric transmembrane glycoprotein with a structure analogous to that of immunoglobulin, featuring both variable and constant domains. The αβ TCR binds to antigens in association with MHC class I or MHC class II molecules. The TCR is involved in both priming and activation of antigen-specific T cells.

TAP (transporter). It is a transporter associated protein found in the membrane of the endoplasmic reticulum. It is responsible for transporting small peptides from the cytoplasm into the ER where they can associate with MHC Class I molecules. The TAP structure is formed of two proteins TAP-1 and TAP-2, and have one hydrophobic and one ATP-binding region each.

Tapasin. Tapasin is a MHC class I antigen-processing molecule present in the lumen of the endoplasmic reticulum, and plays an important role in the maturation of MHC class I molecules in the ER lumen. It is also important in the formation of the complex between MHC class I molecules and TAP transporters.

T-helper cell. A functionally defined subset of thymus-differentiated lymphocytes which are required for the generation of antibody against certain antigens by B cells and the generation of cytotoxic T cells. These cells bear a surface marker called CD4. There are at least 4 phenotypes of T-helper cells based upon "stable" cytokine profiles: Th0, Th1, Th2 and Th?.

Thrombocyte. Thrombocytes are anucleated cells that play a key role in blood clotting. In mammals, thrombocytes are anucleated cell fragments called plateletst.

Thrombopoietin. Cytokine which promotes the proliferation and differentiation of cells of the megakaryocyte lineage.

Thromboxanes. A group of inflammatory mediators derived from arachidonate metabolism via the cyclooxygenase pathway, and one of the major metabolites formed by platelets.

Thymocyte. Usually refers to immature T lymphocyte, but can mean any T cell in the thymus. Cortical thymocytes usually exhibit a range of maturational states, and medullary thymocytes are usually mature but naive.

Thymocyte activating factor. Mainly, interleukin-1 β.

TNF binding protein (TNF-BP). TNF-BP is also referred to as "sTNF R," for "soluble TNF receptor." A soluble homologue of the TNF receptor which is believed to exert negative regulatory effects by binding TNF-α.

Tolerance. A phenomenon discovered by Felton, 1942, which is defined as the immunologic 'unresponsiveness' to a given antigen; ie., the lack of a measurable response. Either high or low antigen doses appear to be able to induce tolerance to that antigen.

Toll - like receptors (TLR). Toll and Toll-like receptors are "pattern-recognition" molecules which mediate a signaling cascade using the NF-κB pathway and is essential for immune responses in adult Drosophila and humans. There are any TLRs in which TLR-2 stimulation results in the production of inducible nitric oxide synthase, interleukin-12, and eventual apoptosis.

Transforming growth factor (TGF). TGF-β is a multifunctional, 25 kD peptide which stimulates osteoblasts, and fibroblasts and inhibits osteoclasts, epithelial cells, and most immune cells. Receptors for TGF-β are found in almost all cells. There are at least 5 members of the TGF-βfamily which have been designated TGF-β1, TGF-β2, TGF-β3, TGF-β4, and TGF-β5.

Transferrin. A 75-86 kD serum glycoprotein (serum concentration is 2.6 mg/mL), known for it's iron binding capacity.

T-switch cells. T cells which induce B cell isotype switching; especially, in the Peyer's Patches, where they induce IgM-to-IgA isotype switching. They do not induce IgA secretion.

T-suppressor cell. A functionally defined subpopulation of T cells which reduces the immune responses expressed by B cells and cytotoxic T cells. In man, these cells bear the CD8 surface antigen.

Tumor necrosis factor. TNF is the prototype member of the TNF family of molecules which are important in costimulation and apoptosis. TNF is also known as cachectin, in reference to its ability to produce profound metabolic wasting associated with chronic infection and neoplastic disease. It may be TNF-α and TNF-β.

Ubiquitin. A highly conserved, 8 kD (76 amino acid) cytosolic polypeptide which tags proteins and thereby controls their trafficking within the cell. It activate transcription and assist with DNA repair.

Vaccination. Vaccination is repeated inoculation of a host with an antigen in order to confer specific immunity to that antigen. Term is derived from the cow pox virus (Vaccinia) which was used to confer immunity against small pox (Variola), which is also an example of the cross-reactivity between the two viruses.

Variable domain. The domain structure of an antibody or TCR which confers antigenic specificity. One variable domain exists on each chain (both light and heavy) of an antibody or a TCR.

Vascular cell adhesion molecule-1 (VCAM-1). VCAM-1 is a cell adhesion molecule of 90-110 kD expressed constitutively by bone marrow stromal cells and is important in the interaction of stromal cells with maturing B cells.

Vascular endothelial growth factor (VEGF). It induces proliferation of endothelial cells through the binding of a kinase domain receptor and the Fms-like tyrosine kinase receptor. It also causes endothelium to become more permeable to solutes and stimulates the transendothelial migration of monocytes.

VDJ recombination. The elimination of intervening sequences between various V, D, and J gene segments and ligation of those segments together. These recombination of the germline segments result in antibody and TCR idiotypic diversity and occur prior to antigen exposure.

Zone of equivalence. That point at which antibody paratopes and antigenic epitopes are equal. If the interacting molecules are multivalent, a precipitin will form from the "immune complexes" at the zone of equivalence.

Index

O

P

R

S

T

U

V

X